Kenji Tasaka

New Advances in
Histamine Research

With 192 Figures, Including 1 in Color

Springer-Verlag
Tokyo Berlin Heidelberg
New York London Paris
Hong Kong Barcelona Budapest

Kenji Tasaka, M.D., PH.D.
Chairman and Professor,
The Department of Pharmacology in the Faculty of Pharmaceutical Sciences,
Okayama University,
1-1-1, Tsushima-Naka, Okayama, 700 Japan

ISBN-13:978-4-431-68265-3   e-ISBN-13:978-4-431-68263-9
DOI: 10.1007/978-4-431-68263-9

Preface

In 1910 histamine was first isolated from ergotinum dialysatum and its hypotensive effect and uterine-stimulation action were discovered by Sir Henry Dale. Since then, its physiological roles and pathological significance in various organs and diseases have been elucidated to a great extent. Some representative contributions carried out in the subsequent 80 years are listed below. However, many other important studies on the physiology and pharmacology of histamine have been carried out in different areas.

- —Discovery of histamine by the decarboxylation of histidine (Windaus and Vogt, 1907)
- —Isolation of histamine from biological materials. Ergot preparation (Barger and Dale, 1910)
- —Pharmacological actions akin to those produced by "shock" and to certain anaphylactic reactions (Dale and Laidlaw, 1910-1911). Effect on capillaries (Dale and Richards, 1918; Lewis, 1924-1927)
- —Gastric secretory effect (Popielski, 1920)
- —Histamine as a natural constituent of some organs (Best, Dale, Dudley and Thorpe, 1927)
- —Liberation of histamine in anaphylactic shock (Dragstedt et al., 1932-1936; Bartosch, Feldberg and Nagel, 1932)
- —Development of antihistaminics (Bovet and Staub, 1937)
- —Histamine inactivation by histaminase (Zeller, 1951; Schayer, 1952)
- —Histamine in mast cells (Riley and West, 1953)
- —Compound 48/80, as a potent histamine liberator (Feldberg, Paton, and Schachter, 1951)
- —Fluorometric histamine assay (Shore, 1959)
- —Concept of H_1- and H_2-receptor (Advent of H_2 receptor blocking agent) (Ash and Schild, 1966; Black et al., 1972)

Recently there have been many important findings in relation to the physiological effects of histamine on the central nervous system, and systematic studies of great importance are now appearing regularly. One of the new trends in histamine research was investigated by Schwartz and his associates: They clearly showed the existence of an H_3

receptor in the brain, thereby stimulating much research to clarify the function of histamine in the brain. One remarkable outcome was the discovery that the principal role of histamine in the brain may be excitatory. Concurrently with this, Tasaka and his associates presented the exciting new finding that histamine exerts an excitatory effect on the arousal system. Further experimentation revealed that histamine also exerts a stimulatory effect on memory retention, the acquisition of learning, and memory recollection.

Furthermore, it recently became clear that histamine plays a very critical role in the differentiation and proliferation of neutrophil progenitor cells in the bone marrow. Although the mechanism of leukocyte release from bone marrow is not yet clearly understood, it has been shown that histamine plays a critical role in efficiently releasing mature neutrophils from the bone marrow. Recent progress in molecular biology has allowed the mechanism of histamine action on bone marrow cells to be successfully analyzed at the molecular level. In related work, the interaction between histamine and cytokines has drawn the keen attention of both histaminologists and hematologists.

Since histamine release in the antigen-antibody reaction was discovered in 1930, the various stages in the mechanism leading to histamine release is still one of the main areas of histamine research. The methodology in this field has progressed rapidly, stimulating research carried out on the basis of molecular pharmacology and refined immunoelectron-microscopy. The study of histamine release from mast cells is also progressing rapidly.

This book covers many new findings, including the role of histamine in the arousal system, learning, and memory. It also features research into the role of histamine in neutrophil differentiation and the mechanism of histamine release using new techniques such as molecular pharmacology and refined immunoelectronmicroscopy.

Enormous endeavors have been made at the Department of Pharmacology, Okayama University in these new areas of histamine research. This book summarizes the research done in this department, and the author sincerely hopes that it will be useful in understanding the new advances that have been made in histamine research.

KENJI TASAKA

The author

Kenji Tasaka, M.D. Ph.D., is chairman and professor of the Department of Pharmacology in the Faculty of Pharmaceutical Sciences at Okayama University. He was also the Dean of the Faculty of Pharmaceutical Sciences from 1981 to 1986 and from 1990 to 1994 (for a total of 4 terms).

Professor Tasaka graduated from Okayama University Medical School in 1952, obtaining the degree of M.D. In 1953 he became a staff member of the Pharmacology Department at Okayama University and received his Ph.D. in 1958. After he completed 2 years of postdoctoral research in the Department of Physiology at Mayo Clinic (Rochester, Minnesota), he was appointed to Research Instructor of the Department of Medicine at Virginia Medical College. He became Associate Professor of Pharmacology in 1970 at Okayama University Medical School. In 1971 he was promoted to Professor of Pharmacology in the Faculty of Pharmaceutical Sciences of the same university.

Professor Tasaka is a member of the Japanese Pharmacological Society (Auditor, Councilor), the Pharmaceutical Society of Japan (Board of Governors, Councilor) the Japanese Society of Allergology (Councilor), the European Histamine Research Society, the American Academy of Allergy and Immunology, the European Academy of Allergology and Clinical Immunology, and is a Fellow of the Royal Microscopical Society. He was the President of the following conferences: The Japanese Pharmacological Society (Kinki Branch) in 1977, the International Histamine Symposium in 1981, and the Asian Histamine Research Society in 1989 and 1992, the latter were both held in Okayama, Japan.

He received awards from the European Histamine Research Society in 1985, 1988, 1990, and 1992. In 1993 he received the Mochida Memorial Scientific Award and the Chugoku Scientific Award, and in 1994 he was given the Pharmaceutical Education Award from the Pharmaceutical Society of Japan.

At present his main interests are histamine release from mast cells, the histamine-induced differentiation of neutrophil progenitor cells and the excitatory effect of histamine on the contral nervous system.

Table of contents

Chapter 1

Excitatory effect of histamine on the arousal system

1. Introduction 1

2. EEG power spectral analysis 2

3. Changes in EEG patterns and EEG spectral powers recorder at the frontal cortex (FCOR) and the nucleus ventralis thalami (VE) after electrical stimulation to the RF................................. 4

4. Influence of the stimulation frequency and voltage 5

5. Changes in EEG spectral powers after stimulation to the certain areas of the RF 6

6. Effects of histamine and its related compounds on EEG spectral powers 7

7. Effects of H_1 And H_2 antagonists on histamine-induced decrease in EEG spectral powers 10

8. Effects of H_3 agonists and antagonist on EEG spectral powers 12

9. The pathway responsible for EEG synchronization and effect of histamine on this system 16

10. Conclusion 22

References 23

Chapter 2

The role of histamine on learning and memory

1. Introduction 27

2. One-way active avoidance and passive avoidance 28

3. Age-related changes in acquisition and retention 29

4. Recovery of memory deficit in relation with a long interruption
   of training in old rats... 31

5. Role of endogenous histamine in learning and memory............... 39

6. Learning and memory deficits induced by hippocampal lesions.......... 45

7. Effects of certain H_1 antagonists on active avoidance response 51

8. Conclusion.. 62

References 63

Chapter 3

Role of brain histamine on corticosteroid release

1. Introduction 69

2. Effect of parenteral administration of histamine on plasma ACTH
   and corticosteroid concentrations 70

3. Effect of intracerebroventricular injection of histamine on plasma
   ACTH and corticosteroid concentrations 73

4. Effects of H_1 and H_2 agonists on plasma ACTH and corticosteroid
   concentrations 75

5. Effects of H_1 and H_2 antagonists on the elevation of plasma ACTH and corticosteroid concentrations induced by histamine or 4-methylhistamine 77

6. Effects of H_3 agonist and antagonist on plasma ACTH and corticosterone concentrations in rats 80

7. Effect of CRF on plasma ACTH and corticosteroid concentrations...... 80

8. Effect of histamine on plsama ACTH and corticosteroid concentrations in hypophysectomized animals 81

9. Histamine depletion in rat brain and its influence on corticosterone release 82

10. Influence of lesions or stimulation to the posterior hypothalamus 84

11. Interaction between the posterior hypothalamus and the adrenal gland .. 89

12. Effects of some neurotransmitters on steroidogenesis in dog adrenocortical cells 92

13. Conclusion 93

References 94

Chapter 4

Role of Ca^{2+} and cAMP in histamine release from mast cells

Introduction 97

1. Intracellular Ca^{2+} release induced by histamine releasers and its inhibition by some antiallergic drugs.................................. 97

2. Inhibition of intracellular Ca mobilization, Ca uptake and histamine release induced by some antiallergic drugs 103

3. Sequential analysis of histamine release and intracellular Ca^{2+} release
   from murine mast cells 111

4. Ca uptake and Ca releasing properties of the endoplasmic reticulum
   in rat peritoneal mast cells 114

5. Role of endoplasmic reticulum, an intracellular Ca^{2+} store, in histamine
   release from mast cell 121

6. Histamine release from β-escin-permeabilized rat peritoneal mast cells
   and its inhibition by intracellular Ca^{2+} blockers, calmodulin inhibitors
   and cAMP 128

7. Role of ATP and activation of protein kinase C in Ca^{2+}-dependent histamine
   release from permeabilized rat mast cells 139

8. Ca^{2+}-induced translocation of protein kinase C during Ca^{2+}-dependent
   histamine release from β-escin-permeabilized mast cells 146

9. Phosphorylation of smg p21B in rat peritoneal mast cells in association
   with histamine release inhibition by dibutyryl-cAMP 154

References 160

Chapter 5

Role of cytoskeleton in the histamine release from mast cells

1. Role of microfilaments in the exocytosis of rat peritoneal mast cells 169

2. Microfilament-associated degranulation 178

3. Role of microtubules on Ca^{2+} release from the endoplasmic reticulum and
   associated histamine release from rat peritoneal mast cells 184

4. Identification of vimentin in rat peritoneal mast cells and its phosphorylation
   in association with histamine release 192

References 199

Chapter 6

Histamine release induced by basic peptides and proteins

Introduction 203

1. Substance P-induced histamine release from rat peritoneal mast cells
   and its inhibition by antiallergic agents and calmodulin inhibitors 203

2. Histamine release induced by histone and related morphological changes
   in mast cells 212

3. Guinea pig eosinophil major basic protein as a potent histamine
   releaser 215

References 228

Chapter 7

Inhibitory effect of interleukin-2 on the histamine release from rat peritoneal mast cells in association with the production of lipocortin-I

1. Introduction 233

2. Inhibitory effects of interleukin-2 on histamine release from rat peritoneal
   mast cells induced by compound 48/80 and concanavalin A 233

3. Effect of IL-2 on ^{45}Ca uptake and IP$_3$ production in rat peritoneal
   mast cells 235

4. Effect of IL-2 on cAMP-protein kinase A system in rat mast cells 236

5. [^{3}H]-Leucine uptake into rat mast cells elicited by IL-2 237

6. Determination of the protein synthesis elicited by IL-2 238

7. Western blotting analysis of lipocortin-I induced by IL-2.............. 238

8. Participation of tyrosine kinase in histamine release inhibition due to
   IL-2 239

9. Discussion 241

References 243

Chapter 8

Histamine-induced leukocytosis

1. Introduction 245

2. Histamine-induced leukocytosis in experimental animals 245

3. Effect on neutrophil precursors..................................... 252

4. Histamine and cell proliferation 257

5. Histamine-induced differentiation of HL-60 cells: Analysis of the
   mechanism of action of histamine 258

6. Effect of histamine on IL-1 production in bone marrow stromal cells 264

7. Histamine-induced externalization of G-CSF receptor 273

References 287

Chapter 9

Pharmacology of newly developed H_1 antagonists: antiallergic profile of H_1 antagonists

1. Introduction 293

2. Determination of H_1 antagonistic activity (isolated guinea pig ileum) 293

3. Effect on histamine release and SRS-A release 295

4. Effect on ^{45}Ca uptake and intracellular Ca mobilization 298

5. The role of cyclic nucleotide for inhibition of histamine release 301

6. Effect on bronchoconstriction in guinea pigs......................... 304

7. Effect on cutaneous reaction in rats 306

8. Effect on experimental allergic rhinitis 308

9. Effect on experimental allergic conjunctivitis........................ 309

10. Effect on LTs- or PAF-induced contractions of isolated guinea
    pig trachea 312

11. Effect on rabbit platelet aggregation induced by PAF 313

12. Effect on the central nervous system 313

13. conclusion 318

References 319

Subject index 325

Chapter 1
Excitatory effect of histamine on the arousal system

1. Introduction

It has been reported that histamine appears to act as a neurotransmitter in the mammalian brain (Schwartz, 1975). In 1954, Feldberg and Sherwood demonstrated that intracerebroventricular (i.c.v.) injection of histamine ($200\,\mu$g) elicited a decrease in spontaneous activity and sleepiness in cats. In mice and rats, it was also found that high doses of histamine (i.c.v., 20-100 μg) produced sedation, ptosis and drowsy patterns in spontaneous EEGs characterized by high amplitude and low frequency EEG waves (Kamei et al., 1981). Moreover, i.c.v. injection of histamine (at doses of $10\,\mu$g or more) elicited an inhibition of the pinna reflex (Kamei et al., 1984) as well as a prolongation of thiopental sleeping time in mice (Kamei et al., 1986). These findings seem to suggest that high doses of histamine exerts an inhibitory influence on the central nervous system (CNS). By contrast, Monnier et al. (1970, 1977) reported that intravenous (i.v.) and i.c.v. injection of histamine induced an EEG arousal response characterized by decreased delta activity in the cerebral cortex. It has also been shown that histamine (250 ng) induced a significant increase in spontaneous motor activity in the conscious rat, including increased grooming and exploratory behavior, indicating that histamine may be involved in modulating behavioral arousal (Kalivas, 1982). In accordance with this view, H_1 antagonists have caused drowsy patterns in spontaneous EEGs and sedative effects in humans (Carruthers et al., 1978) and experimental animals (Heinrich, 1953; Tasaka et al., 1986).

On the other hand, the presence of the presynaptic H_3 receptor in the CNS has become apparent since the first, specific antagonist, thioperamide, and agonist, (R)-α-methylhistamine, have been developed (Arrang et al., 1983, 1987). Lin et al. (1990) recently reported that oral administration of (R)-α-methylhistamine caused a significant increase in deep slow wave sleep in cats while that of thioperamide enhanced wakefulness. Although both thioperamide and (R)-α-methylhistamine were reported to be effective even through oral administration (Lin et al., 1990), the absorption and metabolism of the drugs as well as the extent of the penetration into the CNS should be clarified sufficiently in order to understand the function of H_3 receptors in the CNS. In fact, it has been reported that (R)-α-methylhistamine (6.3 mg/kg) and thioperamide (2 mg/kg) caused no significant changes in the histamine content of the mouse brain when they were administered by i.p. (Oishi et al., 1989). I.c.v. injection of these drugs seems to be the most expedient for assessing the effect which appears via the H_3 receptors in the brain.

In this chapter, the effects of histamine and its related compounds on the arousal system are reviewed on the basis of the findings obtained from experiments carried out by means of low frequency stimulation to the midbrain reticular formation (RF), which inevitably induces EEG synchronization. Furthermore, the pathway responsible for this EEG synchronization and the effect of histamine on this system are also described.

2. EEG power spectral analysis

2.1. Hardware

To analyze EEGs with a personal computer (NEC, PC-9801F2), signals were amplified with an EEG amplifier (voltage gain, 1×10^4) and the analogue signals were converted into digital values by means of a multichannel A-D converter (Micro-Science, DAS-1298BPC-16) with a 20 msec interval (τ). Using these instruments, analog signals between -5.12 and $+5.12$ volts (V) were linearly converted into 12 bits of binary digital values (1 digit $= 2.5$ mV) at the conversion rate of 25μsec/channel with a voltage resolution of 0.25μV. To enable real time analysis of the EEGs, the calculating speed of the computer was usually augmented with a numerical data processor (Intel, i8087-2).

2.2. Software

In order to analyze EEG signals, EEG analysis software was programmed (Fig. 1). In addition to converting analog EEG signals into digital values which were stored in the computer's memory, the program allowed computation of the fast Fourier transform (FFT) in real time.

The main routine of this program was written in BASIC and compiled by a BASIC compiler (NEC) using MS-DOS (Microsoft). The subroutines, used for high-speed calculations, such as A-D conversion and FFT, were programmed in C-language and compiled to obtain machine language programs by means of Microsoft-C (Microsoft). Programming of the FFT subroutine was carried out according to the algorithm of Cooley and Tukey (1965) and the power spectrum was calculated from digitized EEG data ($\tau = 20$ msec). The time schedule for the measurement of EEG power spectra before and after drug administration is shown in Fig. 2. To obtain each tracing of EEG power, data were collected for 2.56 sec and a subsequent 2.6 sec was spent calculating the FFT and EEG power. Since the data sampling was performed at a rate of 50 Hz (1/20 msec) for 2.56 sec, the number of samples was 128. Under these conditions, the minimum frequency resolution was 0.39 Hz (50/128). The values of the power spectra were displayed as compressed array; each tracing was depicted every 5.16 sec. The power density values collected during about 2 min (5.16 sec $\times$ 24), including 61.4 sec for data sampling, were averaged and distributed into the four frequency bands: delta (0-6 Hz), theta (6-10 Hz), alpha (10-16 Hz) and beta waves (16-25 Hz). All calculated data were transferred to minifloppy disks displayed on a cathode ray tube (CRT) screen and printed by a dot-matrix printer with the least possible delay (Tasaka et al., 1989).

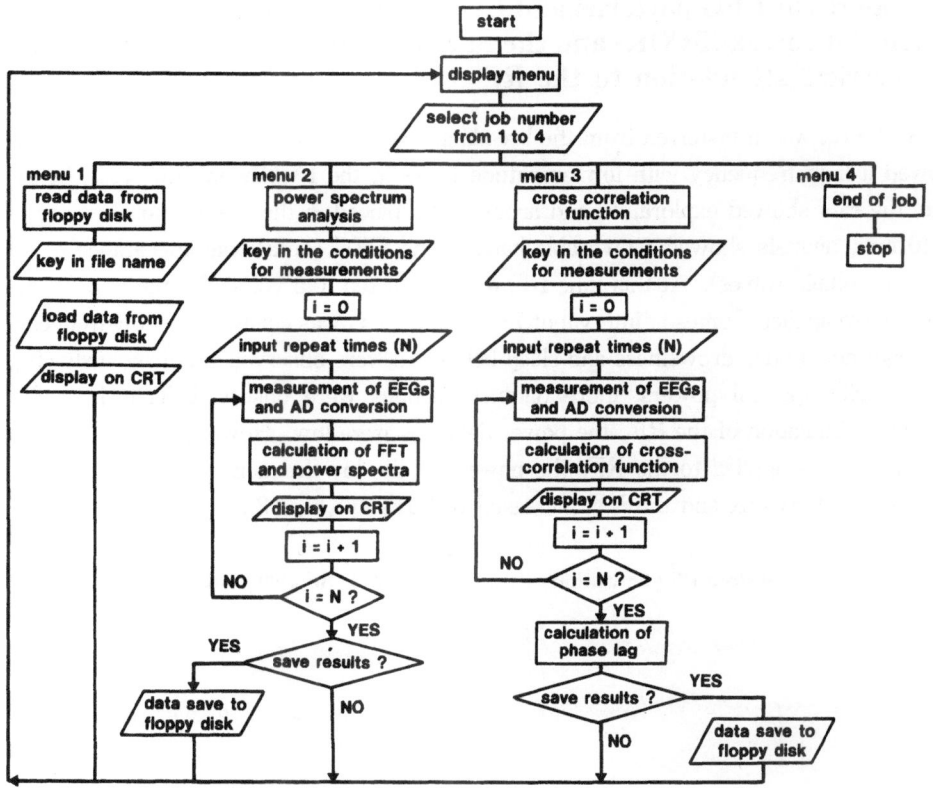

Fig. 1. A flow chart for the EEG analysis program.

menu 1: Read data from floppy disk. The analyzed data stored on the floppy disk, is loaded and displayed on the CRT screen.

menu 2: Power spectrum analysis. In this menu, the power spectrum of digitized EEG data are calculated. After selection of this menu, one must input the suitable conditions for measurement, such as sampling interval (τ), sampling number, channel selection, etc., and must indicate how many times (N) FFT should be calculated. Compressed arrays of power spectra are displayed in a series on the screen at fixed intervals.

menu 3: Cross correlation function.

menu 4: End of job. If this menu is selected, all the data in the memory of the computer are erased and the job is finished.

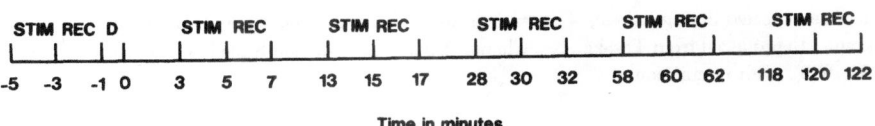

Fig. 2. Time schedule for EEG spectral analysis. Stimulation (STIM) and recording (REC) refer to RF stimulation and EEG recording, respectively. D indicates drug application time. In some cases, EEG analysis was performed without drug application.

3. Changes in EEG patterns and EEG spectral powers recorded at the frontal cortex (FCOR) and the nucleus ventralis thalami (VE) after electrical stimulation to the RF

When the rat was transferred from the breeding cage to the observation box, the animal showed a high frequency with low amplitude EEG at the FCOR and the VE. At this time, the rats showed exploration and arousal. As placed in the observation box for 30 to 40 min, animals showed a favorable background EEG (occasional appearance of 5-8 Hz high voltage waves). At this time RF was stimulated (0.5 volts, 0.1 msec, 3 Hz, for 10 sec were applied 7 times at interval of 7 sec), the EEG was changed to a low frequency with high amplitude, drowsiness and sleep behaviors were observed simultaneously (Fig. 3). In EEG spectral powers, more distinct changes were observed. That is, before electrical stimulation of the RF, the power densities were low. However, after the same stimulation was applied to the RF, the powers in the low frequency bands (delta wave region) of the FCOR and the VE increased remarkably (Fig. 3).

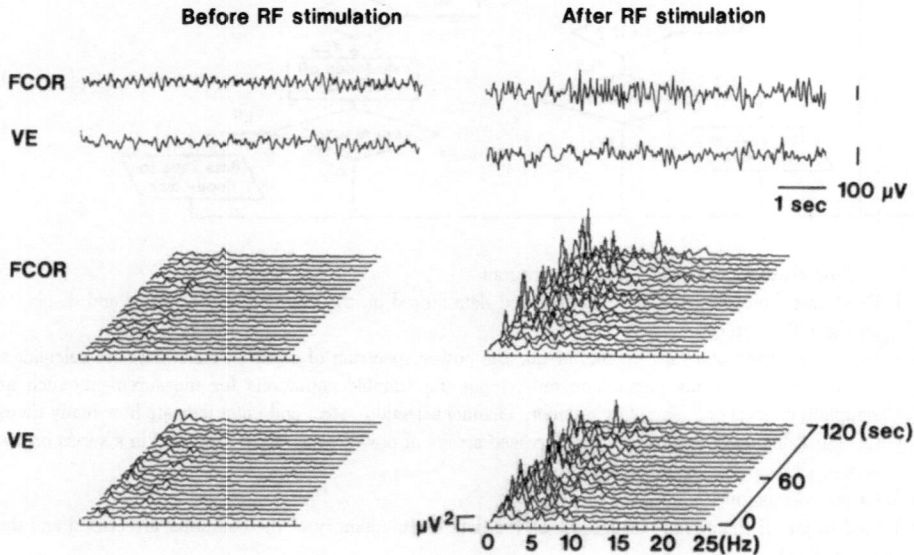

Fig. 3. Changes in the EEG patterns and EEG spectral powers induced at the FCOR and VE after stimulation to the RF. FCOR: Frontal cortex; VE: Nucleus ventralis thalami. Each tracing was calculated from tha data collected during 2.6 sec. The tracings were displayed sequentially from the bottom to the top of the graph. Reproduced from Tasaka, K., Chung, Y.H., Mio, M. and Kamei, C.: Brain Res. Bull., 32, 365-371 (1993), with permission.

4. Influence of the stimulation frequency and voltage

Changing the stimulation voltage from 0.1 to 5 V with a fixed stimulation frequency of 3 Hz (0.1 msec) or changing the frequency from 0.2 to 200 Hz with a fixed voltage of 0.5 V (0.1 msec) increased the powers in the low frequency bands (delta wave) of the FCOR and the VE spectra significantly (Fig. 4). However, the power densities in the higher frequency bands (theta, alpha and beta waves) of the FCOR and VE were not altered markedly.

As shown in Fig. 4, low frequency stimulation to the RF elicited EEG synchronization as illustrated by the marked increase in power density in the low frequency band (delta). It is clear that repeated electrical stimulation (3 Hz, 0.5 V) to the RF causes the transition of EEG waves from low voltage-fast waves to high voltage-slow waves. Moruzzi and Magoun (1949) reported that the lowest stimulation frequency to effectively induced EEG desynchronization was 50 Hz and this response became evident by increasing the stimulation frequency up to 300 Hz. Although they stated that low voltage, high

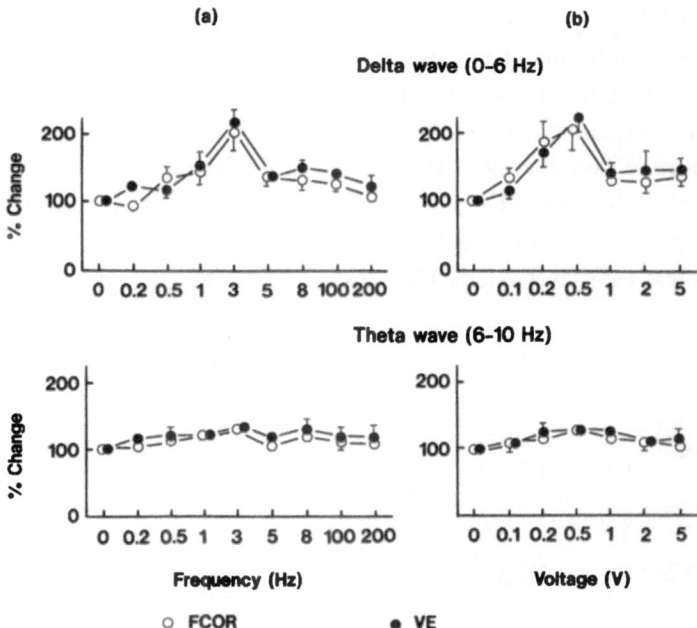

Fig. 4. Influence of the stimulation frequency and voltage on the EEG power spectra recorded from the FCOR and VE. The power densities determined before stimulation were taken as 100%. (a) Changes in power in relation to frequency changes. Stimulation of 0.5 V, 0.1 msec for 10 sec was applied 7 times at 7 sec of intervals. (b) Changes in power in association with voltage. The stimulation conditions were exactly the same as (a) except for voltage. The power densities determined before stimulation were taken as 100% (N = 5). From Tasaka, K., Chung, Y.H., Sawada, K. and Mio, M.: Brain Res. Bull., 22, 271-275 (1989), with permission.

frequency stimulation is suitable for EEG arousal, the voltages they employed were 1.0 to 3.0 V. According to Monnier and Herkert (1977), however, experimental arousal was induced when they stimulated the RF with 0.2 V at a frequency of 150 Hz. In our experiment, when the RF was stimulated at 3 Hz (0.1 msec for 10 sec × 7) with various voltages (0.1 to 5 V), the maximal increase in EEG power, especially in delta region, was induced at 0.5 V. These observations seem to indicate that the frequency is more important in eliciting an increase in EEG power than the strength of the stimulating voltages. Monnier et al. (1970) used the increase in spontaneous delta activity of the EEG as the criterion of cortical activity; such an increase indicates reduction of wakefulness or sleepiness, while a decrease is a useful signal for arousal or alertness. In accordance with these findings, the increase in low frequency power (delta) seems to indicate the reduction of wakefulness (sleepiness).

5. Changes in EEG spectral powers after stimulation to the certain areas of the RF

In order to study whether or not some special areas are essential to inducing an increase in the FCOR and VE power densities after stimulation to the RF, bipolar electrodes were implanted into the various RF sites as depicted in Fig. 5. As summarized in Table 1, no obvious differences were observed in the EEG spectral powers recorded from the FCOR and VE, even though electrical stimulation was applied at several different sites in the RF. It became clear that no particular area in the RF induces EEG synchronization. This indicates that the RF may be organized as one functional unit and that stimulation applied to any part of the RF always induced the same response. The EEG synchronization was extensively studied by Rossi (1965). He cleary described that EEG

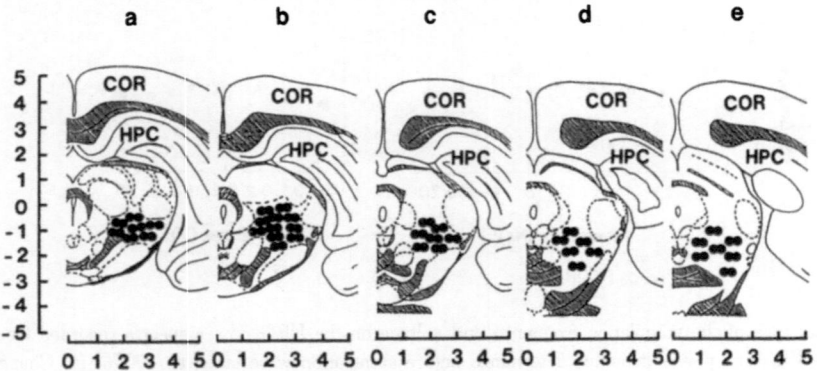

Fig. 5. Localization of the stimulating electrodes in the RF. Each pair of dots represents the location of electrode tips. RF: Reticular formation; COR: Cortex; HPC: Hippocampus. a, b, c, d, e: See Table 1. From Tasaka, K., Chung, Y.H., Mio, M. and Kamei, C.: Brain Res. Bull., 32, 365-371 (1993), with permission.

synchronization was dependent on the low rate (4-12 Hz) stimulation to the RF. When they situated 48 stimulating electrodes in the brain stem starting from the dorsal and ventral tegmentum in the midbrain to nucleus parvo-cellularis and nucleus ventralis in the medulla oblongata and tested the EEG synchronization, 43 gave a good response while in the remaining 5, no synchronization was detected (Favale et al., 1961). As shown in Fig. 5, all 38 paired electrodes showed EEG synchronizing signs in the present study. Furthermore, they noticed that the appearance of synchronization is dependent on the proper general conditions of the animal at the moment of the stimulation (relaxed wakefulness).

Table 1. Changes in EEG spectral powers of the FCOR and VE after stimulation of various areas of the RF

Sites of RF	FCOR	VE
a	211 ± 8**	217 ± 16**
b	215 ± 4**	225 ± 15**
c	223 ± 13**	217 ± 10**
d	222 ± 11**	224 ± 6**
e	214 ± 8**	212 ± 11**

a, A: 2.2, L: 1.8~3.1, H: −1.3~−1.8
b, A: 1.8, L: 1.4~2.8, H: −1.2~−2.2
c, A: 1.4, L: 1.4~2.6, H: −1.5~−2.1
d, A: 1.0, L: 1.3~2.3, H: −1.6~−2.6
e, A: 0.6, L: 1.1~2.3, H: −1.6~−2.6

RF: Reticular formation; FCOR: Frontal cortex; VE: Nucleus ventralis thalami. The power densities determined before stimulation were taken as 100%. **: Significantly different from the values measured before stimulation with $P < 0.01$ (N = 6−11). From Tasaka, K., Chung, Y.H., Mio, M. and Kamei, C.: Brain Res. Bull., 32, 365-371 (1993), with permission.

6. Effects of histamine and its related compounds on EEG spectral powers

6.1. Intracerebroventricular (i.c.v.) injection of histamine

At a dose of 1 μg, histamine caused a significant decrease in the powers of the FCOR and VE after low frequency stimulation to the RF (Fig. 6) which was markedly observed in the low frequency band (0-6 Hz). At doses higher than 1 μg, histamine had a significant and dose-dependent inhibitory effect. At doses of 2 and 5 μg, a significant effect was observed from 5 to 30 min after injection of the drug. Kalivas (1982) demonstrated that i.c.v. injection of histamine at doses of 0.25-2.5 μg significantly increased spontaneous motor activity compared with a saline-treated group 20-40 min after histamine injection. This finding may correspond to those described above, though the parameters in measuring the histamine-induced arousal response were quite different.

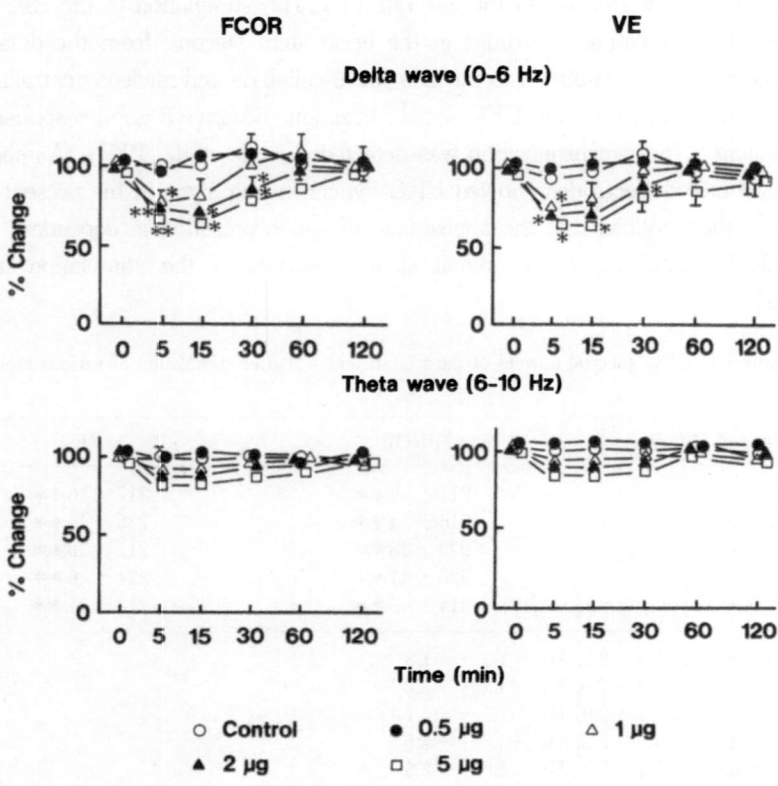

Fig. 6. Effect of histamine (i.c.v.) on EEG spectral power recorded from the FCOR and VE after stimulation to the RF. *,**: Significantly different from control group with $P < 0.05$ and $P < 0.01$, respectively (N = 6). Reproduced from Tasaka, K., Chung, Y.H., Sawada, K. and Mio, M.: Brain Res. Bull. 22, 271-275 (1989), with permission.

6.2. Intraperitoneal (i.p.) injection of histidine

I.p. injection of histidine caused no significant influences on EEG powers at doses of 20 and 100 mg/kg. However, at a dose of 200 mg/kg it elicited a significant reduction from 15 min to 60 min after drug injection (Fig. 7). It is well known that i.p. injection of histidine induces an increase in brain histamine contents. Schwartz et al. (1972) reported that i.p. injection of histidine at doses of 250 and 500 mg/kg elevated brain histamine levels by 17 and 51%, respectively. Therefore, the effect observed with histidine can be ascribed to an increase in brain histamine contents.

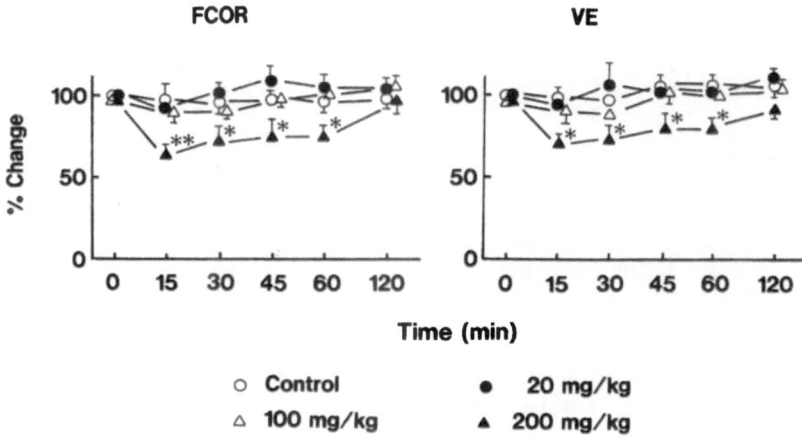

Fig. 7. Effect of histidine (i.p.) on EEG spectral power (delta band) recorded from the FCOR and VE after stimulation to the RF. **∗,∗∗** : Significantly different from control group with P < 0.05 and P < 0.01, respectively (N = 5).

6.3. Intracerebroventricular (i.c.v.) injection of 2-methylhistamine and 4-methylhistamine

2-Methylhistamine and 4-methylhistamine have been generally used as pharmacological tools for specific H_1 and H_2 receptor agonists, respectively. 2-Methylhistamine at a dose of $5\,\mu$g caused a marked decrease in the EEG powers of the low frequency band (delta) indicating that 2-methylhistamine induces an arousal effect as does histamine. 4-Methylhistamine, however, was not effective in altering EEG powers, even at a dose of $30\,\mu$g (Fig. 8). It became apparent that not only histamine but also 2-methylhistamine, a typical H_1 agonist, decreased the EEG powers in both the FCOR and VE. It seems therefore that the arousal effect of histamine is exerted via the H_1 receptors with no relation to the H_2 receptors.

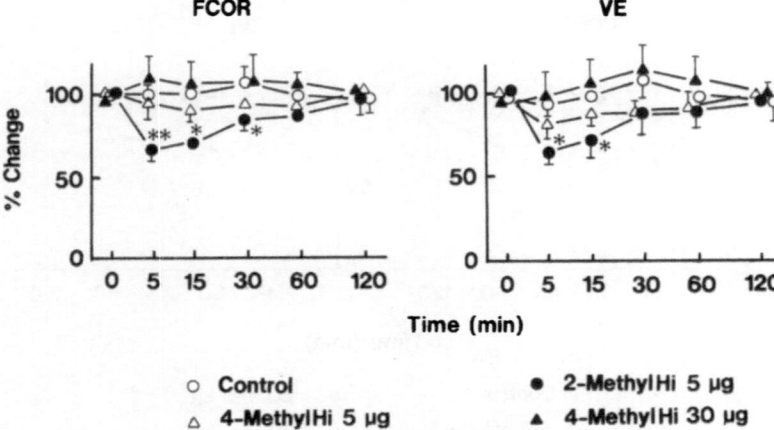

Fig. 8. Effects of 2-methylhistamine and 4-methylhistamine on EEG spectral power (delta band) recorded from the FCOR and VE after stimulation to the RF. ✱, ✱✱: Significantly different from control group with P < 0.05 and P < 0.01, respectively (N = 5). Reproduced from Tasaka, K., Chung, Y.H., Sawada, K. and Mio, M.: Brain Res. Bull., 22, 271-275 (1989), with permission.

7. Effects of H_1 and H_2 antagonists on histamine-induced decrease in EEG spectral powers

7.1. Effects of pyrilamine and diphenhydramine

It is known that pyrilamine and diphenhydramine cause drowsy EEG patterns when administered either i.c.v., i.v. or p.o. Therefore, it is necessary to determine the dose which causes no influence on EEG spectral power after a single i.c.v. injection in order to investigate the antagonistic effect of H_1 antagonists on the histamine-induced decrease in EEG powers. It was found that pyrilamine at a dose of $5.1\,\mu g$ and diphenhydramine at $4.6\,\mu g$ caused no significant effect on EEG power density in either the FCOR or VE (Fig. 9). When pyrilamine or diphenhydramine administered at doses of 5.1 or $4.6\,\mu g$ simultaneously with $2\,\mu g$ of histamine, the decrease in powers induced by histamine was prevented almost completely (Fig. 9); the doses of pyrilamine $(5.1\,\mu g)$ and diphenhydramine $(4.6\,\mu g)$ corresponded to $2\,\mu g$ of histamine in molar base. The arousal effect of histamine was antagonized by simultaneous application of H_1 antagonists which cause no significant effect when administered separately. It is well known that CNS depression is the most apparent side effect induced by H_1 antagonists, nevertheless, how H_1 antagonists produce their depressant effect was not known (Douglas, 1985). Now, it has become clear that the depressant effect on the CNS caused by H_1 antagonists is a result of the blocking of the H_1 receptors. Of course, in the case of acute H_1 antagonist poisoning, various side effects, which may be unrelated to histamine receptors, may appear.

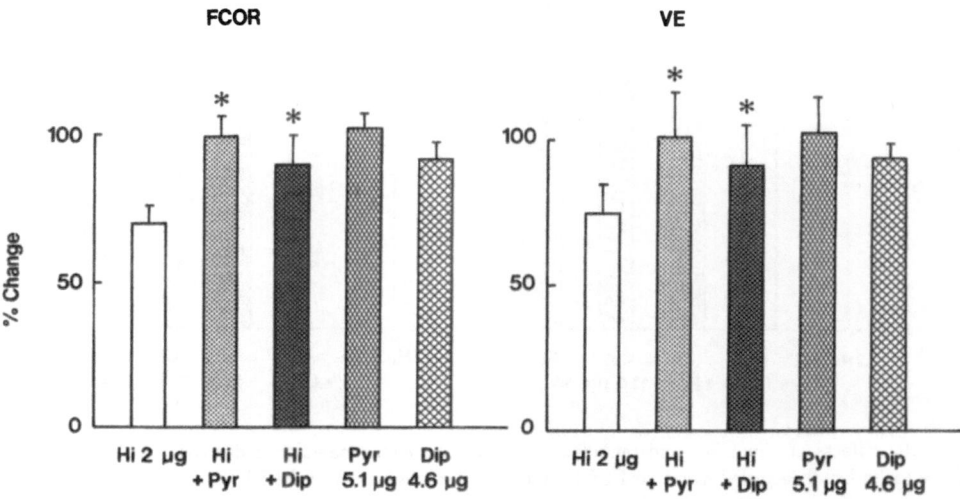

Fig. 9. Effects of pyrilamine and diphenhydramine on the histamine-induced decrease in EEG spectral power (delta band) recorded from the FCOR and VE after stimulation to the RF. Hi: Histamine; Pyr: Pyrilamine; Dip: Diphenhydramine. ✳: Significantly different from histamine-treated group with P < 0.05 (N = 5). Reproduced from Tasaka, K., Chung, Y.H., Sawada, K. and Mio, M.: Brain Res. Bull., 22, 271-275 (1989), with permission.

7.2. Effects of cimetidine and ranitidine

Cimetidine at a dose of 11.4 μg caused no significant changes in the powers, and when the same dose of cimetidine was administered with histamine, the effect of the latter was unaltered. The dose of cimetidine employed was equivalent to 5 μg of histamine in molar base, as was that of ranitidine (14.2 μg). The latter was effective neither in single nor in simultaneous administrations with histamine (Fig. 10).

It has often been reported that the administration of cimetidine causes somnolence, lethargy, confusion and hallucination, especially in patients with hepatic or renal diseases (Douglas, 1985). However, neither cimetidine nor ranitidine effectively inhibited the effect of histamine on EEG powers. The effect of histamine was antagonized by simultaneous administration of H_1 antagonists but not H_2 antagonists. Thus, the arousal effect of histamine is definitely exerted via H_1 receptors.

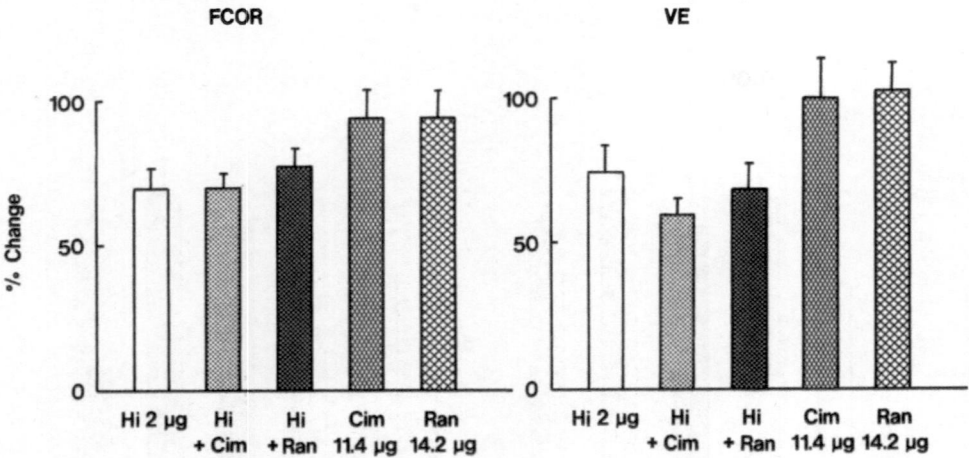

Fig. 10. Effects of cimetidine and ranitidine (i.c.v.) on the histamine-induced decrease in EEG spectral power (delta band) recorded from the FCOR and VE after stimulation to the RF (N = 5). Hi: Histamine; Cim: Cimetidine; Ran: Ranitidine. Reproduced from Tasaka, K., Chung, Y.H., Sawada, K. and Mio, M.: Brain Res. Bull., 22, 271-275 (1989), with permission.

8. Effects of H₃ agonist and antagonist on EEG spectral powers

8.1. Effect of thioperamide

I.v. injection of thioperamide caused a dose-related decrease in the powers of the FCOR and VE. Dose of 0.5 mg/kg caused no significant changes in the EEG power spectra. However, doses of 1 and 2 mg/kg elicited a significant decrease in power densities in the delta bands of both the FCOR and VE. The maximum effect in both areas was achieved 60 min after administration (Fig. 11). When thioperamide was administered i.c.v. at doses of 5 and 10 μg, EEG spectral power also decreased significantly 15 min after injection in the FCOR and VE (Fig. 13).

8.2. Effect of (R)-α-methylhistamine

I.v. injection of (R)-α-methylhistamine at doses of 1-5 mg/kg caused a dose-related increase in EEG power densities in the FCOR and VE, though no stimulation of the RF was performed. At a dose of 1 mg/kg, no significant increase was observed, however, at doses of 2 and 5 mg/kg, a significant effect was noted. The maximum effect was achieved 30 min after injection of (R)-α-methylhistamine (Fig. 12). Nearly the same effect was also observed 15 min after i.c.v. injection of (R)-α-methylhistamine; a significant effect was observed at doses of 10 and 20 μg both in the FCOR and VE (Fig. 13).

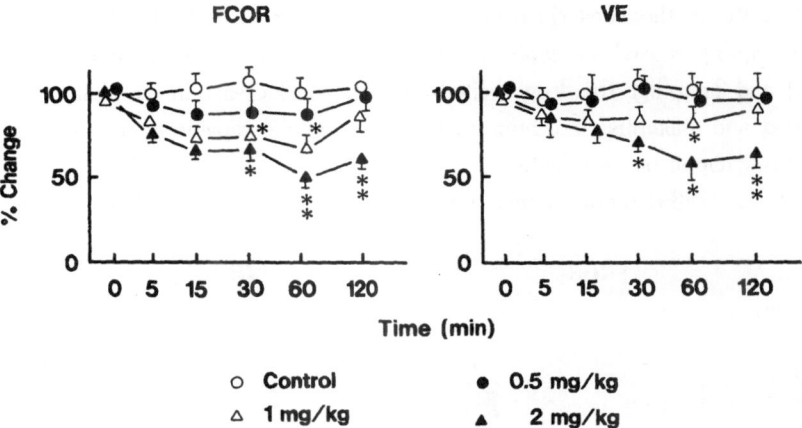

Fig. 11. Effect of thioperamide on EEG spectral power (delta band) recorded from the FCOR and VE after stimulation to the RF. *,**: Significantly different from control group with P < 0.05 and P < 0.01, respectively (N = 5).

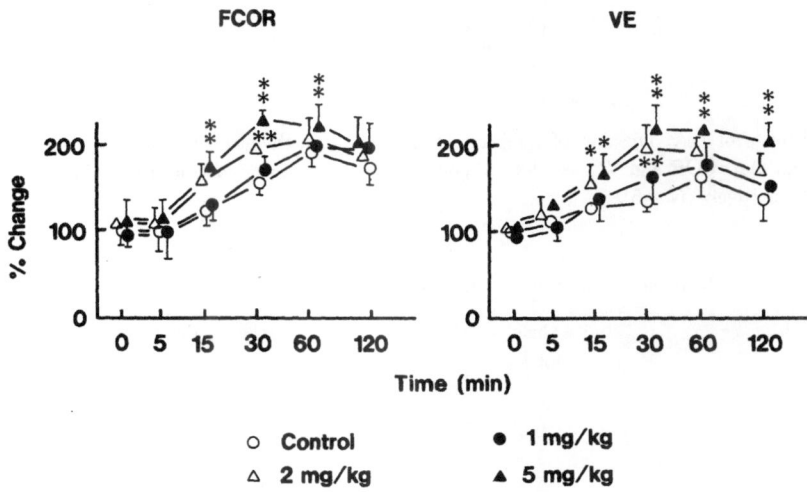

Fig. 12. Effect of (R)-α-methylhistamine on EEG spectral power (delta band) recorded from the FCOR and VE. *,**: Significantly different from control group with P < 0.05 and P < 0.01, respectively (N = 5).

8.3. Changes in brain histamine content after injection of thioperamide

Thioperamide at doses of 0.5-2 mg/kg, i.v. caused a dose-related decrease in the histamine content of the cortex and thalamus. Histamine contents were measured according to the method described previously (Kamei et al., 1993). Thioperamide at a dose of 0.5 mg/kg caused no significant decrease in the histamine content. However, at doses of 1 and 2 mg/kg, the drug elicited a significant decrease in the histamine content of the cortex and thalamus, indicating that thioperamide is capable of releasing histamine from the presynaptic nerve (Table 2).

Oishi et al. (1989) reported that thioperamide (2 mg/kg, i.p.) caused no significant

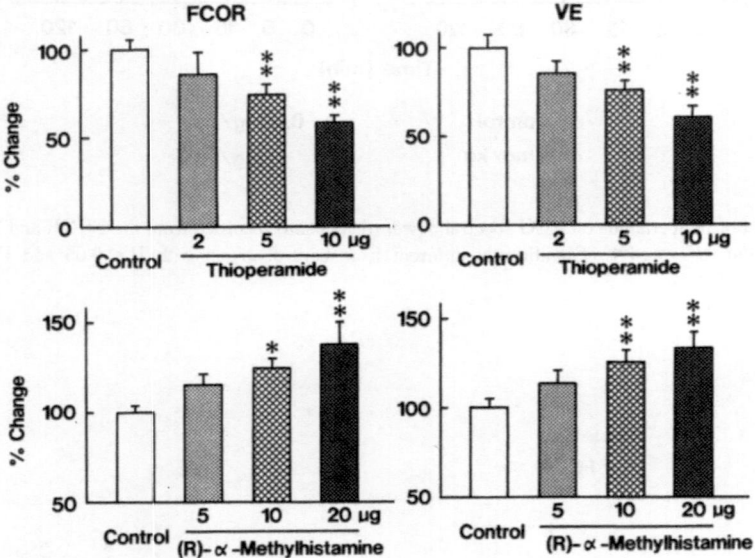

Fig. 13. Effects of i.c.v. injection of thioperamide and (R)-α-methylhistamine on EEG spectral power recorded from the FCOR and VE. *,** : Significantly different from control group with P < 0.05 and P < 0.01, respectively (N = 5).

Table 2. Effect of thioperamide (i.v.) on histamine content of the cortex and thalamus

Drug	Dose (mg/kg)	Histamine contents (ng/g tissue) Cortex	Thalamus
Control	–	26.6 ± 1.9	114.3 ± 8.2
Thioperamide	0.5	24.5 ± 1.2	112.6 ± 2.7
	1	20.0 ± 2.5 *	84.1 ± 7.5 *
	2	17.6 ± 2.7 **	77.0 ± 9.7 **

Histamine contents were measured 60 min after injection of the thioperamide. *,** : Significantly different from control group with P < 0.05 and P < 0.01, respectively (N = 5).

influence on the histamine content of the brain as a whole in mice. On the contrary, Sakai et al. (1992) found that thioperamide at a dose of 25 mg/kg (i.p.) elicited a significant decrease in the histamine content of the mouse brain from 2 to 8 hr after injection. Garbarg et al. (1989) also reported that thioperamide (5 mg/kg, p.o.), caused a significant and long-lasting decrease in the histamine content of rat cerebral cortex. As shown in Table 2, 1 and 2 mg/kg, i.v. injection of thioperamide caused a significant decrease in the histamine content. At the same doses, thioperamide caused a significant decrease in EEG spectral power as shown in Fig. 11.

8.4. Antagonistic effect of thioperamide on the increase in power densities induced by (R)-α-methylhistamine

The decrease in power density elicited by i.c.v. injection of thioperamide (10 μg) was significantly antagonized by an i.c.v. injection of 430 ng (a dose equivalent to 1/10 of thioperamide in molarity) and 4300 ng (a dose equivalent to 10 μg of thioperamide in molarity) of (R)-α-methylhistamine in the FCOR and VE. However, 43 ng (a dose equivalent to 1/100 of thioperamide in molarity) of (R)-α-methylhistamine had no significant influence (Fig. 14).

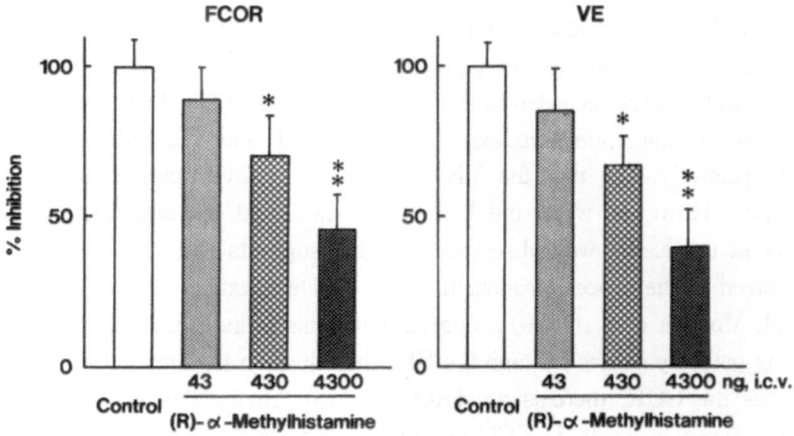

Fig. 14. Inhibitory effect of (R)-α-methylhistamine on the decrease in EEG spectral power induced by thioperamide (10 μg, i.c.v.) recorded from the FCOR and VE. $*$, $**$: Significantly different from control group with $P < 0.05$ and $P < 0.01$, respectively ($N = 5$).

Lin et al. (1990) reported that (R)-α-methylhistamine at doses of 10-20 mg/kg, p.o. caused a significant increase in deep slow wave sleep in cats, while thioperamide (2-10 mg/kg, p.o.) enhanced wakefulness in a dose-dependent manner. The arousal effect of thioperamide (2 mg/kg, p.o.) was prevented by pretreatment with (R)-α-methylhistamine (20 mg/kg, p.o.). This finding was almost identical to the results shown previously, although the species and the administration route were different. They also found that the arousal effects of thioperamide were prevented by pretreatment with pyrilamine, an H_1 receptor antagonist, indicating that the released histamine after thioperamide injection may cause the arousal response and the histamine-induced arousal may be blocked by pyrilamine treatment as described previously.

9. The pathway responsible for EEG synchronization and effect of histamine on this system

9.1. Changes in EEG spectral powers after electrical stimulation to the RF, VE or CM (Nucleus medialis centralis thalami)

To investigate the pathway accountable for EEG synchronization, changes in the EEG spectral powers recorded at the FCOR, VE, CM and RF after electrical stimulation to the RF, VE or CM were observed. EEG power density of the delta region recorded at the L-FCOR, L-VE, R-VE and CM increased significantly after stimulation of the L-RF, L-VE or CM (Table 3). When the same stimulation was applied to the L-VE or CM, EEG synchronization recorded at the L-FCOR was more evident than that seen after L-RF stimulation. These results suggest that either the VE or CM is more directly connected to the FCOR than the RF. Furthermore, while the EEG spectral powers recorded at the L-RF were also significantly increased by stimulation of the L-VE, no significant increase was induced by CM stimulation. L-RF stimulation also markedly increased spectral powers recorded in the ipsilateral VE (L-VE). These results seem to indicate that direct interconnections exist between the RF and VE, and between the VE and CM, respectively, and that the VE and CM may play important roles in EEG synchronization. However, when the L-RF was stimulated, no significant increase in EEG power in the L-RF was observed, and this suggests that the RF may not be directly involved in the process leading to EEG synchronization.

Although Monnier et al. (1970) proposed that in histamine-induced central activation the ascending pathway passes through the RF, then through the intralaminar nuclei, and finally reaches the COR, there is no direct evidence proving the existence of such a pathway. As shown in Table 3, EEG synchronizations of the contralateral VE and the CM were also observed after RF stimulation. These results suggest that the electrical signals eliciting EEG synchronization in the FCOR by RF stimulation were relayed not only at the ipsilateral VE but also at the contralateral VE and the CM. Supporting these findings was the fact that stimulation to the VE and CM induced significant EEG synchronization of the FCOR. These responses seem to be related to the augmenting and recruiting responses, respectively (Jasper, 1949; Mancia et al., 1971; Purpura et al., 1964; Tanaka et al., 1984). EEG synchronization was also found in the bilateral VE and CM when electrical stimulation was applied to the VE and CM. This suggests that a dense neuronal connection exists between the VE and CM. On the contrary, after electrical stimulation to the RF, EEG synchronization was not detected in the RF. These findings seem to indicate that the thalamus, not the RF, is essential for EEG synchronization. In association with this, it is known that electrical stimulation to the medial thalamus (recruiting response) (Arduini, 1958) or to the specific nuclei of the thalamus (augmenting response) (Purpura et al., 1964) induces EEG synchronization.

9.2. Changes in EEG spectral powers after the electrocoagulation of certain areas in the thalamus

To determine the most important area for inducing the EEG synchronization of the FCOR when electrical stimulation was applied to the RF, electrocoagulation of certain areas in the thalamus was performed. A decrease in EEG power densities recorded at the L-FCOR induced by RF stimulation was inhibited significantly after electrocoagulation of either the L-VE or the L&R-VE (Table 4). However, no discernible changes were observed in EEG powers of the L-FCOR when the L-RF was stimulated after lesioning the R-VE and CM. Also no changes were found in the EEG powers recorded at the CM when the R-VE was lesioned and the L-RF was stimulated. The EEG power densities recorded from the L-VE were not much influenced by lesioning of the R-VE and CM (Table 4). These results suggest that the ipsilateral VE play a crucial role in EEG synchronization when electrical stimulation is applied to the RF.

Recently, the reticulo-thalamo-cortical projection has been widely investigated by means of localized injection of horseradish peroxidase (HRP) and its modification (Menétrey, 1985; Nakano et al., 1985; Velayos and Reinoso-Suarez, 1982). Jones and

Table 3. Changes in EEG spectral powers recorded from the L-FCOR, L-VE, R-VE, CM and L-RF in association with stimulation to either the L-RF, L-VE or CM

Recording sites	Stimulation sites		
	L-RF	L-VE	CM
L-FCOR	$212 \pm 11**$	$251 \pm 12**$	$258 \pm 12**$
L-VE	$223 \pm 14**$	$232 \pm 17**$	$244 \pm 15**$
R-VE	$219 \pm 7**$	$217 \pm 7**$	$215 \pm 7**$
CM	$206 \pm 8**$	$204 \pm 12**$	$213 \pm 16**$
L-RF	133 ± 14	$147 \pm 5*$	134 ± 15

L-FCOR: Left frontal cortex; L-VE: Left nucleus ventralis thalami; R-VE: Right nucleus ventralis thalami; CM: Nucleus medialis centralis thalami; L-RF: Left reticular formation. The power density determined before electrical stimulation was taken as 100% and changes in EEG powers were expressed as percent changes. *, **: Significantly different from the values measured before stimulation with $P < 0.05$ and $P < 0.01$, respectively. From Tasaka, K., Chung, Y.H., Mio, M. and Kamei, C.: Brain Res. Bull., 32, 365-371(1993), with permission.

Yang (1985) studied autoradiographic study of anterograde axonal transport and found that when they were injected [³H] leucine into the brain stem (the reticularis mesence-phali), the greatest amount of radioactivity was transported into the forebrain and next greatest in the thalamus. Within the brain stem, fibers crossed the midline from the injection site to contralateral RF. Further, long ascending projections coursed ipsilateral-ly from the RF into the diencephalon with a smaller homologous projection on the contralateral side. This observations are almost in agreement with our results that electrical signals started from the RF were transferred mainly through the ipsilateral VE and synchronizing pulses are transferred to the FCOR.

Table 4. Changes in EEG power densities induced by RF stimulation after electrocoagulation was performed in certain areas of the thalamus

Lesion sites	Recording sites		
	L-FCOR	L-VE	CM
Control	212 ± 11	223 ± 14	227 ± 13
L-VE	103 ± 2**	—	99 ± 4**
R-VE	193 ± 5	215 ± 7	206 ± 8
L&R-VE	113 ± 4**	—	109 ± 4**
CM	197 ± 12	205 ± 27	—

—: Not tested; L-FCOR: Left frontal cortex; L-VE: Left nucleus ventralis thalami; R-VE: Right nucleus ventralis thalami; L&R-VE: Left and right nucleus ventralis thalami; CM: Nucleus medialis centralis thalami. The power densities determined before stimulation were taken as 100%. **: Significant-ly different from control animals with P < 0.01. From Tasaka, K., Chung, Y.H., Mio, M. and Kamei, C.: Brain Res. Bull., 32, 365-371 (1993), with permission.

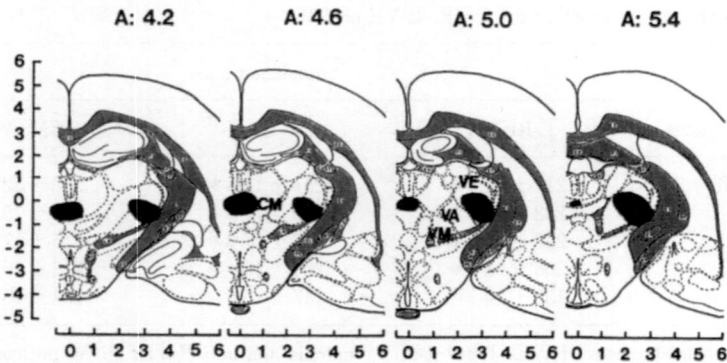

Fig. 15. Electrocoagulated areas in the VE and CM. Under pentobarbital anesthesia, electrocoagulation was made monopolarly. The black areas correspond to the injured regions. VE: Nucleus ventralis thalami; CM: Nucleus medialis centralis thalami. From Tasaka, K., Chung, Y.H., Mio, M. and Kamei, C.: Brain Res. Bull., 32, 365-371 (1993), with permission.

9.3. Effect of histamine in thalamic-lesioned rats

To clarify the importance of the ipsilateral VE in inducing EEG synchronization after stimulation to the RF, i.c.v. injection of histamine was performed in VE- and CM-lesioned rats. Fig. 15 represents the electrically lesioned areas in the VE and CM.

In VE-lesioned rats, no synchronization was elicited after RF stimulation. Furthermore, when i.c.v. injection of histamine at a dose of $2\,\mu g$ was performed, no significant changes in EEG power of the FCOR were induced by RF stimulation. In CM-lesioned rats, however, EEG synchronization was elicited by RF stimulation and i.c.v. injection of histamine caused a significant decrease in the EEG powers as that seen in normal rats (Fig. 16).

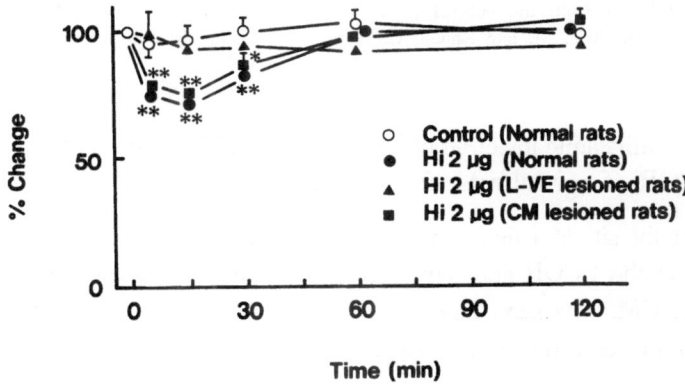

Fig. 16. Effect of histamine on EEG spectral power recorded at the FCOR after RF stimulation in rats lesioned at certain thalamic sites. $*, **$: Significantly different from the control group with $P < 0.05$ and $P < 0.01$, respectively. From Tasaka, K., Chung, Y.H., Mio, M. and Kamei, C.: Brain Res. Bull., 32, 365-371 (1993), with permission.

When histamine was injected with the equivalent dose of pyrilamine ($5.1\,\mu g$) in molar ratio, the desynchronization effect of histamine was completely abolished, though no influence was observed when the same dose of pyrilamine alone was i.c.v. injected. Simultaneous i.c.v. injection of histamine and cimetidine ($4.6\,\mu g$, corresponding to $2\,\mu g$ of histamine in molar ratio) induced desynchronization, as did the single i.c.v. injection of histamine (Fig. 17). As shown in Fig. 16, it was found that histamine was not effective in decreasing EEG power densities of the FCOR when the VE was lesioned, while CM-lesioned rats displayed the same EEG synchronization as seen in normal rats. These findings suggest that histamine-induced decrease in EEG spectral power is due to activation of the specific thalamic system and that the nonspecific thalamic system may not be involved.

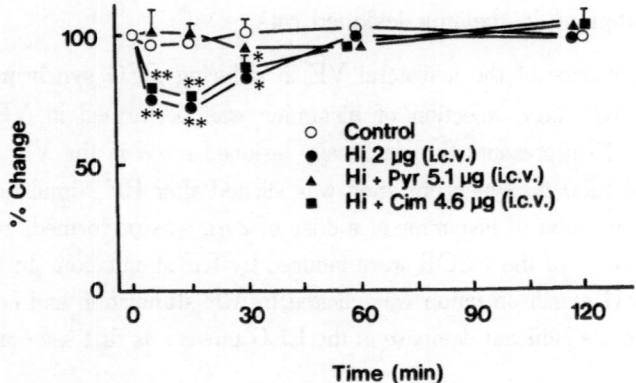

Fig. 17. Effects of pyrilamine and cimetidine on the histamine-induced decrease in EEG spectral power induced by RF stimulation in CM-lesioned rats. Hi: Histamine; Pyr: Pyrilamine; Cim: Cimetidine. *, **: Significantly different from the control group with $P < 0.05$ and $P < 0.01$, respectively. From Tasaka, K., Chung, Y.H., Mio, M. and Kamei, C.: Brain Res. Bull., 32, 365-371 (1993), with permission.

9.4. Effect of intrathalamic injection of histamine on EEG spectral powers recorded at the FCOR of normal rats after stimulation to the RF

In order to locate the site of action at which histamine causes a decrease in EEG spectral power recorded at the FCOR after stimulation to the RF, histamine was injected into either the VE or CM. The i.c.v. injection of histamine into normal rats at a dose of 1 μg caused, as previously reported, a significant decrease in the FCOR EEG power

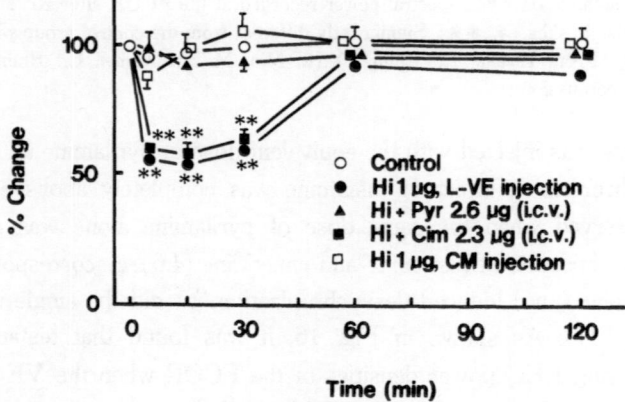

Fig. 18. Effect of intrathalamic injection of histamine on EEG spectral power recorded at the FCOR after RF stimulation and the effects of pyrilamine and cimetidine on EEG synchronization. L-VE: Left nucleus ventralis thalami. **: Significantly different from the control group with $P < 0.01$. From Tasaka, K., Chung, Y.H., Mio, M. and Kamei, C.: Brain Res. Bull., 32, 365-371 (1993), with permission.

induced by RF stimulation. At the same dose, histamine also caused a marked decrease in EEG power density in the FCOR when it was injected into the L-VE. The inhibitory effect lasted more than 30 min. As seen in the control animals, simultaneous injection of histamine (1 μg) and pyrilamine (2.6 μg) into the L-VE produced EEG synchronization after L-RF stimulation (Fig. 18). However, simultaneous injection of cimetidine (2.3 μg) did not influence the histamine effect. In addition, when the same dose of histamine was injected into the CM, no appreciable change was observed in the EEG powers recorded at the FCOR (Fig. 18).

9.5. Effect of intrathalamic injection of histamine on EEG spectral powers recorded at the FCOR after stimulation to the VE

When the same electrical stimulation was applied to the L-VE instead of the RF, a similar cortical synchronization was elicited. Histamine (1 μg) injection into the L-VE significantly decreased EEG power densities in the FCOR. As shown previously, when the equivalent dose of pyrilamine (2.6 μg) was injected simultaneously with histamine, pyrilamine abolished the histamine effect, while the simultaneous injection of cimetidine (2.3 μg) caused no significant changes in the histamine effect (Fig. 19). Furthermore, when the same dose of histamine was injected into the CM, no changes were observed in EEG power in the FCOR after VE stimulation (Fig. 19).

EEG synchronization in the FCOR induced by stimulation to the RF and the ipsilateral VE disappeared when histamine was injected into the VE, whereas histamine injection into the CM was not effective in decreasing EEG synchronization. These findings support our proposal that a histamine-induced decrease in EEG spectral power occurs through the activation of the specific thalamic system.

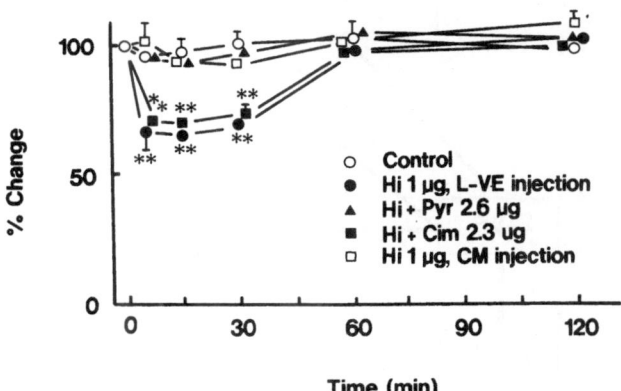

Fig. 19. Effect of intrathalamic injection of histamine on EEG spectral power recorded at the FCOR after VE stimulation and the effects of pyrilamine and cimetidine on EEG synchronization. **: Significantly different from the control group with P < 0.01. From Tasaka, K., Chung, Y.H., Mio, M. and Kamei, C.: Brain Res. Bull., 32, 365-371(1993), with permission.

10. Conclusion

To clarify whether or not the sedative effect of H_1 antagonists is exerted in relation to H_1 receptors in the brain, EEG processing was performed by the FFT method and displayed as compressed spectral arrays. When a train of low frequency electrical stimulation was applied to the RF of conscious rats, EEG spectral powers recorded at the FCOR and VE were increased, especially in the low frequency bands (0-6 Hz). The i.c.v. administration of histamine suppressed the increase in EEG power, and this inhibition was antagonized by simultaneous administration of pyrilamine or diphenhydramine, but not in combination with cimetidine or ranitidine. As in the case of histamine, the administration of 2-methylhistamine, an H_1 agonist, decreased power in the slow wave region, while administration of 4-methylhistamine, an H_2 agonists, did not. It was assumed that the arousal effect of histamine is exerted via H_1 but not related to H_2 receptors. Adverse effects of H_1 antagonists, such as sedation and drowsiness, may be induced by their inhibitory effect on histamine-induced arousal response which appears through H_1 receptors.

Thioperamide, an H_3 antagonist, caused a similar effect to that of histamine when the drug was administered either i.v. or i.c.v. On the other hand, (R)-α-methylhistamine, a representative H_3 agonist, caused an increase in EEG power as in the case of pyrilamine and diphenhydramine. An increase in powers elicited by (R)-α-methylhistamine was dose-dependently antagonized by thioperamide. It is certain that the released histamine elicited by thioperamide caused an excitatory effect on the arousal system via the H_1 receptors.

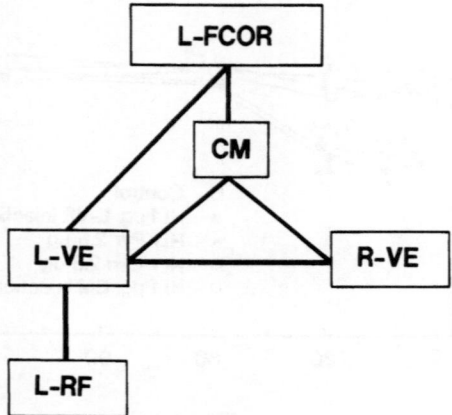

Fig. 20. Scheme of the EEG synchronization induced by RF stimulation. L-RF: Left reticular formation; L-FCOR: Left frontal cortex; L-VE: Left nucleus ventralis thalami; R-VE: Right nucleus ventralis thalami; CM: Nucleus medialis centralis thalami. From Tasaka, K., Chung, Y.H., Mio, M. and Kamei, C.: Brain Res. Bull., 32, 365-371(1993), with permission.

Electrical stimulation (3 Hz, 0.5 V) to the RF of conscious rats induced significant increase of EEG power densities (synchronization) recorded at the FCOR, VE or CM. Significant synchronization was also observed in the FCOR when electrical stimulation was applied to the VE and CM. When ipsilateral and bilateral VEs were electrocoagulated, no EEG synchronization was observed in the FCOR and CM. I.c.v. administration of histamine caused a marked suppression of FCOR EEG synchronization in both CM-lesioned and normal rats through H_1 receptors. EEG synchronization in FCOR was not induced in ipsilateral or bilateral VE-coagulated rats after RF stimulation. When histamine (1 μg) was injected into the VE of normal rats, EEG synchronization of FCOR was markedly reduced after RF or VE stimulation. No such changes were induced when histamine was injected into the CM. These findings may further support the hypothesis that there is a direct connection between the RF and VE or between the VE and FCOR, but not between the RF and CM. It can be assumed that RF stimulation is mainly projected to the VE, after which the synchronization pulses are transferred to the FCOR. However, in the other pathway, the CM is stimulated by pulses that pass through the L-VE, after which the pulses are transferred to the FCOR through an indirect pathway (Fig. 20).

References

Arduini, A.: Enduring potential changes evoked in the cerebral cortex by stimulation of brainstem reticular formation and thalamus. In: Jasper, H.H. et al., (ed) Reticular formation of the brain. Boston: Little Brown and Co., pp. 333-351, 1958

Arrang, J.-M., Garbarg, M. and Schwartz, J.-C.: Auto-inhibition of brain histamine release mediated by a novel class (H_3) of histamine receptor. Nature, 302, 832-837 (1983)

Arrang, J.-M., Garbarg, M. and Schwartz, J.-C.: Autoinhibition of histamine synthesis mediated by presynaptic H_3-receptors. Neuroscience, 23, 149-157 (1987)

Carruthers, S.G., Shoeman, D.W., Hignite, C.E. and Azarnoff, D.L.: Correlation between plasma diphenhydramine level and sedative and antihistamine effects. Clin. Pharmacol. Ther., 23, 375-382 (1978)

Cooley, J.W. and Tukey, J.W.: An algorithm for the machine calculation of complex Fourier series. Math. Comput., 19, 297-301 (1965)

Douglas, W.W.: Histamine and 5-hydroxytryptamine (serotonin) and their antagonists. In: Gilman, A.G., Goodman, L.S., Rall, T.W., Murad, F. (ed) Goodman and Gilman's The Pharmacological Basis of Therapeutics, 7th ed. New York, Macmillan Publishing Co., pp. 605-638, 1985

Favale, E., Loeb, C., Rossi, G.F. and Sacco, G.: EEG synchronization and behavioral signs of sleep following low frequency stimulation of the brain stem reticular formation. Acta. Ital. Biol., 99, 1-22 (1961)

Feldberg, W. and Sherwood, S.L.: Injections of drugs into the lateral ventricle of the cat. J. Physiol., 123, 148-167 (1954)

Garbarg, M., Trung Tuong, M.D., Gros, C. and Schwartz, J.C.: Effects of histamine H_3-receptor ligands on various biochemical indices of histaminergic neuron activity in rat brain. European J. Pharmacol., 164, 1-11 (1989)

Heinrich, M.A. Jr.: The effects of antihistaminic drugs on the central nervous system in rats and mice. Arch. int. Pharmacodyn., 92, 446-463 (1953)

Jasper, H.: Diffuse projection systems: The integrative action of the thalamic reticular system. EEG Clin. Neurophysiol., 1, 405-420 (1949)

Jones, B.E. and Yang, T.-Z.: The efferent projections from the reticular formation and the locus coeruleus studied by anterograde and retrograde axonal transport in the rat. J. Comp. Neurol., 242, 56-92 (1985)

Kalivas, P.W.: Histamine-induced arousal in the conscious and pentobarbital-pretreated rat. J. Pharmacol. Exp. Ther., 222, 37-42 (1982)

Kamei, C., Chung, Y.H., Dabasaki, T. and Tasaka, K.: Species differences elicited by intraventricular injection of histamine on EEGs and behavior. Japan. J. Pharmacol., 31 (suppl), 86p (1981)

Kamei, C., Dabasaki, T. and Tasaka, K.: Effect of intraventricular injection of histamine on the pinna reflex in mice. Japan. J. Pharmacol., 35, 193-195 (1984)

Kamei, C., Akahori, H. and Tasaka, K.: Influence of histamine and related compounds on the hypnotic effect of thiopental in mice. J. Pharmacobio-Dyn., 9, 112-116 (1986)

Kamei, C., Okumura, Y. and Tasaka, K.: Influence of histamine depletion on learning and memory recollection in rats. Psychopharmacology, 111, 376-382 (1993).

Lin, J.-S., Sakai, K., Vanni-Mercier, G., Arrang, J.-M., Garbarg, M., Schwartz, J.-C. and Jouvet, M.: Involvement of histaminergic neurons in arousal mechanisms demonstrated with H_3-receptor ligands in the cat. Brain Res., 523, 325-330 (1990)

Mancia, M., Avanzini, G., Caccia, M. and Rocca, E.: Recruiting responses following splitting of the brain-stem in cats. EEG Clin. Neurophysiol., 31, 259-268 (1971)

Menétrey, D.: Retrograde tracing of neural pathways with a protein-gold complex. I. Light microscopic detection after silver intensification. Histochemistry, 83, 391-395 (1985)

Monnier, M., Sauer, R. and Hatt, A.M.: The activating effect of histamine on the central nervous system. Int. Rev. Neurobiol., 12, 265-305 (1970)

Monnier, M. and Herkert, B.: Variations of histamine concentration in sleep and arousal hemodialysates (seasonal, thalamic and reticular influences). Arch. int. Pharmacodyn., 228, 39-49 (1977)

Moruzzi, G. and Magoun, H.W.: Brain stem reticular formation and activation of the EEG. EEG Clin. Neurophysiol., 1, 455-473 (1949)

Nakano, K., Kohno, M., Hasegawa, Y. and Tokushige, A.: Cortical and brain stem afferents to the ventral thalamic nuclei of the cat demonstrated by retrograde axonal transport of horseradish peroxidase. J. Comp. Neurol., 231, 102-120 (1985)

Oishi, R., Itoh, Y., Nishibori, M. and Saeki, K.: Effects of the histamine H_3-agonist (R)-α-methylhistamine and the antagonist thioperamide on histamine metabolism in the mouse and rat brain. J. Neurochem., 52, 1388-1392 (1989)

Purpura, D.P., Shofer, R.J. and Musgrave, F.S.: Cortical intracellular potentials during augmenting and recruiting responses. II. Patterns of synaptic activities in pyramidal and nonpyramidal tract neurons. J. Neurophysiol., 27, 133-151 (1964)

Rossi, G.F.: Brain stem facilitating influences on EEG synchronization. Experimental findings and observations in man. Acta Neurochir., 13, 257-288 (1965)

Sakai, N., Sakurai, A., Sakurai, E., Yanai, K., Maeyama, K. and Watanabe, T.: Effects of the histamine H_3 receptor ligands thioperamide and (R)-α-methylhistamine on histidine decarboxylase activity of mouse brain. Biochem. Biophys. Res. Commun., 185, 121-126 (1992)

Schwartz, J.C., Lampart, C. and Rose, C.: Histamine formation in rat brain in vivo: Effects of histidine loads. J. Neurochem., 19, 801-810 (1972)

Schwartz, J.-C.: Histamine as a transmitter in brain. Life Sci., 17, 503-518 (1975)

Tasaka, M., Kojima, H. and Akashi, A.: An electroencephalographical study on timiperone, a new antipsychotic drug. Folia pharmacol. japon., 84, 213-219 (1984)

Tasaka, K., Kamei, C., Katayama, S., Kitazumi, K., Akahori, H. and Hokonohara, T.: Comparative study of various H_1-blockers on neuropharmacological and behavioral effects including 1-(2-ethoxyethyl)-2-(4-methyl-1-homopiperazinyl) benzimidazole difumarate (KB-2413), a new antiallergic agent. Arch. int. Pharmacodyn., 280, 275-291 (1986)

Tasaka, K., Chung, Y.H., Sawada, K. and Mio, M.: Excitatory effect of histamine on the arousal system and its inhibition by H_1 blockers. Brain Res. Bull., 22, 271-275 (1989)

Tasaka, K., Chung, Y.H., Mio, M. and Kamei, C.: The pathway responsible for EEG synchronization and effect of histamine on this system. Brain Res. Bull., 32, 365-371 (1993)

Velayos, J.L. and Reinoso-Suarez, F.: Topographic organization of the brainstem afferents to the mediodorsal thalamic nucleus. J. Comp. Neurol., 206, 17-27 (1982)

Chapter 2
The role of histamine on learning and memory

1. Introduction

It is well known that the central cholinergic system is intimately related to the modulation of learning (acquisition) and memory processes (retention and retrieval) in humans and animals (Haroutunian et al., 1985). Actually, there are many reports about the ameliorating effects of cholinergic drugs on memory impairment induced by transient cerebral ischemia (Yamamoto et al., 1987), electroshock (Sakurai et al., 1989) or anticholinergic drug administration (Yamamoto and Shimizu., 1987). On the other hand, it has been reported that classic H_1 antagonists such as diphenhydramine, pyrilamine and promethazine provided potent depressant actions on the central nervous system (CNS) in many different situations including learning and memory (Winter and Flataker, 1951). Kamei et al. (1981a) demonstrated that diphenhydramine and promethazine induced not only a drowsy pattern in EEGs characterized by high voltage and slow waves but also an inhibition of the EEG arousal response induced by electrical stimulation of the midbrain reticular formation. Tasaka et al. (1985) reported that diphenhydramine, pyrilamine and promethazine caused a potent suppression of the two-way conditioned avoidance response (retardation of memory retrieval) in rats. In connection with this, de Almeida and Izquierdo (1988) found that immediate posttraining after intracerebroventricular (i.c.v.) administration of histamine at doses of 1 and 10 ng facilitated retention performance of step-down inhibitory avoidance behavior (passive avoidance response) measured 24 hr after drug administration in rats. In addition, this histamine-induced effect was inhibited by the simultaneous administration of promethazine (1000 ng) and cimetidine (1000 ng), indicating that histamine-induced facilitation of memory retention is mediated via both H_1 and H_2 receptors. Bhattacharya (1990) also reported that histamine at doses of 1, 5 and 10 μg produced a dose related dual effect on learning acquisition and retention of memory, with the lower two doses facilitating and high dose retarding the memory paradigms. In these two studies, however, normal animals were used for estimating the drug effect. This chapter provides information about the role of histamine on learning and memory not only in normal but also in some amnesic animals.

Recently, it has been reported that newly developed H_1 antagonists such as azelastine, oxatomide and astemizole provided potent suppression of allergic reactions, especially type I hypersensitivity (Tasaka, 1986). These drugs are generally recognized to cause little CNS depressant actions different from classic H_1 antagonists. The reason why newer H_1 antagonists showed less CNS depressant activity can be ascribed to their possessing less affinity to the brain tissue (Laduron et al., 1983; Ahn and Barnett, 1986) or their being less liable to cross the blood-brain barrier (Soda et al., 1981).

In this chapter, the effects of certain H_1 antagonists including newly developed drugs on learning and memory are also discussed.

2. One-way active avoidance and passive avoidance

In animal studies, to investigate the function of learning and memory, active avoidance and passive avoidance procedures have been commonly used. One-trial step-through type passive avoidance response has been widely employed for evaluating the effect of cerebral metabolic activators (Yamazaki et al., 1984; Ader et al., 1972; Rigter et al., 1974). A one-way active avoidance response has the following characteristic properties which differ from a one-way passive avoidance procedure. Firstly, when rats were placed in a lighted room to test passive avoidance every 1 hr over 6 trials, 3 to 4 out of 10 rats did not move into the dark compartment at the 5th or 6th trial, even after the limited time of 300 sec. This can be ascribed to adaptation. On the contrary, in active avoidance procedure, when the rats were placed in the dark compartment initially, none of them moved spontaneously into the lighted compartment even after a number of trials, since rats prefer to stay in a dark place. This clearly indicates that no adaptation takes place in active avoidance test. Secondary, in one-way active avoidance task, the maximum avoidance response can be achieved after about 10-15 training sessions in all of the test rats, suggesting that since the task is not only simple but also straightforward, it can be quickly and completely trained compared with two-way shuttle box avoidance response (Oka and Shimizu, 1975). On the other hand, the training (acquisition) trial in testing for passive avoidance response was a single session, therefore the task was very simple and quickly acquired. However, the question has been raised as to whether or not this task accurately reflects the learning process and memory of the animals. Based on these findings, a one-way active avoidance procedure was used in the study.

The apparatus was divided into 2 compartments, a $19 \times 15 \times 14$ cm lighted room and a $33 \times 20.5 \times 14$ cm dark room, linked by a guillotine door (5×5 cm). The lighted compartment was painted white, had a flat floor and was illuminated by a lamp. The floor of the dark compartment consisted of a series of copper rods (3 mm in outside diameter), arranged side by side, 1.0 cm apart, through which electroshock can be delivered (40 V, 0.2 mA). When the animals were trained, rats were placed in the dark compartment and unless the animals moved into the lighted side within 5 sec the guillotine door was closed and an electric shock was delivered to the feet for 5 sec. The acquisition training was repeated once a day for 10-15 consecutive days. To evaluate drug effect, the latency before rat moves into the lighted side was measured (Kamei et al., 1990a).

3. Age-related changes in acquisition and retention

In aged animals, impaired acquisition has been recognized in several learning procedures such as passive avoidance, shuttle box avoidance and maze learning (Ishikawa et al., 1981; Inami et al., 1990; Rüthrich et al., 1982). Similarly, in the one-way active avoidance, as shown in Fig. 1, the latencies measured in 6- and 12-month-old rats were longer than those in 2-month-old rats during the acquisition process, indicating that 6- and 12-month-old rats showed a lower acquisition than 2-month-old rats. In younger rats, the latency decreased even in the 2nd trial and it became shorter, rather rapidly reaching a minimum within 10 days. A marked delay in acquisition of active avoidance response was seen in 12-month-old rats, and the amount of time required for acquisition in 6-month-old rat was between that of 2-month-old rat and 12-month-old rat groups. However, there is no significant difference between the 6-month-old rat group and the 12-month-old rat group (Fig. 1).

Nakagawa et al. (1989) reported that in the shuttle box active avoidance procedure, aged rats demonstrated the escape failure that is associated with age-related motor disturbances. On the other hand, Ikeda and Matsushita (1985) demonstrated that aged rats showed a lowering in learning behavior regardless of motor function impairment. Wallace et al. (1980) reported that complex and demanding levels of motor and reflexive functioning behaviors, such as suspension from a horizontal wire, descent of a wire mesh pole, traversal of an elevated platform and rotarod performance declined markedly with an increase in age, although little or no change was seen in simple reflexive behaviors, such as placing, hopping and negative geotaxis, surface righting and mid-air righting

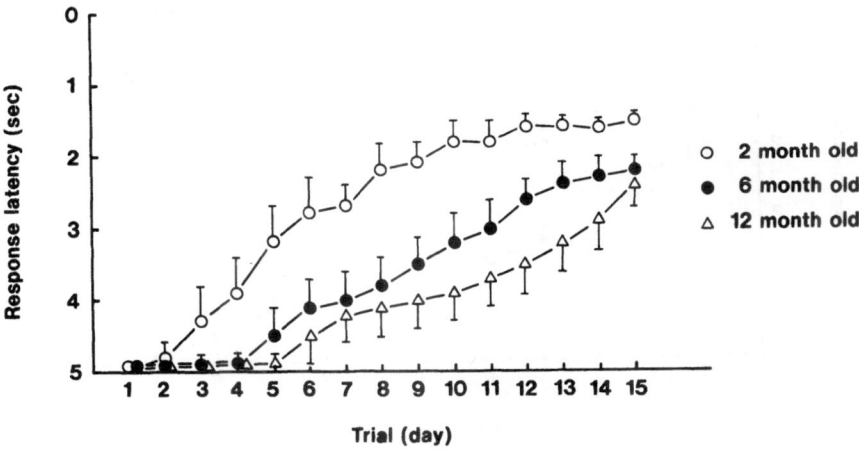

Fig. 1. Influence of aging on the acquisition of the active avoidance response in rats. From Kamei, C., Tsujimoto, S. and Tasaka, K.: J. Pharmacobio-Dyn., 13, 772-777 (1990a), with permission.

reactions during the course of aging in Fischer 344 albino rats. In the one-way active avoidance task, there are no significant differences among the 3 different age groups in the ability of alley progression and sensitivity to the electroshock. That is, the time which was required for the rats to move into the dark room when they were placed in the lighted room was the same, regardless of age (alley progression: 2-month-old rats, 11.5 ± 1.8 sec; 6-month-old rats, 12.0 ± 2.7 sec; 12-month-old rats, 13.0 ± 2.0 sec), and when rats were received the electroshock in the dark room (40 V, 0.2 mA), all the animals in the 3 different age groups moved to the lighted room immediately.

To study how durably information is stored after acquisition was completed with 15 trials, the active avoidance test was repeated once a week (retention test). Response latency was prolonged in all groups as time passed, especially in the older groups, indicating retention of the active avoidance response was disturbed in old rats. Significant difference exists between younger rat group (2-month-old rats) and older rat group (6-month-old and 12-month-old rats) in the prolongation of response latency. There is, however, no significant difference between 6-month-old rats and 12-month-old rats (Fig. 2). Similar results were obtained in the brightness discrimination test. (Rüthrich et al., 1982). Lippa et al. (1980) also reported that in the passive avoidance task, retention was considerably diminished as a result of aging in rats. In the passive avoidance procedure, when memory retention was disturbed, response latency decreased differently from the active avoidance procedure. When 24 hr retention latencies were measured in rats of various ages, response latencies on the test day were inversely correlated with age; the decrease in latency was first observed at approximately 15 months of age with maximal effects occurring after 20 months of age.

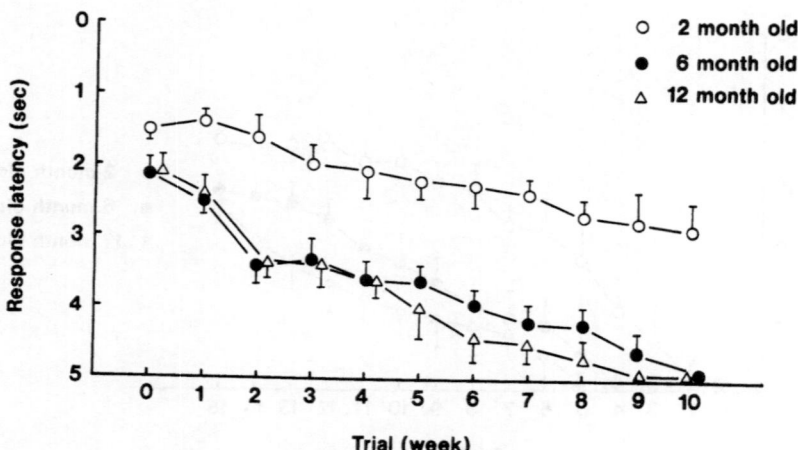

Fig. 2. Influence of aging on the retention of the active avoidance response in rats. From Kamei, C., Tsujimoto, S. and Tasaka, K.: J. Pharmacobio-Dyn., **13**, 772-777 (1990a), with permission.

4. Recovery of memory deficit in relation with a long interruption of training in old rats

4.1. Effect of acetylcholine

For this purpose, rats 12 months of age were used after 6 months of interruption from the last acquisition trial. When rats were placed in the dark compartment, some moved into the lighted side after 5 to 10 sec, while some loitered around the border line (guillotine door). I.c.v. injection of acetylcholine (ACh) into old rats at doses of 5 and 10 ng caused a decrease, though not significant, in response latency. At doses of 20 and 50 ng, i.c.v. the drug elicited a significant shortening of response latency, indicating that ACh remarkably facilitates memory retrieval (Fig. 3).

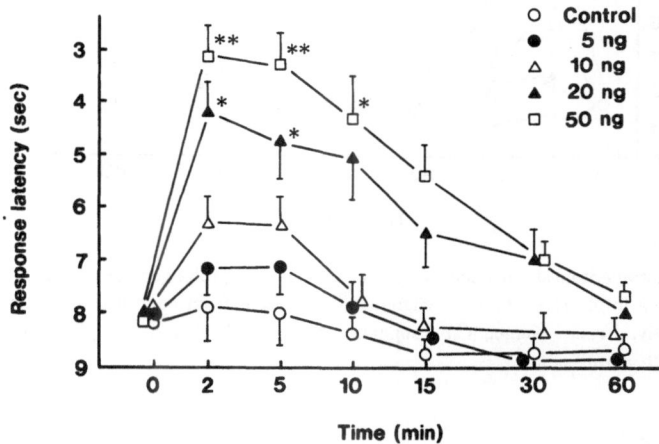

Fig. 3. Effect of intracerebroventricular injection of acetylcholine on prolonged latency induced by a long interruption of training in old rats. *, **: Significantly different from control group with $P < 0.05$ and $P < 0.01$, respectively. From Kamei, C. and Tasaka, K.: Biol. Pharm. Bull., 16, 128-132 (1993b), with permission.

4.2. Effects of cholinergic drugs

It has been shown that physostigmine at a dose of 0.01 mg/kg, i.p. caused no appreciable changes. However, the drug showed a significant facilitation of memory retrieval at doses of 0.02 and 0.05 mg/kg. At a dose of 0.02 mg/kg, significant effect was noted only transiently 15 min after administration, but at 0.05 mg/kg, significant effect was observed 15 and 30 min after drug administration. At a dose of 0.1 mg/kg, no improvement in memory retrieval was observed (Fig. 4). Arecoline also had a facilitation of memory retrieval. At doses of 0.2 and 0.5 mg/kg, i.p., the drug caused a significant effect, while at a dose of 1.0 mg/kg, the drug caused no significant effect (Fig. 5).

Haroutunian et al. (1985) reported that physostigmine (0.03 mg/kg, i.p.) and arecoline (1.0 mg/kg, i.p.) enhanced memory retention for 72 hr of rats tested in passive avoidance response. Egashira et al. (1990) reported that the number of muscarinic receptors, the V_{max} values of acetylcholinesterase, choline acetyltransferase and choline uptake in different brain regions of aged rats (24-month-old), were significantly dimini-

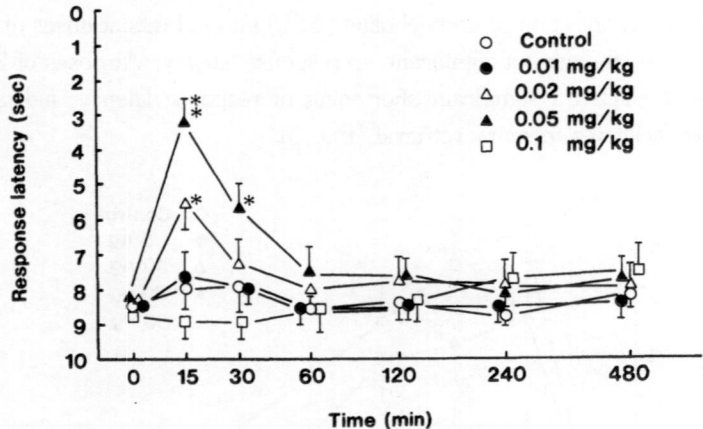

Fig. 4. Effect of intraperitoneal injection of physostigmine on prolonged latency induced by a long interruption of training in old rats. *, **: Significantly different from control group with $P < 0.05$ and $P < 0.01$, respectively. From Kamei, C., Tsujimoto, S. and Tasaka, K.: J. Pharmacobio-Dyn., 13, 772-777 (1990a), with permission.

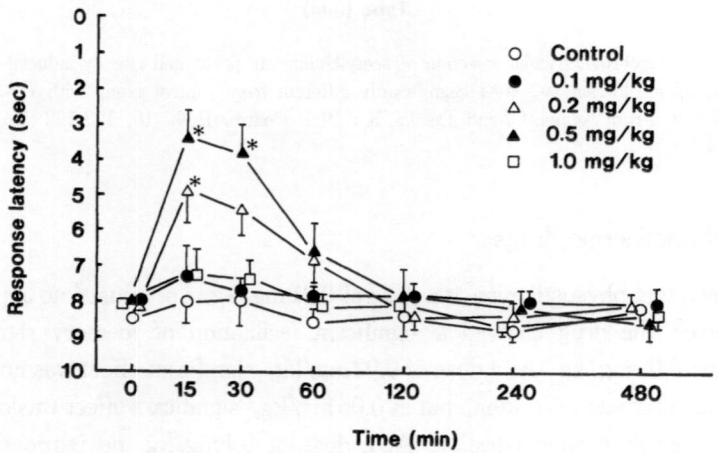

Fig. 5. Effect of intraperitoneal injection of arecoline on prolonged latency induced by a long interruption of training in old rats. *: Significantly different from control group with $P < 0.05$. From Kamei, C., Tsujimoto, S. and Tasaka, K.: J. Pharmacobio-Dyn., 13, 772-777 (1990a), with permission.

shed compared with those found in younger rats (2-month-old). Decreased activities of choline acetyltransferase and acetylcholinesterase of the cortex, degeneration of the cholinergic cells of the basal forebrain nucleus of Meynert, and degeneration of the cholinergic cells of the medial septal nucleus have been observed in clinical research on Alzheimer's disease (Whitehouse et al., 1982; Coyle et al., 1983; Perry et al. 1981). These findings supported this result that ACh (i.c.v.) and cholinergic drugs caused a remarkable facilitation of memory retrieval in old rats. These pharmacological findings strongly suggest that there is an intimate relationship between the cholinergic mechanism and age-related memory deficiency. As shown in Figs. 4 and 5, the dose-response curves for physostigmine and arecoline were bell-shaped. Similarly, other investigators have observed same dose-response relation for some cholinergic drugs, though no clear explanation is given for such complicated relationships (Yamamoto et al., 1987; Sakurai et al., 1989). However, Haroutunian et al. (1985) reported that their cholinergic effects on the peripheral nervous system and the resulting tremors may be the reasons for such relationships.

4.3. Effects of cerebral metabolic activators

Recently, some cerebral metabolic activators have been developed, and used in clinics to treat cerebral vascular damage, head injury, Alzheimer's disease and senile dementia (Yamamoto et al., 1987; Yamazaki et al., 1984). Therefore, effects of some cerebral metabolic activators on this memory deficit were investigated. As shown in Table 1, hopantenate calcium significantly diminished the response latency at a dose of 100 mg/kg (p.o.) 60 and 90 min after administration. Oral administration of idebenone was not effective even at a dose of 200 mg/kg, but significant reduction in response latency was elicited at doses of 20 and 50 mg/kg when the drug was injected i.p. Indeloxazine had some effect at a dose of 50 mg/kg (p.o.), but no such effect was observed at lower doses. Nefiracetam, a cognition enhancer (Tanaka et al., 1990) also caused a significant decrease in response latency at a dose of 30 mg/kg (p.o.).

Nakahiro et al. (1985) reported that calcium hopantenate facilitates cholinergic function in the CNS; the drug prevented scopolamine- and atropine-induced locomotor activities in mice and the binding of $[^3H]$ muscimol to rat brain membrane. Furthermore, the drug dose-dependently enhanced K^+-induced release of $[^3H]$ ACh from slices of the cerebral cortex and hippocampus. As for indeloxazine, Yamamoto et al. (1987) described how the drug antagonized amnesia induced by scopolamine with a bell-shaped dose-response curve similar to that of physostigmine. Nefiracetam also improved scopolamine-induced amnesia in the step-through passive avoidance task when administered before the training trial (Sakurai et al., 1989). Moreover, Kakihana et al. (1984) found that in ischemic rats, a significant decrease in ACh and a marked increase in choline were observed in the cerebral cortex, hippocampus, striatum and diencephalon. They also

Table 1. Effects of some metabolic activators on the prolonged latency induced by a long interruption of training in old rats

Drugs	Dose (mg/kg, p.o.)	Response latency (sec)						
		0	30	60	90	120	180	240
Control	–	8.5±0.5	7.8±0.5	7.7±1.0	8.0±0.7	8.4±0.3	8.4±0.2	8.0±0.5
Hopantenate	50	8.0±0.9	7.7±1.1	6.1±1.8	7.7±1.2	8.2±0.7	8.5±0.4	8.4±0.5
calcium	100	7.9±0.4	7.6±0.7	5.4±1.1*	5.3±1.4*	6.1±1.8	7.4±1.2	7.7±1.3
	200	8.1±0.3	7.4±0.9	6.3±1.2	6.5±1.3	7.5±0.9	8.0±0.3	7.8±0.6
	500	8.0±0.4	8.6±0.1	8.6±0.1	8.6±0.1	8.6±0.1	8.6±0.1	8.6±0.1
Idebenone	200	8.0±0.6	7.7±0.9	8.7±0.1	8.7±0.1	7.2±1.3	7.1±1.4	7.1±1.4
Indeloxazine	10	8.0±0.6	7.4±1.3	8.4±0.3	7.9±0.8	8.4±0.3	7.6±1.0	8.0±0.8
	20	8.0±0.5	7.2±1.0	7.0±1.0	7.0±1.0	7.6±0.9	8.2±0.7	8.1±0.9
	50	7.8±0.2	7.2±0.6	6.6±0.5*	6.3±0.6*	6.9±0.6	7.5±0.3	7.8±0.2
Nefiracetam	3	8.1±0.9	7.8±1.4	7.0±1.5	7.0±1.4	8.0±0.3	8.4±1.0	8.7±0.4
	10	8.0±0.5	6.8±1.0	6.3±1.2	7.7±0.5	7.6±0.6	8.2±0.3	8.0±0.5
	30	7.8±0.4	6.5±0.9	5.4±1.1*	6.9±0.8	7.5±0.4	8.1±0.9	8.0±0.5

Drugs	Dose (mg/kg, i.p.)	Response latency (sec)						
		0	15	30	60	90	120	240
Control	–	8.0±0.6	8.1±0.8	8.6±0.8	8.4±0.5	8.6±0.3	8.8±0.2	8.2±0.4
Idebenone	10	8.5±0.1	7.8±0.7	6.4±1.6	7.3±1.6	8.4±1.2	8.5±0.1	8.5±0.1
	20	8.3±0.6	7.0±0.4	5.8±1.2*	6.5±1.2	7.5±1.0	8.4±1.2	8.4±1.2
	50	8.5±0.8	6.3±1.4	4.9±1.0*	6.4±1.0	7.4±0.8	8.6±0.4	8.7±0.3

*: Significantly different from the control group with $P < 0.05$. From Kamei, C., Tsujimoto, S. and Tasaka, K.: J. Pharmacobio-Dyn., 13, 772-777 (1990a), with permission.

found that pretreatment with idebenone inhibited a decrease in ACh and an increase in choline in the forebrain regions, although in normal rats the drug did not alter the concentrations of ACh and choline in those brain regions. It seems likely that all the drugs tested demonstrate cholinergic activities. Therefore, drugs demonstrating a facilitating effect of memory retrieval in old rats may have some cholinergic properties. However, the question has been arisen whether or not this ameliorating effect of the drugs may be due to the central stimulant activity of the tested compounds. However, it has not been shown that cerebral metabolic activators used in this study caused a stimulant action on the CNS such as increase in locomotor activity.

4.4. Effects of histamine and histidine

In the case of histamine, i.c.v. injection at doses of 10 and 20 ng reduced the response latency moderately but not significantly. However, at doses of 50 and 100 ng the drug elicited a significant shortening of the response latency (Fig. 6). I.p. injection of histidine at doses of 200 and 500 mg/kg, also caused a shortening of the response latency. At a dose of 500 mg/kg, a significant effect was observed from 15 to 90 min after the drug injection (Fig. 7).

It is assumed that when histamine acts to facilitate the memory retrieval, the excitation of the arousal system may be useful. Tasaka et al. (1989a) reported that histamine at a dose of 1 μg, i.c.v. is significantly effective in stimulating the arousal system. Furthermore, when histamine was injected i.c.v. at doses larger than 50 ng, the

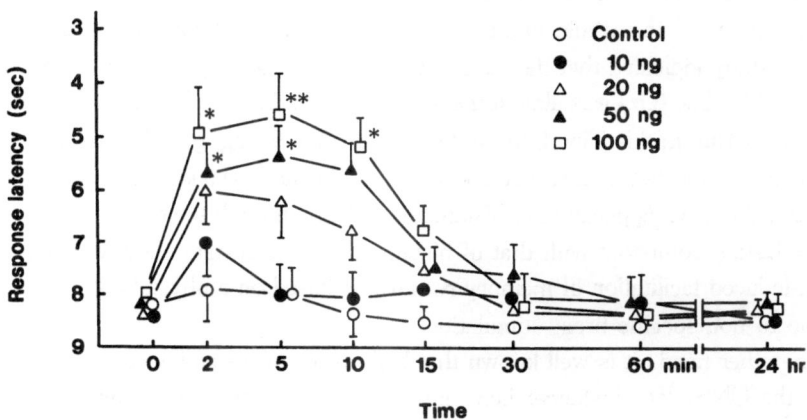

Fig. 6. Effect of intracerebroventricular injection of histamine on the prolonged latency induced by a long interruption of training in old rats. *, **: Significantly different from control group with $P < 0.05$ and $P < 0.01$, respectively. From Kamei, C. and Tasaka, K.: Biol. Pharm. Bull., 16, 128-132 (1993b), with permission.

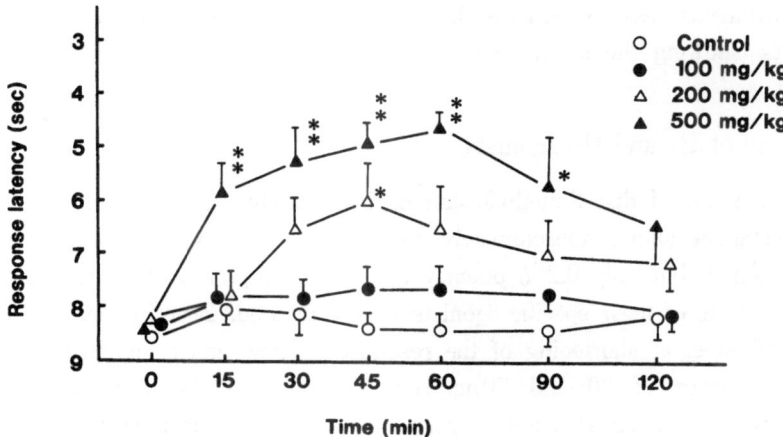

Fig. 7. Effect of intraperitoneal injection of histidine on the prolonged latency induced by a long interruption of training in old rats. *, **: Significantly different from control group with $P < 0.05$ and $P < 0.01$, respectively. From Kamei, C. and Tasaka, K.: Biol. Pharm. Bull., 16, 128-132 (1993b), with permission.

regional cerebral blood flow (rCBF) of the hippocampus was significantly increased (Kamei and Tasaka, 1993b). It is generally recognized that cerebral ischemia may lead to amnesia (Yamazaki et al., 1984). Conversely, an increase in hippocampal rCBF may contribute to the facilitation of memory retrieval. However, when histamine at doses of 50 and 100 ng was applied i.c.v. to rats, no behavioral excitation was observed. This may indicate that the facilitation of memory retrieval induced by histamine is not affected by non-specific excitatory influence on the CNS.

So far it has been reported that histidine at a dose of 500 mg/kg increased the histamine content of the whole brain to a level of 60 ng/g (Schwartz et al., 1972). A preliminary study indicated that the histamine content of the hippocampus after histamine (100 ng) and histidine (i.p.) was almost the same (control, 22.6 ± 3.6 ng/g; histamine 100 ng, 67.7 ± 14.4 ng/g; histidine 500 mg/kg, i.p. 61.3 ± 2.9 ng/g). The results shown in Fig. 7 seems to indicate that histidine injected i.p. (500 mg/kg) had almost the same effect as seen after the i.c.v. application of histamine (100 ng), though the effect of histidine was more long-lasting compared with that of histamine. These results show very clearly that histamine-induced facilitation of memory retrieval is based on a physiological process and is not due to non-specific brain stimulation.

On the other hand, it is well known that high doses of histamine caused a depressive effect on the CNS. For instance, i.c.v. injection of histamine ($10 \mu g$ or more) in mice resulted in an occurrence of catalepsy (Kamei et al., 1983) and an inhibition of pinna reflex (Kamei et al., 1984). In rats and cats, drowsy pattern in EEG, sedation and ptosis were observed at doses of 100 and $20 \mu g$, respectively (Kamei et al., 1981b). Feldberg and Sherwood (1954) also reported that histamine (i.c.v.) at doses of 100-200 μg caused a muscle weakness, decrease of spontaneous activity and sleepiness in cats. In active avoidance task, when large doses of histamine ($5\text{-}10 \mu g$, i.c.v.) was injected to young rats, showing the latency within 2 sec, the response latency was significantly prolonged.

4.5. Effects of H_1 and H_2 agonists

It has been reported that 2-methylhistamine is a selective H_1 receptor agonist, while 4-methylhistamine shows approximately 40% of the potency of histamine as an H_2 receptor agonist, but only 0.2% potency as an H_1 receptor agonist (Ganellin, 1982). When the effects of these specific agonists were compared, 2-methylhistamine caused a dose-related effect in shortening of the response latency, and a significant effect was obtained at doses of 20 and 50 ng, i.c.v. At doses of 50 and 100 ng, i.c.v. 2-thiazolylethylamine (2-TEA), another specific H_1 agonist, also significantly shortened the response latency, while at a dose of 20 ng some improvement was observed but not significantly (Fig. 8). The $ED_{50}S$ for 2-methylhistamine and 2-TEA were 32.6 (25.4-46.6) ng and 67.6 (49.0-117.8) ng, respectively (Fig. 9). No significant shortening of response latency was observed after 4-methylhistamine application at doses of 50-200 ng, i.c.v. in old rats. Also impromidine (a potent H_2 agonist) (Durant et al., 1978) effected no appreciable changes in response latency, even at a dose of 200 ng, i.c.v. (Fig. 9).

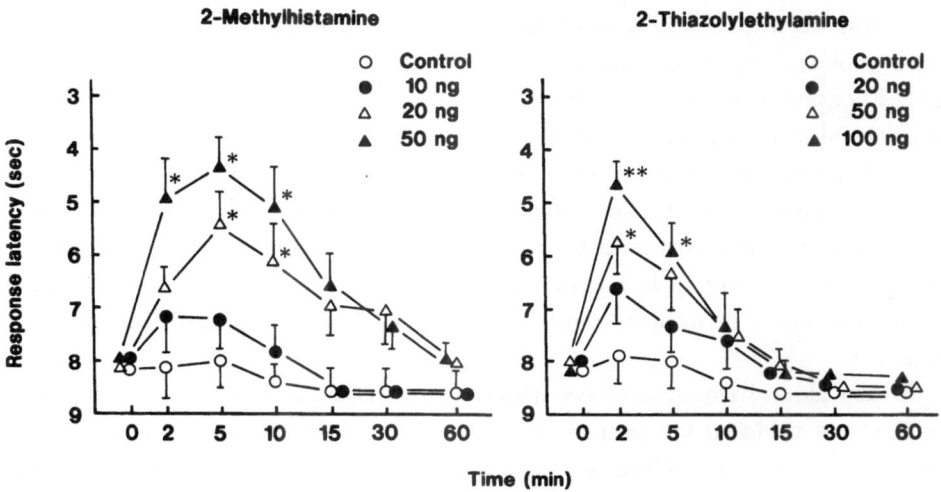

Fig. 8. Effects of intracerebroventricular injection of 2-methylhistamine and 2-thiazolylethylamine on the prolonged latency induced by a long interruption of training in old rats. *, **: Significantly different from control group with $P < 0.05$ and $P < 0.01$, respectively. From Kamei, C. and Tasaka, K.: Biol. Pharm. Bull., 16, 128-132 (1993b), with permission.

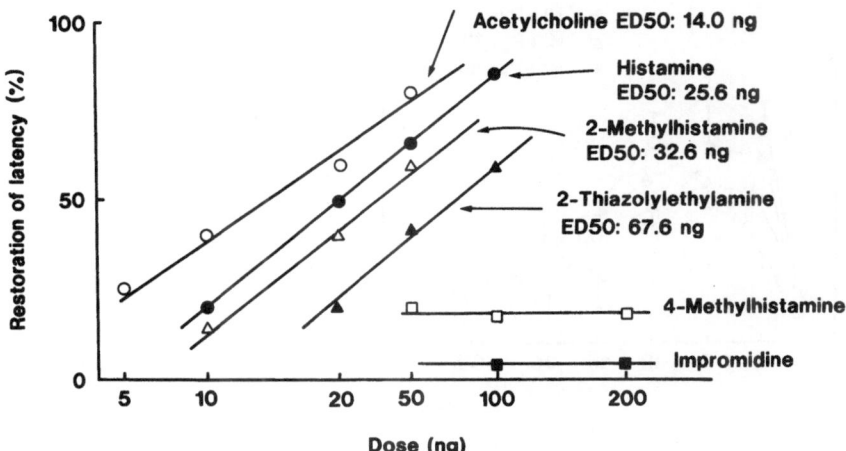

Fig. 9. Dose-response lines for shortening of response latency in active avoidance response in old rats. From Kamei, C. and Tasaka, K.: Biol. Pharm. Bull., 16, 128-132 (1993b), with permission.

4.6. Effects of H_1 and H_2 antagonists on histamine-induced facilitation of memory retrieval

When simultaneous i.c.v. application of pyrilamine, a representative H_1 antagonist, and histamine (100 ng) was carried out, pyrilamine abolished the histamine-induced shortening of response latency. At a dose of 2.6 ng, i.c.v. (a dose equivalent to $1/100$ of histamine dose in molarity), pyrilamine did not antagonize significantly. However, at doses of 13 ng ($1/20$ of histamine dose) and 26 ng ($1/10$ of histamine dose) it significantly inhibited histamine-induced shortening of response latency. On the other hand, cimetidine, a representative H_2 antagonist had no appreciable influence on histamine-induced shortening of response latency even at a dose of 220 ng, i.c.v. (a dose equivalent to histamine dose in molarity) (Fig. 10). Neither pyrilamine nor cimetidine affected the response latency when these drugs were applied separately. Based on these studies, it became apparent that the facilitation of memory retrieval due to histamine is intimately related to H_1 receptor.

de Almeida and Izquierdo (1988) demonstrated that histamine-induced facilitation of retention was mediated via both H_1 and H_2 receptors. However, in the study of de Almeida and Izquierdo, 1000 ng of antagonists (promethazine and cimetidine) was needed to antagonize 1 ng histamine action. Therefore, the question has been raised whether or not H_2 receptor is truly related to histamine-induced facilitation of memory retention. Bhattacharya (1990) found that 2-methylhistamine induces facilitation and 4-

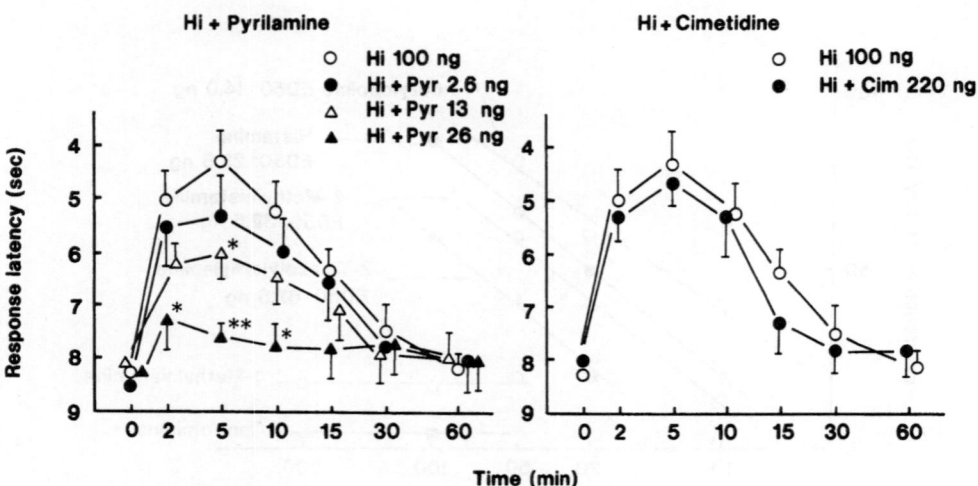

Fig. 10. Effects of pyrilamine and cimetidine on shortening of the response latency induced by histamine in old rats. *, **: Significantly different from histamine-treated group with $P < 0.05$ and $P < 0.01$, respectively. From Kamei, C. and Tasaka, K.: Biol. Pharm. Bull., 16, 128-132 (1993b), with permission.

methylhistamine attenuates learning and memory in the passive avoidance response in rats. This finding is almost identical to the present study which found that the facilitation of memory retrieval caused by histamine may be related to only the H_1 receptor.

4.7. Effects of H_3 agonist and antagonist

Recently, Arrang et al. (1983) demonstrated that the presence of novel class histamine receptors, namely H_3 receptors which regulate through a presynaptic negative feedback process. Highly potent, selective and brain-penetrating ligands to H_3 receptors, i.e., (R)-α-methylhistamine, a chiral agonist, and thioperamide, a competitive antagonist, were also developed (Arrang et al., 1987). Therefore, the effects of both drugs on the memory deficit in old rats.

However, (R)-α-methylhistamine (H_3 agonist) showed no significant changes in response latency at a dose of 5 mg/kg, i.p. Thioperamide (H_3 antagonist) caused a slight shortening of response latency, but it was not significantly. It has been reported that (R)-α-methylhistamine at a dose of 6.3 mg/kg, i.p. and thioperamide at a dose of 2 mg/kg, i.p. caused no significant changes in histamine content of the mouse brain, however, the content of *tele*-methylhistamine, a histamine metabolite, was significantly increased or decreased by administration of thioperamide or (R)-α-methylhistamine, respectively (Oishi et al., 1989). In addition, (R)-α-methylhistamine (3.2 mg/kg, i.p. in rats) inhibited the pargyline-induced *tele*-methylhistamine accumulation by 66-89% in eight brain regions, and thioperamide (2 mg/kg, i.p. in rats) enhanced it by 49-117%, indicating that (R)-α-methylhistamine inhibits histaminergic activity in the brain whereas thioperamide markedly enhances it (Oishi et al., 1989). From these findings, it may be the reason why thioperamide caused no significant effect on this memory deficit different from i.c.v. application of histamine.

5. Role of endogenous histamine in learning and memory

5.1. Effect of α-fluoromethylhistidine

Since a discovery of Kollonitsch et al. (1978), α-fluoromethylhistidine (α-FMH), a potent inhibitor of histidine decarboxylase, has been used for studying functional role of the brain histamine. Garbarg et al. (1980) found that the histamine contents of the cerebral cortex, hypothalamus and striatum in mice decreased by 60-70% after i.p. injection of α-FMH (10 mg/kg). In rats, Onodera et al. (1992) reported that α-FMH at a dose of 100 mg/kg, i.p. caused a decrease in the histamine level by 40-50% in the hypothalamus, thalamus and midbrain. Although α-FMH has no drastic effects on the CNS, it caused a decrease in locomotor activity in mice (Onodera et al., 1992), slow-wave sleep in rats (Monti et al., 1988) and a prolongation of thiopental-induced sleep in mice (Kamei et al., 1986).

When α-FMH at a dose of 10 mg/kg was injected i.p., it caused a moderate but not significant prolongation of the response latency. However, at a dose of 20 mg/kg, retardation of the response onset became apparent and a significant prolongation of the response latency was noticed 1 hr after drug administration; the effect lasted for 6 hr. At doses of 50 and 100 mg/kg, the effect was more marked than that observed at 20 mg/ kg in response of both latency and duration. A significant prolongation of the response latency was observed even 8 hr after the high dose injection (50 and 100 mg/kg) (Fig. 11). There was no motor disturbance, such as ataxia or muscle relaxation, even at a dose of 100 mg/kg.

When α-FMH was injected i.c.v., almost the same effect as i.p. injection was observed; i.e., at doses of 5-50 μg, i.c.v. of α-FMH caused a dose-related prolongation of the response latency. Although the drug showed no significant effect at a dose of 5 μg, at doses of 10-50 μg the drug caused a significant prolongation of the response latency. At a dose of 50 μg a significant prolongation was observed from 30 min to 6 hr after the α-FMH injection (Fig. 12). As shown in Figs. 11 and 12, α-FMH prolonged response latency after both i.p. and i.c.v. injections. This indicates that α-FMH caused a retardation of memory retrieval. It is generally recognized that an increase or decrease in general activity is responsible for facilitation or retardation of memory retrieval, respectively. However, no sign of the motor disturbance was observed even after 50 μg of i.c.v. injection of α-FMH. In addition, locomotor activity and rearing behavior in open-field test were not influenced by i.c.v. injection of α-FMH even at a dose of 50 μg (Kamei et al., 1993a). Therefore, it is unlikely that retardation of active avoidance

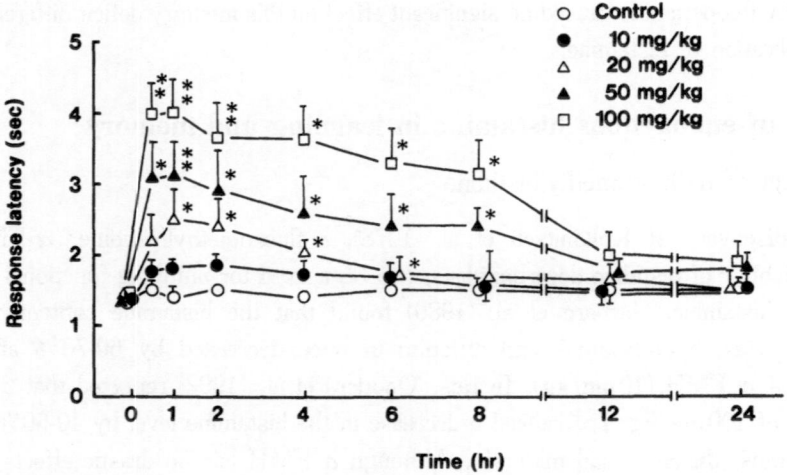

Fig. 11. Effect of intraperitoneal injection of α-fluoromethylhistidine on active avoidance response in rats. *, **: Significantly different from control group with $P < 0.05$ and $P < 0.01$, respectively. From Kamei, C., Okumura, Y. and Tasaka, K.: Psychopharmacology, 111, 376-382 (1993a), with permission.

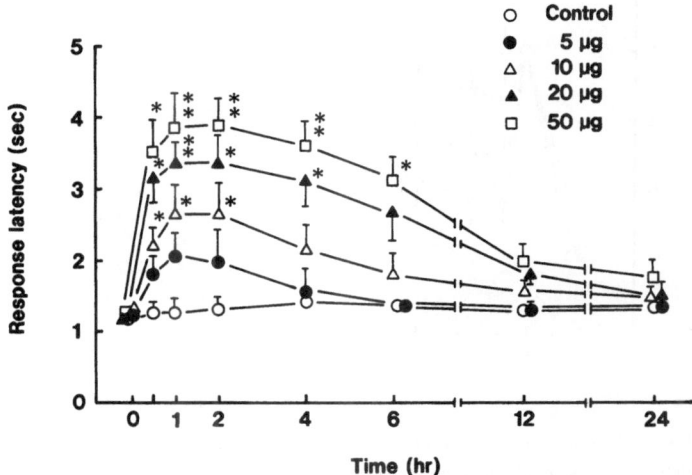

Fig. 12. Effect of intracerebroventricular injection of α-fluoromethylhistidine on active avoidance response in rats. *, **: Significantly different from control group with $P < 0.05$ and $P < 0.01$, respectively. From Kamei, C., Okumura, Y. and Tasaka, K.: Psychopharmacology, 111, 376-382 (1993a), with permission.

response due to α-FMH can be ascribed to the secondary effects of the drugs such as the motor disturbance or behavioral changes.

On the other hand, with regard to the effect of α-FMH on locomotor activity, contradictory results have been reported: a marked decrease of motor activity was observed in mice (Sakai et al., 1992) and a slight decrease of motor activity in rats (Cacabelos and Alvarez, 1991). However, Alvarez and Banzán (1986) reported that locomotor activity increased after α-FMH when it was injected into the hippocampus. Recently Onodera et al. (1992) demonstrated that α-FMH at a dose of 100 mg/kg i.p. caused no significant effect on locomotor activity and rearing behavior, though at a dose of 200 mg/kg i.p., it caused a significant decrease. The difference in the results of these experiments can be ascribed in some way to the species difference or the differences in the administration route.

5.2. Effect of histamine on α-fluoromethylhistidine-induced memory deficit

As shown in Fig. 11, prolongation of response latency was observed after i.p. injection of α-FMH. To investigate whether or not the effect of α-FMH can be reversed by histamine, the effect of i.c.v. application of histamine on α-FMH-induced memory deficit was studied.

When histamine was administered by i.c.v. injection at a dose of 5-20 ng to the rats in which response latency was significantly prolonged after i.p. injection of α-FMH (100 mg/kg), the latency became markedly shorter (Fig. 13). In this case, α-FMH was

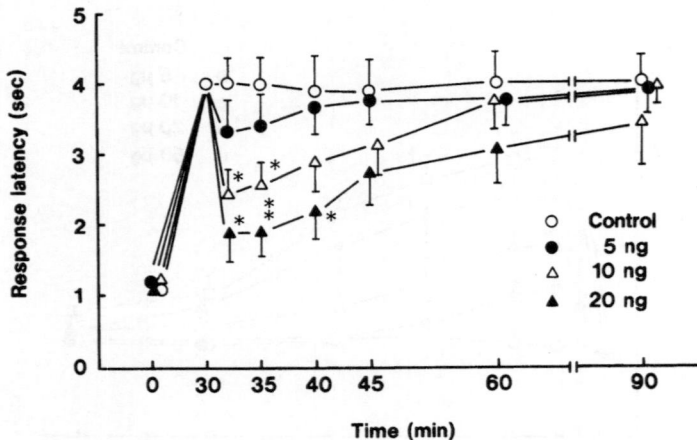

Fig. 13. Effect of intracerebroventricular injection of histamine on the prolonged latency induced by α-fluoromethylhistidine. α-Fluoromethylhistidine (100 mg/kg) was injected intraperitoneally at time 0. *, **: Significantly different from control with $P < 0.05$ and $P < 0.01$, respectively. From Kamei, C., Okumura, Y. and Tasaka, K.: Psychopharmacology, 111, 376-382 (1993a), with permission.

given 30 min prior to the histamine injection. A significant difference was observed at doses of 10 ng (2 and 5 min) and 20 ng (2, 5 and 10 min) after histamine application. As shown in Fig. 6, histamine at doses of 50 and 100 ng elicited shortening of prolonged latency caused by a long interruption of training in old rats. The dose of histamine which was effective in this study was extremely low.

5.3. Effect of α-fluoromethylhistidine on brain histamine contents

α-FMH elicited a decrease in histamine content in the brain both by i.p. and i.c.v. injection. I.p. injection of α-FMH (100 mg/kg) caused a significant decrease in the brain histamine content from 1 to 8 hr after drug injection. A remarkable decrease in histamine content was noticed in the cerebral cortex, hippocampus, thalamus and hypothalamus. A moderate decrease in histamine content of the midbrain and striatum was also noted. Twenty-four hr after α-FMH injection, decreased brain histamine contents were almost restored to the control level (Table 2). When the drug was injected i.c.v. (50 μg), almost the same results were obtained. The histamine contents of the brain was significantly decreased in the cerebral cortex, hippocampus, thalamus and hypothalamus at 1, 2, 4 and 6 hr after i.c.v. injection of α-FMH injection. The decrease in the histamine content of the hippocampus and the hypothalamus was particularly marked. The histamine contents of the striatum and midbrain were also decreased, but not significantly. In all the areas tested, the decrease in histamine content was restored within 24 hr after α-FMH administration (Table 3).

It is known that endogenous neuronal histamine is depleted by i.p. injection of α-FMH without affecting the contents of other monoamines, such as norepinephrine,

Table 2. Effect of intraperitoneal injection of α-fluoromethylhistidine (100 mg/kg) on histamine content of the different regions of the rat brain

Brain regions	Histamine contents (ng/g tissue)						
	Control	1	2	4	8	12	24 hr
Cerebral cortex	26.2 ± 1.9	16.9 ± 1.3**	17.4 ± 1.4**	17.5 ± 1.3**	20.0 ± 2.1*	23.5 ± 1.4	24.6 ± 1.7
Hippocampus	26.4 ± 2.3	14.5 ± 0.4**	15.2 ± 1.5**	15.5 ± 0.4**	20.1 ± 2.1*	22.6 ± 1.5	25.6 ± 1.5
Thalamus	110.6 ± 8.6	78.3 ± 3.2**	80.0 ± 3.2**	79.8 ± 7.3**	82.9 ± 6.2*	102.4 ± 4.6	107.2 ± 3.7
Hypothalamus	239.7 ± 17.9	115.0 ± 11.3**	122.5 ± 21.3**	130.9 ± 14.2**	149.4 ± 27.6*	206.9 ± 11.8	218.4 ± 12.6

*, **: Significantly different from control group with $P < 0.05$ and $P < 0.01$, respectively. From Kamei, C., Okumura, Y. and Tasaka, K.: Psychopharmacology, 111, 376-382 (1993a), with permission.

Table 3. Effect of intracerebroventricular injection of α-fluoromethylhistidine (50 μg) on histamine content of the different regions of the rat brain

Brain regions	Histamine contents (ng/g tissue)						
	Control	1	2	4	6	12	24 hr
Cerebral cortex	24.3 ± 1.9	16.3 ± 1.1*	14.0 ± 1.8**	17.1 ± 2.4*	18.5 ± 0.9*	22.2 ± 1.0	24.5 ± 2.6
Hippocampus	22.9 ± 3.3	9.3 ± 0.5**	9.1 ± 3.1**	10.3 ± 2.4**	13.3 ± 1.4*	19.5 ± 2.3	22.4 ± 3.3
Thalamus	112.1 ± 6.0	82.2 ± 5.3**	79.6 ± 11.8*	80.7 ± 4.6**	83.0 ± 6.6*	96.1 ± 8.1	109.1 ± 10.3
Hypothalamus	220.6 ± 13.3	91.8 ± 9.7**	92.6 ± 4.0**	94.2 ± 6.0**	107.9 ± 10.6*	172.7 ± 17.0	210.2 ± 21.9

*, **: Significantly different from control group with $P < 0.05$ and $P < 0.01$, respectively. From Kamei, C., Okumura, Y. and Tasaka, K.: Psychopharmacology, 111, 376-382 (1993a), with permission.

dopamine and serotonin in the brain (Yamatodani and Watanabe, 1991). Garbarg et al. (1980) have shown that the blockade of histidine decarboxylase activity by i.p. injection of α-FMH decreased brain histamine content only in the neuronal pool and does not affect the non-neuronal sauces (mainly in the mast cells). Maeyama et al. (1983) found almost complete depletion of brain histamine by α-FMH in W/W^V mice, which are devoid of mast cells. Therefore, it is reasonable to assume that a substantial amount of histamine which remained in various regions of the brain after α-FMH administration may be located in the mast cells. From these findings, it seems likely that prolongation of the response latency induced by i.p. injection of α-FMH may be simply due to the decrease in neuronal histamine contents. However, as reported by Garbarg et al. (1980), histidine decarboxylase activity was decreased not only in various regions of the brain but also in the gastric tissue after i.p. injection of α-FMH. Sakurai et al.(1990) reported that α-FMH concentrations in the peripheral tissues were much higher than that determined in the brain after intravenous injection of α-FMH. To avoid any systemic effect, we have also studied the effect of α-FMH when injected i.c.v.

5.4. Correlation between a memory deficit and a decrease in histamine content induced by α-fluoromethylhistidine

The prolongation of response latency and the depletion of the histamine content of the brain was observed by means of both i.p. and i.c.v. injection of α-FMH (100 mg/kg, i.p.; 50 μg, i.c.v.). Therefore, the relationship between two parameters was calculated. As a result (Fig. 14), the regression line of the prolongation of response latency and the decrease in the histamine content of the hippocampus is $Y = 0.191X - 0.068$ ($r = 0.9896$). Also a high correlation coefficient ($r = 0.9827$) was obtained when the histamine contents of the hypothalamus was taken into account. This can be concluded that endogenous histamine may play an important role in learning and memory retrieval in rats, and further support that the hippocampus and hypothalamus are important for learning and memory. In accordance with this, it has been reported that both acquisition and retention of avoidance response were impaired after hippocampectomy in rats (McNew and Thompson, 1966; Papsdorf and Woodruff, 1970). Also, Grossman (1970) and Asdourian et al. (1977) reported that bilateral damage of the lateral hypothalamus impaired active avoidance in rats.

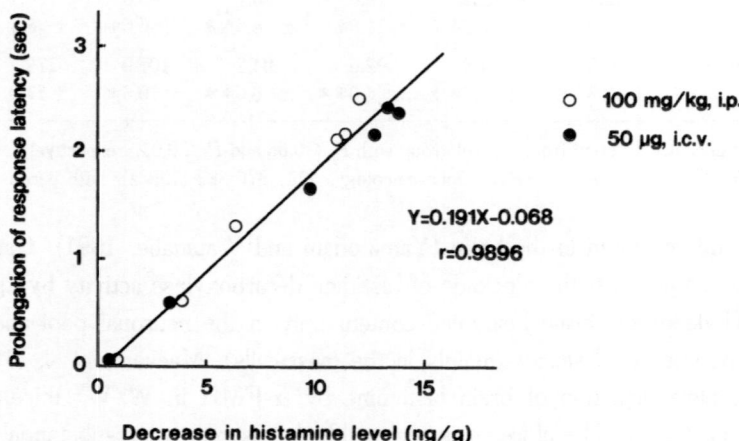

Fig. 14. Correlation between a prolongation of response latency and a decrease in histamine contents of the hippocampus induced by α-fluoromethylhistidine. Reproduced from Kamei, C., Okumura, Y. and Tasaka, K.: Psychopharmacology, 111, 376-382 (1993a), with permission.

5.5. Effects of α-fluoromethylhistidine and histamine on learning acquisition

α-FMH was effective in the prolongation of response latency, indicating that the drug caused a retardation of memory retrieval in acquired memory. The effect of α-FMH on learning acquisition in comparison with histamine was also studied.

Acquisition trial was continued for 12 consecutive days at 1 hr after injection of α-FMH (50 μg) or 5 min after histamine (100 ng) injection. The retarded acquisition of active avoidance response was observed after repeated administration of α-FMH. A significant difference was observed throughout the whole injection period between the control group and α-FMH treated group. On the other hand, histamine was effective in inducing a significant facilitation of the acquisition of active avoidance response (Fig. 15). These findings suggest that α-FMH caused a significant suppression not only in the memory retrieval but also in the acquisition learning of active avoidane response.

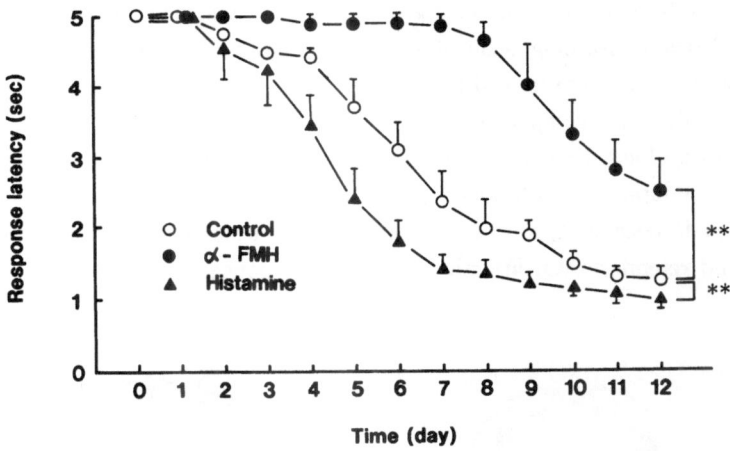

Fig. 15. Effect of α-fluoromethylhistidine or histamine on the acquisition of active avoidance response in rats. Acquisition test was done 1 hr (α-fluoromethylhistidine) or 5 min (histamine) after drug injection. α-Fluoromethylhistidine and histamine were injected intracerebroventricularly at doses of 50 μg and 100 ng, respectively. **∗∗**: Significantly different from saline-treated control group with $P < 0.01$. From Kamei, C., Okumura, Y. and Tasaka, K.: Psychopharmacology, 111, 376-382 (1993a), with permission.

6. Learning and memory deficits induced by hippocampal lesions

6.1. Effect of hippocampal lesions on acquisition and retention

There has been no divergence of opinion as to the hippocampus plays a significant role for learning and memory. However, in active avoidance procedure, contradictly findings were obtained both in acquisition learning and memory retention from those observed in passive avoidance task (Isaacson et al. 1961; Rich and Thompson, 1965). Therefore, the influence of hippocampal lesions on learning and retention were investigated in active avoidance response.

A bilateral lesion of the dorsal hippocampus (A: 3.0, L: 2.0, H: 2.0) was made under sodium pentobarbital anesthesia (35 mg/kg, i.p.) by the passing through of a 5 mA anodal current (D.C.) for 15 sec according to the atlas of de Groot (1959). Sham-operated rats were subjected to a similar surgical procedure, without delivery of the current. At least 10 days were allowed for recovery from the surgery. At the end of the experiments, the brains of all animals were examined histologically. Fig. 16 shows the typical lesions of the hippocampus. The black areas represent the regions of the injury. The greater part of the dorsal hippocampus including the CA1 field and dentate gyrus (CA4) (Paxinos and Watson, 1986) was lesioned in all experimental animals. However, the hippocampus located ventrally in the brain was left intact.

Five days after the hippocampus was lesioned electrically, the acquisition trial was started, and continued for 12 days. The latency measured in hippocampal-lesioned group was longer than those in sham-operated group during the entire period of time, indicating that severe deficit in learning acquisition was produced after hippocampal lesions. In retention test, the rats were trained to move in the lighted room within 2 sec, when they were placed in the dark compartment after training. Response latency of hippocampal-lesioned group prolonged soon after hippocampal lesions and significant difference was noticed between hippocampal-lesioned group and control group (Fig. 17). From the present study, it became apparent that hippocampal lesions caused a deficit of both acquisition and retention even in active avoidance response.

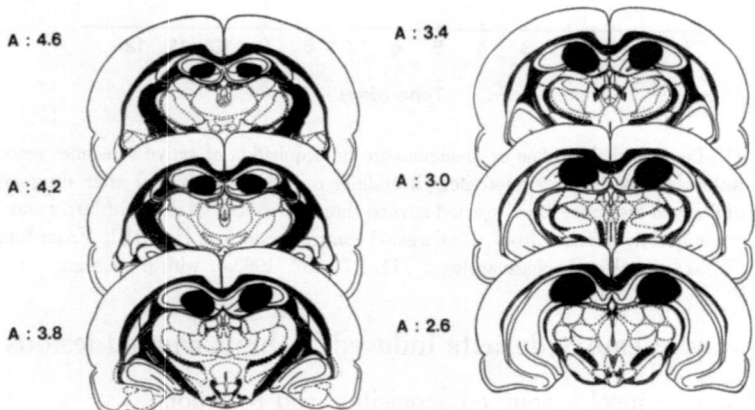

Stereotaxic atlas of de Groot

Fig. 16. Schematic drawing of the representative lesions of the dorsal hippocampus. The black areas indicate lesions.

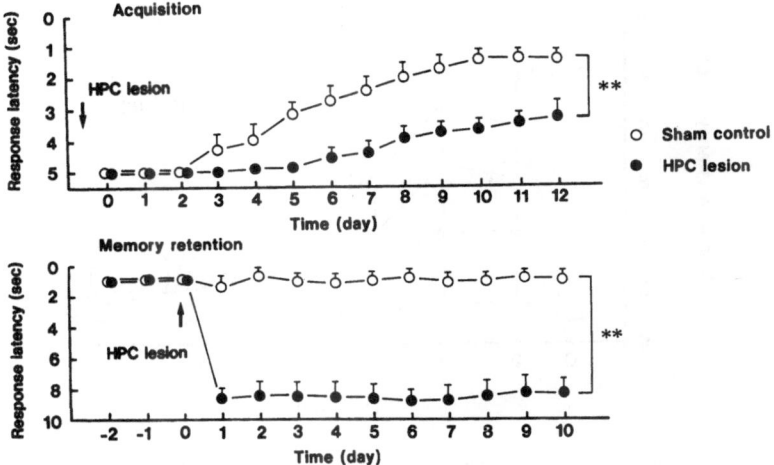

Fig. 17. Effect of hippocampal lesions on acquisition and retention of active avoidance response in rats. ∗∗: Significantly different from the control group with $P < 0.01$.

6.2. Effects of histamine and histidine

The effects of histamine and histidine on the prolongation of response latency in hippocampal-lesioned rats were investigated. Histamine at a dose of 20 ng (i.c.v.), it caused a shortening of response latency, but not significantly. However, at doses higher than 50 ng histamine elicited a dose-related and significant shortening of the response latency. At a dose of 100 ng, significant effect continued from 2 to 10 min after injection (Fig. 18). Histidine at a dose of 100 mg/kg, i.p. showed a moderate shortening of response latency, but not significantly. At doses of 200 and 500 mg/kg, however, histidine significantly shortened the response latency. At a dose of 500 mg/kg, significant effect continued until 90 min after injection (Fig. 19).

Tagami et al. (1984) and Sunami and Tasaka (1991) reported that histamine produced an excitatory effect on CA3 pyramidal cells in hippocampal slices of guinea pigs. Nearly the same results were reported by Haas (1984) when the CA1 pyramidal cells were stimulated in rats. These findings indicate that histamine may exert an excitatory influence on the hippocampus and this may be related with histamine which plays an active part in the facilitating of memory retrieval. On the other hand, it was found also that the locomotor activity and rearing behavior measured by open-field test was significantly increased after hippocampectomy (Kamei and Tasaka, 1992). The similar findings were reported by Muñoz and Grossman (1980) that ambulation (locomotor activity) and rearing was increased by i.c.v. injection of kainic acid, a selective neuronal depleting agent, into the dorsal hippocampus. Therefore, it is undoubtedly that the improvement of learning behavior seen after histamine injection in hippocampal-lesioned rat was not due to an increase in exploratory behavior.

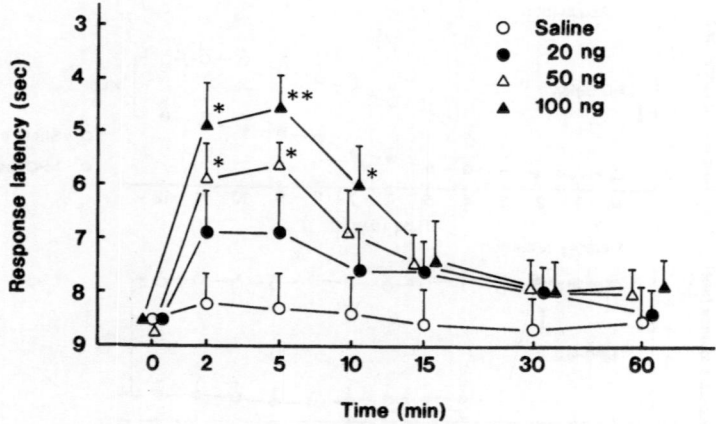

Fig. 18. Effect of intracerebroventricular injection of histamine on prolonged latency induced by hippocampal lesions. *, **: Significantly different from control group with $P < 0.05$ and $P < 0.01$, respectively.

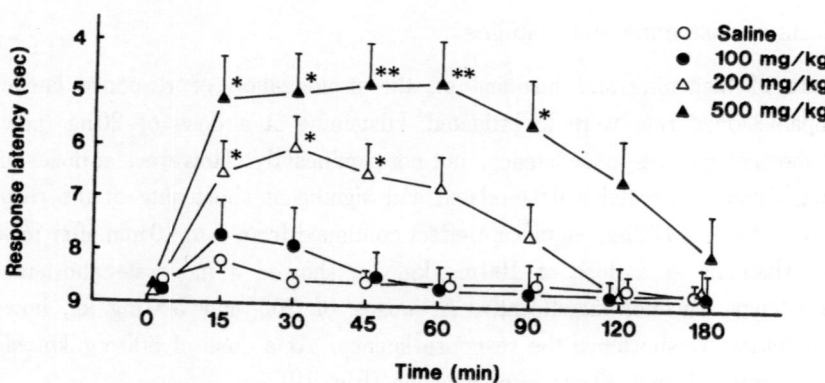

Fig. 19. Effect of intraperitoneal injection of histidine on prolonged latency induced by hippocampal lesions. *, **: Significantly different from control group with $P < 0.05$ and $P < 0.01$, respectively.

6.3. Changes in brain histamine contents after hippocampectomy

When α-FMH was injected, a decrease in the histamine content of the hippocampus and the prolongation of response latency occurred simultaneously, therefore, it seems likely that it is of great value to investigate the histamine content after hippocampectomy, because the prolongation of response latency was observed after hippocampectomy. The hippocampus was lesioned electrically, and the histamine contents of some brain areas were measured 3 weeks after surgery. Although histamine contents in the striatum, thalamus and hypothalamus were not influenced significantly by hippocampal lesions, that

Table 4. Changes in brain histamine contents after hippocampal lesions in rats

Brain areas	Histamine contents (ng/g)	
	Sham-operated rats	Hippocampal-lesioned rats
Cerebral cortex	25.4 ± 2.1	13.0 ± 2.1**
Hippocampus	26.4 ± 2.3	7.3 ± 2.8**
Striatum	28.1 ± 4.3	28.2 ± 6.6
Thalamus	119.5 ± 18.4	121.4 ± 10.2
Hypothalamus	227.7 ± 14.8	234.1 ± 14.4

Hippocampus was lesioned electrically and 3 weeks after surgery, the brain histamine content was measured. **: Significantly different from sham-operated group with $P < 0.01$.

in the hippocampus was decreased significantly compared with those of sham-operated rats. Histamine content in the cerebral cortex also decreased, though the extent was less than that in the hippocampus (Table 4).

Panula et al. (1984) found that histamine-containing neurons are located only in a small area of the posterior hypothalamus and these cells are probably the sources of ascending and descending fibers. Based on this finding, it may be that the histaminergic pathway to the cerebral cortex runs through the hippocampus (Kamei and Tasaka, unpublished observation). Burešová et al. (1962) demonstrated that there are connections between the neocortex and hippocampus in both electrophysiological and behavioral experiments. On the other hand, Garbarg et al. (1976) reported that both the histamine level and histidine decarboxylase activity in the cortex decreased after lesions of the medial forebrain bundle. Therefore, it seems likely that a decrease in the histamine level of the hippocampus and cortex may be related to the decrease in histidine decarboxylase activity.

6.4. Changes in histamine content of the hippocampus after histamine and histidine injection

Table 5 shows the histamine contents after histamine (100 ng, i.c.v.) and histidine (500 mg/kg, i.p.) injection in hippocampal-lesioned rats. The decrease in histamine content in the hippocampus induced by hippocampal lesions was restored completely after histamine and histidine treatment (500 mg/kg, i.p.). Not only histamine but also histidine caused an improvement of memory deficient induced by hippocampal-lesioned rats, and the efficacy in memory improvement was almost the same, though the duration was quite different. From the biochemical determinations, the extents of histamine increase in the brain after histamine (100 ng, i.c.v.) application or histidine (500 mg/kg, i.p.) injection was also almost the same. Therefore, it was assumed that the ameliorating effect in response latency induced by histamine was not due to a simple artifact, but related to the increased histamine level in the brain especially in the hippocampal region.

Table 5. Effects of histamine and histidine injection on the decrease in histamine contents of the hippocampus after hippocampal lesions

Drugs	Histamine contents (ng/g)	
	Sham-operated	Hippocampal-lesions
Control	25.4 ± 2.1	7.3 ± 2.8
Histamine (100 ng)	67.7 ± 12.6**	51.2 ± 14.1**
Histidine (500 mg/kg)	60.9 ± 3.7**	46.0 ± 2.9**

Histamine contents were measured 5 min after histamine and 60 min after histidine injection. **: Significantly different from control group with $P < 0.01$.

6.5. Relationship between a facilitation of memory retrieval and an increase in histamine content after histamine injection

As shown in Fig. 18 and Table 5, histamine facilitated the memory retrieval in hippocampus lesioned rats, and the decrease in histamine content of the hippocampus was significantly recovered by histamine application. Therefore, the relationship between the shortening of the response latency and the increase in histamine content of the brain induced by i.c.v. injection of histamine (100 ng) may exist, so that the correlation coefficient (r) was calculated. A high correlation was obtained between an increase in histamine content of the hippocampus and shortening of response latency ($r = 0.9938$), indicating these two findings are intimately related. The same relation was also obseved between the histamine content of the cerebral cortex and the shortening of response latency, however, the correlation coefficient ($r = 0.9109$) was somewhat lower compared with that seen in hippocampus (Fig. 20).

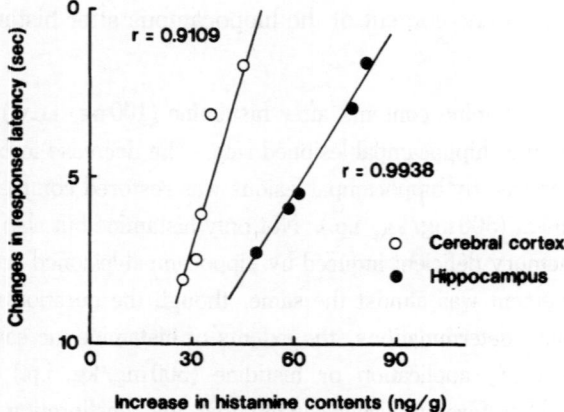

Fig. 20. Relationship between changes in response latency and an increase in histamine content of the brain induced by histamine (i.c.v.).

7. Effects of certain H_1 antagonists on active avoidance response

7.1. Effect of oral administration

Recently non-sedating H_1 antagonists have been developed, and used clinically to control some allergic symptoms (Devlin and Hargrave, 1985). On the other hand, it is well known that classic H_1 antagonists produce a CNS depressant effect clinically (Wyngaarden and Seevers, 1951). In animal studies, these drugs caused a decrease in locomotor activity (Kaneko et al., 1981), a prolongation of thiopental-induced sleep (Tasaka et al., 1990) and an increase in drowsy EEG pattern (Kamei et al., 1981a). In addition, an inhibition of learning behavior was also observed (Tasaka et al., 1986). Therefore, the effects of newer H_1 antagonists on active avoidance response were studied in comparison with those of classic H_1 antagonists.

The effects of H_1 antagonists on active avoidance response were determined after oral administration of each compounds. Pyrilamine at a dose of 5 mg/kg caused a significant prolongation of response latency at 30 min after drug administration as shown in Fig. 21. Response latency increased dose-dependently. Chlorpheniramine also elicited a dose-dependent prolongation of the latency. ED_{50} of the drug was almost the same as those of pyrilamine and diphenhydramine (Table 6). Promethazine appeared to be the most potent drugs, and the effect was 1.3 times more potent than that of diphenhydramine. Ketotifen was not effective up to a dose of 20 mg/kg. However, at 50 mg/kg, it significantly prolonged the response latency and maximum prolongation occurred 1 hr after administration of the drug. The ED_{50} of ketotifen was 3.3 times larger than that of pyrilamine. Astemizole and oxatomide had practically no suppressive effect on avoidance response, though at 50 mg/kg, both drugs protracted latency slightly but significantly; their ED_{50}s were 1.8-3.5 times larger than that of ketotifen. Azelastine prolonged response latency most evidently among the newer H_1 antagonists tested. At doses of 10-50 mg/kg, the drug dose-dependently increased response latency and significant prolongation was observed even at a dose of 10 mg/kg. The ED_{50} value of azelastine was approximately twice that of pyrilamine.

It became evident that classical H_1 antagonists markedly prolonged the response latency of the active avoidance response. This indicates that classical H_1 antagonists caused a retardation of memory retrieval. The potency of these drugs in their depressive action, from greatest to least, was as follows: promethazine, pyrilamine, diphenhydramine and chlorpheniramine. This order is in accordance with that determined using the two-way shuttle box avoidance response in rats (Tasaka et al., 1985). In clinical use, it is well known that these classical H_1 antagonists cause remarkable sedation and sleepiness (Wyngaarden and Seevers, 1951), so that considerable efforts have been made to develop newer H_1 antagonists devoid of side effects on the CNS.

There are several reports that ketotifen had a depressant effect on the CNS in animal

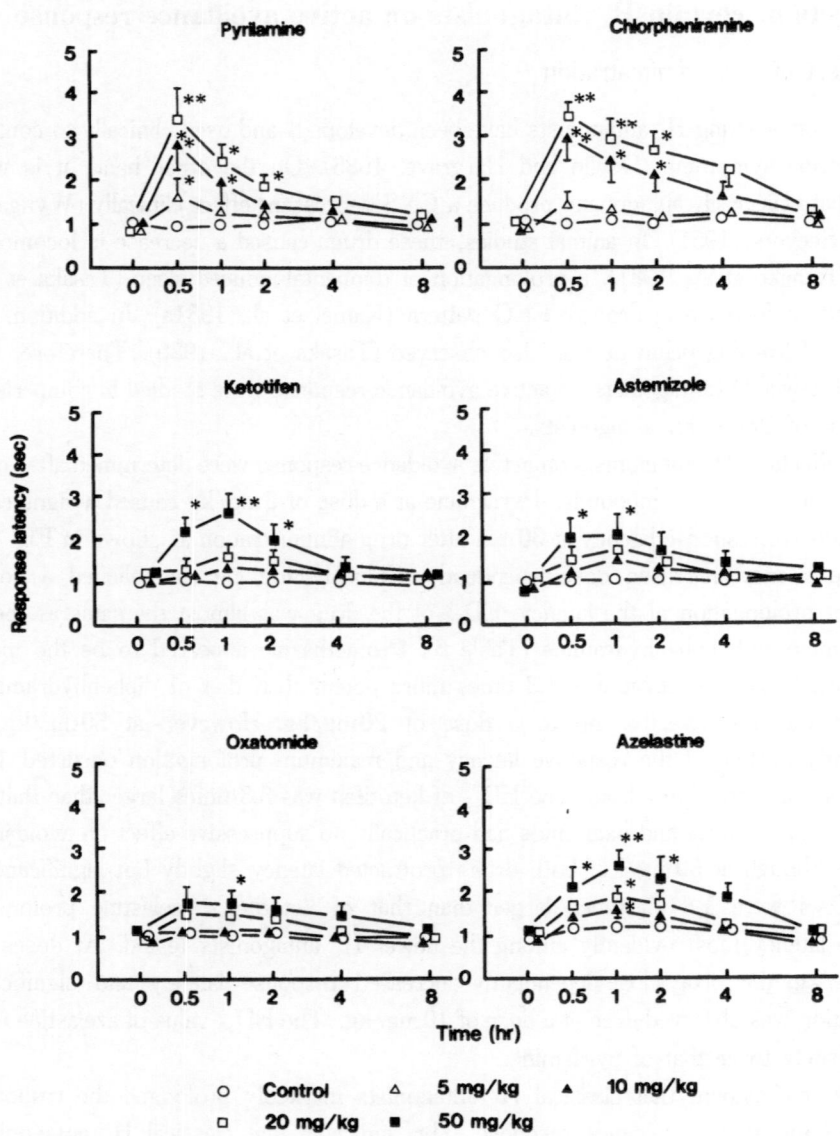

Fig. 21. Effects of oral administration of certain H_1 antagonists on active avoidance response in rats. *, **: Significantly different from control group with $P < 0.05$ and $P < 0.01$, respectively. From Kamei, C., Chung, Y.H. and Tasaka, K.: Psychopharmacology, 102, 312-318 (1990b)

studies. For instance, in a climbing test, ketotifen exerted an inhibitory effect at doses higher than 30 mg/kg (p.o.) in mice (Martin and Römer, 1978). The drug also increased slow-wave sleep in rats (Tasaka et al., 1989b) and decreased REM sleep in dogs at 10 mg/kg (p.o.) (Wauquier et al., 1981). However, a large dose of ketotifen (50 mg/kg)

Table 6. ED_{50} values of H_1 antagonists for the prolongation of the response latency in the active avoidance response in rats

Drugs	ED$_{50}$ (95% Confidence limits)		
	p.o. (mg/kg)	i.v. (mg/kg)	i.c.v. (μg)
Pyrilamine	10.0 (6.71-14.9)	4.83 (4.31-5.42)	29.1 (27.1-31.5)
Diphenhydramine	11.0 (3.67-33.0)	3.95 (2.26-6.91)	27.0 (23.2-32.4)
Promethazine	8.20 (5.16-13.0)	2.50 (1.19-5.25)	19.8 (17.8-22.0)
Chlorpheniramine	12.0 (8.70-16.6)	7.34 (6.94-7.80)	35.6 (31.3-41.9)
Mequitazine	55.6 (33.7-91.7)	21.4 (18.9-25.2)	177 (128 - 305)
Astemizole	74.0 (26.4- 207)	25.6 (23.8-28.1)	135 (134 - 135)
Oxatomide	60.0 (23.1- 156)	12.2 (11.9-12.5)	98.8 (86.2- 117)
Epinastine	23.0 (14.6-49.0)		
Azelastine	21.0 (9.55-46.2)	12.6 (11.1-14.7)	32.7 (30.9-34.7)
Ketotifen	32.5 (19.7-53.6)	approximately 2	14.3 (14.0-14.5)
Levocabastine	111 (101 - 124)		

Reproduced from Kamei, C., Chung, Y.H. and Tasaka, K.: Psychopharmacology, 102, 312-318 (1990b) and Kamei, C. and Tasaka, K.: Japan. J. Pharmacol., 57, 473-482 (1991), with permission.

was required to cause a significant prolongation of latency, suggesting that the inhibitory effect of ketotifen on the avoidance behavior is relatively weak when the drug was administered by p.o. Azelastine among the newer H_1 antagonists was found to have a potent inhibitory effect on the active avoidance response. Kaneko et al. (1981) reported that azelastine at doses of 4-40 mg/kg (p.o.), displayed definite actions on the CNS in mice, such as a decrease in locomotor activity, potentiation of pentobarbital-induced sleep and antagonism of caffeine-induced hyperactivity. However, the clinical dosage of azelastine is much lower than that of the other H_1 antagonists, so that it may cause no noticeable sedative effect in clinical use. Wauquier et al. (1981) found that astemizole at a dose of 10 mg/kg (p.o.) had no influence on the sleep-wakefulness pattern in dogs. Oxatomide was also reported to produce little or no effect on spontaneous and coopera-tive movements in mice at oral doses of 40-100 mg/kg (Ohmori et al., 1983a,b). These findings are in good agreement with the results in Fig. 21 and Table 6.

It is known that the classical H_1 antagonists possess not only anti-histaminic but also anti-muscarinic properties. When these parameters were compared among the classical H_1 antagonists tested, the anti-histaminic property decreased in the following order: pyrilamine ($pA_2 = 9.36$, Ganellin, 1982), chlorpheniramine ($pA_2 = 9.04$, Ganellin, 1982), promethazine ($pA_2 = 8.93$, Ganellin, 1982) and diphenhydramine ($pA_2 = 8.14$, Ganellin, 1982). Regarding the anti-muscarinic properties, pA_2 values decreased in the following order: promethazine (7.62, Reuse, 1948), diphenhydramine (6.04, Ohmori et al., 1983b), chlorpheniramine (5.61, van den Brink and Lien, 1978) and pyrilamine (4.86, Schild, 1947). Since these values were determined using isolated organs, it is rather difficult to make any correlation between these pA_2 values and ED_{50} values determined in prolongation of the active avoidance response. No meaningful correlation seems to exists between these values and the ED_{50} values shown in Table 6.

7.2. Effects of intravenous and intracerebroventricular administration

In oral administration, classic H_1 antagonists inhibited active avoidance response, whereas newer H_1 antagonists caused a weak depressant effect. However, when the drugs were administered by p.o., an absorption from the gastrointestinal tract or distribution of the drug to the CNS affected the potency of the drugs. Therefore, the effects of newer H_1 antagonists were also studied being administered by both i.v. and i.c.v. The effects of pyrilamine, chlorpheniramine, ketotifen, astemizole, oxatomide and azelastine on response latency after i.v. administration were compared and shown in Fig. 22. At a dose of 10 mg/kg, pyrilamine prolonged the response latency at 15 min and a significant effect was observed even 120 min after drug administration. At the same dosage, chlorpheniramine was less potent than pyrilamine. Ketotifen elicited a slight but significant prolongation of the latency even at a dose of 2 mg/kg. Because the LD_{50} of ketotifen by i.v. was 5.3 mg/kg (Martin and Römer, 1978), doses higher than 2 mg/kg were not tested. The effect of astemizole was almost negligible at a dose of 10 mg/kg, but at 20 mg/kg, the drug prolonged the response latency significantly. Oxatomide showed a significant effect at doses of 10 and 20 mg/kg. Promethazine appeared to be the most potent drug, with an ED_{50} of 2.50 (1.19-5.25) mg/kg (Table 6). Mequitazine had low potencies as well as astemizole; the ED_{50} was 21.4 (18.9-25.2) mg/kg.

Fig. 23 and Table 6 show the effects of the same group of drugs tested after i.c.v. administration. At doses of 20 and 50 μg, pyrilamine caused a significant retardation of the response latency. Chlorpheniramine elicited a significant prolongation at doses of 20 and 50 μg. Ketotifen extended the response latency significantly at doses of 10 and 20 μg. However, astemizole elicited no significant effect even at a dose of 50 μg. Both oxatomide and azelastine showed significant effects at 50 or 100 μg and 10, 20 or 50 μg, respectively. The ED_{50} values determined after i.c.v. administration are also included in Table 6. Ketotifen and promethazine were effective in inhibiting the avoidance response (ED_{50} values of 14.3 (14.0-14.5) μg and 19.8 (17.8-22.0) μg, respectively).

Ketotifen caused a relatively potent inhibition of the avoidance response when injected i.c.v., it exerted only a weak influence on this response after oral administration when the ED_{50} values were compared (Table 6). Therefore, it seems likely that either 1) ketotifen may not be easily absorbed from the gastrointestinal tract or 2) the drug may go through the first-pass effect. Martin and Römer (1978) reported that an inhibition of passive cutaneous anaphylaxis (PCA) in rats induced by i.v. injection of ketotifen was 17 times more potent than that induced by oral administration. The discrepancy due to the difference in administration routes was also found in the other drugs used. For instance,

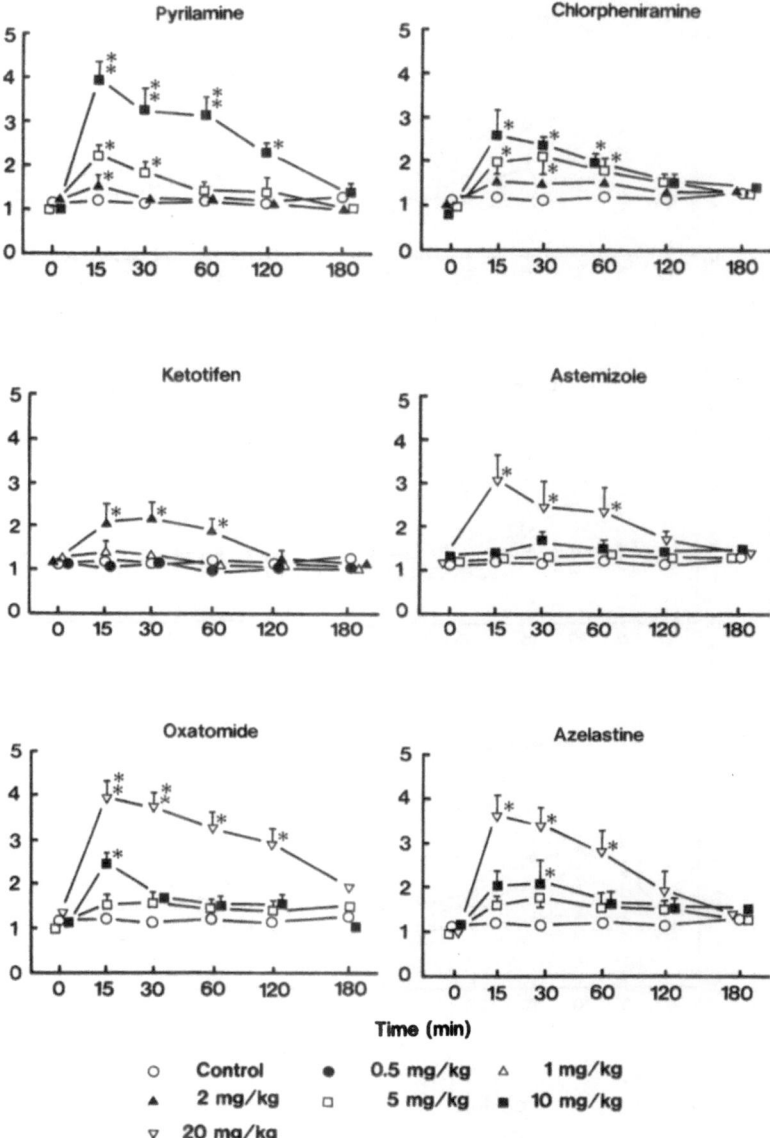

Fig. 22. Effects of intravenous injection of H₁ antagonists on the active avoidance response in rats. *, **: Significantly different from the control group with $P < 0.05$ and $P < 0.01$, respectively. From Kamei, C. and Tasaka, K.: Japan. J. Pharmacol., 57, 473-482 (1991)

the potency of azelastine exerted by i.c.v. injection was slightly stronger than that of chlorpheniramine, while, i.v. injections, azelastine was evidently less effective than chlorpheniramine. This suggests that azelastine may be much less effective than chlorpheniramine in penetrating the CNS. Actually, Tatsumi et al. (1980) reported that the

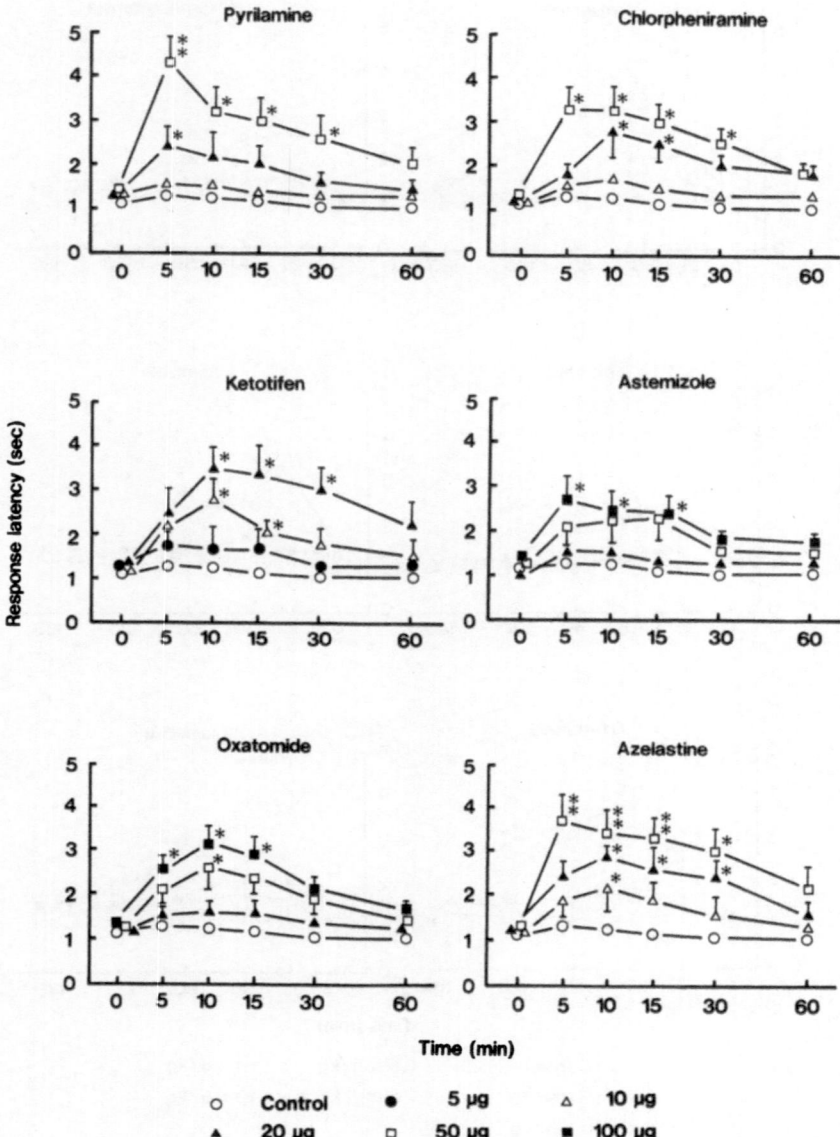

Fig. 23. Effects of intracerebroventricular injection of H_1 antagonists on the active avoidance response in rats. *, **; Significantly different from the control group with $P < 0.05$ and $P < 0.01$, respectively. From Kamei, C. and Tasaka, K.: Japan. J. Pharmacol., 57, 473-482 (1991)

distribution of azelastine in the brain was extremely low when the drug was injected i.v. in rats. Ohmori et al. (1987) found that azelastine caused a marked inhibition of $[^3H]$ pyrilamine binding to guinea pig cerebellum, indicating that the drug is able to bind the CNS efficiently. Based on these findings, it is reasonable to assume that the relatively weak depressant effect on the active avoidance response caused by i.v. injection of azelastine may not be due to its feeble CNS activity but simply to its low capacity for penetrating the CNS. Inhibitory effects of mequitazine and oxatomide were considerably weak compared with those of classical H_1 antagonists when administered both i.v. or i.c.v. It has been reported that the concentration of mequitazine in the brain after i.v. administration was not lower than that determined in the blood (Soda et al., 1981). Similar results were obtained in the case of oxatomide (Shibata et al., 1984). Therefore, it was assumed that mequitazine may have almost negligible affinity for the brain. Uzan et al. (1979) reported that mequitazine is 15-20 times less potent than chlorpheniramine and promethazine in preventing [^{3}H]pyrilamine binding to guinea pig brains. Astemizole has been reported to show a larger Ki value than mequitazine in the [^{3}H]pyrilamine binding test. (Ahn and Barnett, 1986). Oxatomide also caused an inhibition of [^{3}H] pyrilamine binding, although the effect was weaker than that of mequitazine (Ohmori et al., 1987). It seems, therefore, weak depressant effects of mequitazine, astemizole and oxatomide on the avoidance response may be attributable to their poor affinities for brain tissue in addition to the difficulty in penetrating the brain.

7.3. Effect of histamine on H_1 antagonist-induced memory deficit

As shown above, i.c.v. application of histamine caused a facilitation of memory retrieval after a long interruption of training or ablation of the hippocampus in rats via H_1 receptors. Therefore, to determine whether or not H_1 antagonists exert their inhibitory effect by competing with histamine for H_1 receptors in the brain, the effect of i.c.v. application of histamine on H_1 antagonist-induced memory deficit was studied.

I.c.v. injection of histamine at doses of 0.1-1 μg suppressed dose-dependently the prolonged latency in the avoidance response produced by i.v. injection of pyrilamine (9.7 mg/kg, double dose of ED_{50}). Significant effect appeared almost instantaneously after histamine injection at doses higher than 0.5 μg (Fig. 24). The prolongation of the response latency in the avoidance response induced by diphenhydramine (7.9 mg/kg, double dose of ED_{50}) was also shortened after a brief period by i.c.v. injection of histamine. ID_{50} values of histamine on pyrilamine- or diphenhydramine-induced prolongation of latency were 0.52 (0.22-2.99) μg and 0.79 (0.44-3.32) μg, respectively. Histamine alone at a dose of 1 μg caused no significant effect on the active avoidance response.

7.4. Effects of histamine agonists on pyrilamine-induced memory deficit

It is generally recognized that 2-methylhistamine and 4-methylhistamine were widely used as drugs which have H_1 agonist and H_2 agonist activity. I.c.v. injection of 2-methyl-histamine antagonized pyrilamine-induced inhibition of the active avoidance response in the same way as histamine; a significant shortening of the response latency was observed at doses greater than 1 μg. ID_{50} of 2-methylhistamine was 1.40 (1.03-2.53) μg. In

58

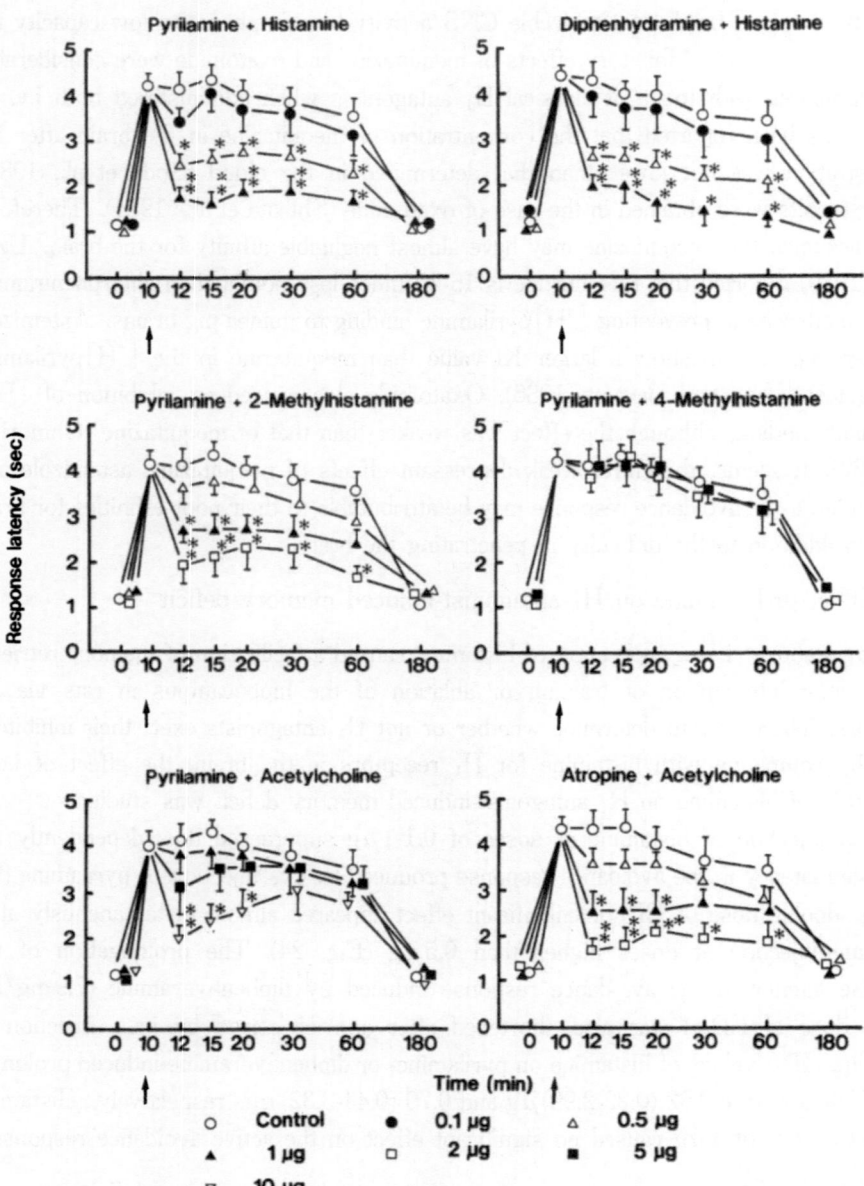

Fig. 24. Effects of intracerebroventricular injection of histamine, 2-methylhistamine, 4-methylhistamine and acetylcholine on the prolonged latency induced by pyrilamine or diphenhydramine or atropine. Pyrilamine, diphenhydramine and atropine were injected intravenously at doses of 9.7 mg/kg, 7.9 mg/kg and 5.2 mg/kg, respectively. Arrows indicate the injection time of saline or drugs. *, **: Significantly different from the control group with $P < 0.05$ and $P < 0.01$, respectively. From Kamei, C. and Tasaka, K.: Japan. J. Pharmacol., 57, 473-482 (1991), with permission.

contrast, 4-methylhistamine did not affect the response latency, even at a dose of $5\,\mu g$ (Fig. 24). This may indicate that the inhibitory effect of H_1 antagonists on the avoidance response may be exerted through the H_1 receptor.

7.5. Effect of acetylcholine on pyrilamine- and atropine-induced memory deficit

It is well known that ACh and some cholinergic drugs have a facilitatory effect on memory retrieval. On the other hand, classic H_1 antagonists including pyrilamine have an anti-cholinergic effect (Reuse, 1948). Therefore, the effect of ACh on pyrilamine and atropine-induced memory deficit was investigated.

I.c.v. injection of ACh at doses of 1 and $5\,\mu g$ did not influence the prolonged latency in the active avoidance response induced by pyrilamine previously administered intravenously. At a dose of $10\,\mu g$, however, ACh produced a significant shortening 2-10 min after ACh administration (Fig. 24). However, the prolongation of latency induced by atropine ($5.2\,mg/kg$, twice of ED_{50} value) (Tasaka et al., 1988) was strongly antagonized by ACh injection; a significant suppression was observed at $1\,\mu g$ of ACh. ID_{50} values for ACh on pyrilamine- and atropine-induced inhibitions of the avoidance response were 11.6 (5.11-65.4) μg and 1.23 (0.75-4.48) μg, respectively. No influence on the response latency was found following a single injection of ACh at a dose of $10\,\mu g$. Based on these findings, it can be assumed that an inhibition of active avoidance due to H_1 antagonists may be exerted via H_1 receptor in the brain.

7.6. Effect of chronic administration of H_1 antagonists on acquisition and retrieval of the memory

Clinically, newer developed H_1 antagonists have often been used in the treatment of chronic allergic diseases such as asthma and atopic dermatitis (Devlin and Hargrave, 1985); in some cases chronic administration is required. Therefore, it is important to investigate the effects of the chronic administration of these drugs on CNS activity. The effects of classical and newer H_1 antagonists on the acquisition of active avoidance response are shown in Figs. 25 and 26. Drugs were administered once daily for 17 consecutive days. Acquisition trial was performed 1 hr after oral administration of the drugs. The dose used for the study was ED_{50} values of each drug (Table 6). Acquisition trials were started on day 8. As shown in Fig. 25, in promethazine- and pyrilamine-treated groups, acquisition of active avoidance response was significantly retarded compared with the control group. Rats treated with diphenhydramine and chlorpheniramine also showed slower acquisition than in the control, though no significant difference were observed. In all of the groups treated with newer H_1 antagonists, acquisition of active avoidance was somewhat delayed in comparison with the control group, although there were no significant differences (Fig. 26). Similar results were obtained when the avoidance test was performed 24 hr after drug administration for all of the drugs tested.

As shown in Fig. 21 and Table 6, both classic H_1 antagonists and newer H_1 antagonists prolonged the response latency, indicating that these drugs inhibited retrieval in the memory by single administration. Chronic administration of H_1 antagonists on memory retrieval was also studied and shown in Fig. 27. Representative classical H_1 antagonists, pyrilamine and promethazine prolonged the response latency after the

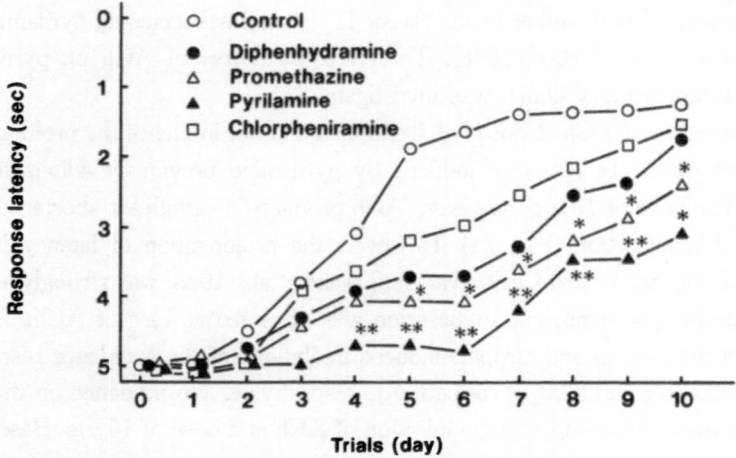

Fig. 25. Effects of chronic administration of classical H_1 antagonists on acquisition of the active avoidance response. *, **: Significantly different from control group with $P < 0.05$ and $P < 0.01$, respectively. From Kamei, C., Chung, Y.H. and Tasaka, K.: Psychopharmacology, 102, 312-318 (1990b), with permission.

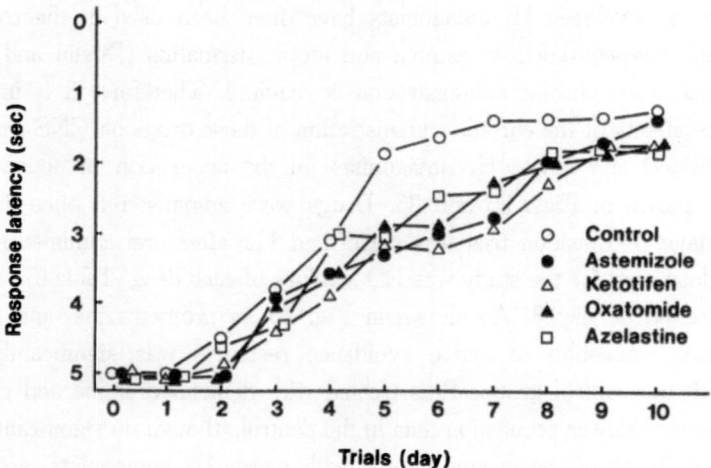

Fig. 26. Effects of chronic administration of newer H_1 antagonists on acquisition of the active avoidance response. From Kamei, C., Chung, Y.H. and Tasaka, K.: Psychopharmacology, 102, 312-318 (1990b), with permission.

initiation of drug administration; the extent of the prolongation of response latency remained almost constant throughout the entire period of drug administration. In the case of chlorpheniramine, significant prolongation of latency was observed until the 7th day but became progressively weaker over days. However, when newer H_1 antagonists such as astemizole, ketotifen, oxatomide and azelastine were administered similarly, the inhibition was transient, latency periods were restored to within the control values on the 3rd

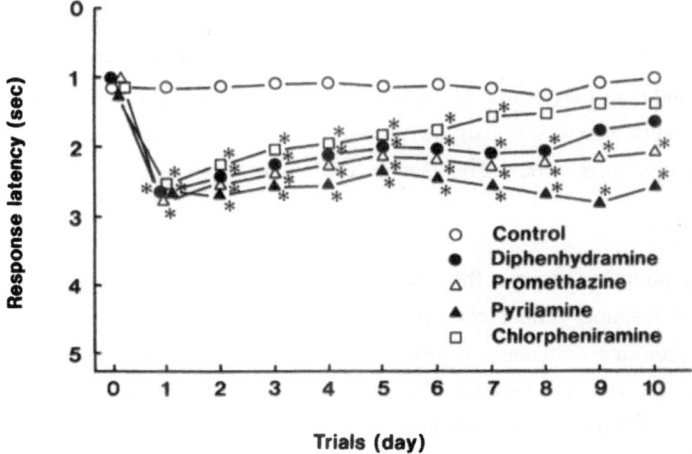

Fig. 27. Effects of chronic administration of classical H_1 antagonists on active avoidance response. ∗: Significantly different from control group with $P < 0.05$. From Kamei, C., Chung, Y.H. and Tasaka, K.: Psychopharmacology, 102, 312-318 (1990b), with permission.

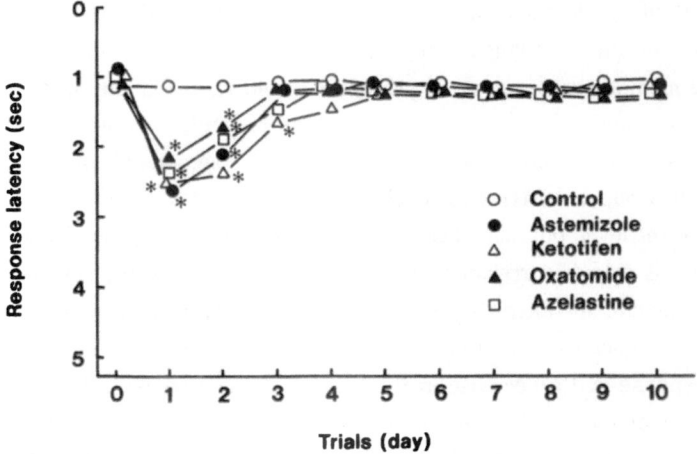

Fig. 28. Effects of chronic administration of newer H_1 antagonists on active avoidance response. ∗: Significantly different from control group with $P < 0.05$. From Kamei, C., Chung, Y.H. and Tasaka, K.: Psychopharmacology, 102, 312-318 (1990b), with permission.

day and no further prolongation was observed in any animals during the rest of the administration periods (Fig. 28).

So far it became apparent that animals given chronic administration of classical H_1 antagonists, especially promethazine and pyrilamine, showed much slower acquisition than those seen in the control group. Chronic administration of oxatomide, astemizole, azelastine and ketotifen had little effect on the acquisition of avoidance response. A significant finding was that transient retardation of memory retrieval caused by chronic administration of newer H_1 antagonist was restored on the 3rd day and no further prolongation of latency was observed during the chronic administration period. These findings may indicate that newer H_1 antagonist caused a lowering of the attentiveness for 2-3 days after initiation of the drug ingestion, however, after that the effects disappeared when these drugs were repeatedly used in the clinic.

8. Conclusion

It seems reasonable to assume that memory impairments in old rats induced by a long interruption of training reflect certain aspects of senile dementia in human beings. In humans, the memory deficiency is probably induced by pathological conditions such as cerebral ischemia or Alzheimer's disease. However, aging is common to all these pathological conditions. A way to improve memory deficiency in old humans is now eagerly sought. Based on the findings shown in Fig. 6, it can be concluded that i.c.v. injection of histamine takes an active part in facilitating memory retrieval in old rats. H_1 agonists effected a dose-related shortening of response latency as seen after histamine application, whereas H_2 agonists failed to prompt the response latency. Simultaneous i.c.v. injection of pyrilamine, abolished the histamine-induced shortening of response latency. Based on these findings, it may be concluded that histamine takes an active part in facilitating memory retrieval via H_1 receptor in old rats. On the other hand, whether or not i.c.v. injection of histamine truly reflect the histamine action exerted in the brain. To answer this question and to elucidate the role of endogenous histamine in learning and in memory retrieval, α-FMH was used. α-FMH caused a significant suppression not only memory retrieval but also learning acquisition of active avoidance response. In addition, there is a high correlation between a prolongation of the response latency and a decrease in histamine content of these brain areas. Therefore, it was concluded that an intimate relation may exist between a prolongation of response latency in the active avoidance response and a decrease in the brain histamine content. It would appear that endogenous histamine located in neuronal cells, especially in the hippocampus and hypothalamus, plays some crucial role in learning and memory of rats in an active avoidance paradigm. In connection with this, it was also found that a severe deficit for learning acquisition and memory retrieval was observed after hippocampectomy. Histamine (i.c.v.) application and histidine (i.p.) injection caused a visible improvement not only in learning acquisition but also in memory retrieval impaired after hippocampal lesions.

As shown above, histamine caused an activation of memory retrieval via H_1 receptor in some memory deficit models, consequently it is reasonable to assume that H_1 antagonists had an inhibition of memory retrieval. Therefore, the effects of some H_1 antagonists on the active avoidance response was studied. Classic H_1 antagonists caused a dose-related depressant effect on the active avoidance response, whereas newer H_1 antagonists showed a relatively weak effect. Chronic administration of astemizole and oxatomide caused only transient suppression of the response. However, classical H_1 antagonists such as promethazine caused sustained inhibition for as long as drug administration was continued. Clinically newer H_1 antagonists have often been used in the treatment of chronic allergic diseases such as asthma and atopic dermatitis; in some cases chronic administration is required for a long period of time. Although these drugs caused only transient suppressive effect, attention should be given when these drugs were administered chronically for a long time.

References

Ader, R., Weijnen, J.A.W.M. and Moleman, P.: Retention of a passive avoidance response as a function of the intensity and duration of electric shock. Psychon. Sci., 26, 125-128 (1972)

Ahn, H.-S. and Barnett, A.: Selective displacement of [^{3}H]mepyramine from peripheral vs. central nervous system receptors by loratadine, a non-sedating antihistamine. European J. Pharmacol., 127, 153-155 (1986)

Alvarez, E.O. and Banzán, A.M.: Histamine in dorsal and ventral hippocampus. II. Effects of H_1 and H_2 histamine antagonists on exploratory behavior in male rats. Physiol. Behav., 37, 39-45 (1986)

Arrang, J.-M., Garbarg, M. and Schwartz, J.-C.: Auto-inhibition of brain histamine release mediated by a novel class (H_3) of histamine receptor. Nature, 302, 832-837 (1983)

Arrang, J.-M., Garbarg, M., Lancelot, J.-C., Lecomte, J.-M., Pollard, H., Robba, M., Schunack, W. and Schwartz, J.-C.: Highly potent and selective ligands for histamine H_3-receptors. Nature, 327, 117-123 (1987)

Asdourian, D., Dark, J.G., Chiodo, L. and Papich, P.S.: Active avoidance in rats with unilateral hypothalamic and optic nerve lesions. Physiol. Behav., 19, 209-211 (1977)

Bhattacharya, S.K.: Central histamine receptors in learning and memory in rats. European J. Pharmacol., 183, 295 (1990)

Burešová, O., Bureš, J., Fifková, E., Vinogradova, O. and Weiss, T.: Functional significance of corticohippocampal connections. Exp. Neurol., 6, 161-172 (1962)

Cacabelos, R. and Alvarez, X.A.: Histidine decarboxylase inhibition induced by α-fluoromethylhistidine provokes learning-related hypokinetic activity. Agents Actions, 33, 131-134 (1991)

Coyle, J.T., Price, D.L. and DeLong, M.R.: Alzheimer's disease: A disorder of cortical cholinergic innervation. Science, 212, 1184-1190 (1983) de Almeida, M.A.M.R. and Izquierdo, I.: Intracerebroventricular histamine, but not 48/80, causes posttraining memory facilitation in the rat. Arch. int. Pharmacodyn., 291, 202-207 (1988)

de Groot, J.: The rat forebrain in stereotaxic coordinates. Verh. K. Ned. Akad. Wet. Natuurkund., 52, 1-40 (1959)

Devlin, J.P. and Hargrave, K.D.: Pulmonary and antiallergic drugs: Design and synthesis. In: Devlin, J.P. (ed) Pulmonary and Antiallergic Drugs. John Wiley & Sons, New York, pp.191-316, 1985

Durant, G.J., Duncan, W.A.M., Ganellin, C.R., Parsons, M.E., Blakemore, R.C. and Rasmussen, A.C.: Impromidine (SK & F 92676) is a very potent and specific agonist for histamine H_2 receptors. Nature, 276, 403-405 (1978)

Egashira, T., Murayama, F., Kimba, Y. and Yamanaka, Y.: Age-related decrements of cholinergic markers in aged rat brain. Japan. J. Pharmacol., 52 (suppl), 74P (1990)

Feldberg, W. and Sherwood, S.L.: Injections of drugs into the lateral ventricle of the cat. J. Physiol., 123, 148-167 (1954)

Ganellin, C.R.: Chemistry and structure-activity relationships of drugs acting at histamine receptors. In: Ganellin C.R. and Parsons, M.E. (ed) Pharmacology of Histamine Receptors. Wright·PSG, Bristol, pp. 10-102, 1982

Garbarg, M., Barbin, G., Bischoff, S., Pollard, H. and Schwartz, J.C.: Dual localization of histamine in an ascending neuronal pathway and in non-neuronal cells evidenced by lesions in the lateral hypothalamic area. Brain Res., 106, 333-348 (1976)

Garbarg, M., Barbin, G., Rodergas, E. and Schwartz, J.C.: Inhibition of histamine synthesis in brain by α-fluoromethylhistidine, a new irreversible inhibitor: In vitro and in vivo studies. J. Neurochem., 35, 1045-1052 (1980)

Grossman, S.P.: Avoidance behavior and aggression in rats with transections of the lateral connections of the medial or lateral hypothalamus. Physiol. Behav., 5, 1103-1108 (1970)

Haas, H.L.: Histamine potentiates neuronal excitation by blocking a calcium-dependent potassium conductance. Agents Actions, 14, 534-537 (1984)

Haroutunian, V., Barnes, E. and Davis, K.L.: Cholinergic modulation of memory in rats. Psychopharmacology, 87, 266-271 (1985)

Inami, T., Tanaka, T., Sakurai, M. and Hayashi, S.: Learning impairment in aged, spontaneously hypertensive rats. Japan. J. Pharmacol., 52 (suppl), 355P (1990)

Ikeda, Y. and Matsushita, K.: The learning, memory and drug effects in aged rats In: Tadokoro, S. (ed) Investigation of Antidementia Drug. Animal Experiments of Learning and Memory. (in Japanese), Seiwa Shoten, Pub. Tokyo, pp. 108-120, 1985

Ishikawa, A., Ikeda, Y., Takato, M., Matsushita, H. and Hata, S.: The light-dark discriminative learning of aged rats. A comparison of the positive and negative reinforcement. Eighth International Congress of Pharmacology, p. 626, 1981

Isaacson, R.L., Douglas, R.J. and Moore, R.Y.: The effect of radical hippocampal ablation on acquisition of avoidance response. J. Comp. Physiol. Psychol., 54, 625-628 (1961)

Kakihana, M., Yamazaki, N. and Nagaoka, A.: Effects of idebenone (CV-2619) on the concentrations of acetylcholine and choline in various brain regions of rats with cerebral ischemia. Japan. J. Pharmacol., 36, 357-363 (1984)

Kamei, C., Kiniwa, S., Ikegami, N. and Tasaka, K.: Effect of 4-(p-chlorobenzyl)-2-[N-methylperhydroazepinyl-(4)]-1-(2H)-phthalazinone hydrochloride (azelastine) on EEGs and behavior in rats. Jpn. J. Clin. Pharmacol. Ther., 12, 297-310 (1981a)

Kamei, C., Chung, Y.H., Dabasaki, T. and Tasaka, K.: Species differences elicited by intraventricular injection of histamine on EEGs and behavior. Japan. J. Pharmacol., 31 (suppl), 86P (1981b)

Kamei, C., Dabasaki, T. and Tasaka, K.: Cataleptic effect of histamine induced by intraventricular injection in mice. Japan. J. Pharmacol., 33, 1081-1084 (1983)

Kamei, C., Dabasaki, T. and Tasaka, K.: Effect of intraventricular injection of histamine on the pinna reflex in mice. Japan. J. Pharmacol., 35, 193-195 (1984)

Kamei, C., Akahori, H. and Tasaka, K.: Influence of histamine and related compounds on the hypnotic effect of thiopental in mice. J. Pharmacobio-Dyn., 9, 112-116 (1986)

Kamei, C., Tsujimoto, S. and Tasaka, K.: Effects of cholinergic drugs and cerebral metabolic activators on memory impairment in old rats. J. Pharmacobio-Dyn., 13, 772-777 (1990a)

Kamei, C., Chung, Y.H. and Tasaka, K.: Influence of certain H_1-blockers on the step-through active avoidance response in rats. Psychopharmacology, 102, 312-318 (1990b)

Kamei, C. and Tasaka, K.: Participation of histamine in the step-through active avoidance response and its inhibition by H_1-blockers. Japan. J. Pharmacol., 57, 473-482 (1991)

Kamei, C. and Tasaka, K.: Effect of intracerebroventricular injection of histamine on memory impairment induced by hippocampal lesions in rats. Japan. J. Pharmacol., 58 (suppl), 55P (1992)

Kamei, C., Okumura, Y. and Tasaka, K.: Influence of histamine depletion on learning and memory recollection in rats. Psychopharmacology, 111, 376-382 (1993a)

Kamei, C. and Tasaka, K.: Effect of histamine on memory retrieval in old rats. Biol. Pharm. Bull., 16, 128-132 (1993b)

Kaneko, T., Kitahara, A., Ozaki, S., Takizawa, K. and Yamatsu, K.: Effects of azelastine hydrochloride, a novel anti-allergic drug, on the central nervous system. Arzneim.-Forsch., 31, 1206-1212 (1981)

Kollonitsch, J. Patchett, A.A., Marburg, S., Maycock, A.L., Perkins, L.M., Doldouras, G.A., Duggan, D.E. and Aster, S.D.: Selective inhibitors of biosynthesis of aminergic neurotransmitters. Nature, 274, 906-908 (1978)

Laduron, P.M., Janssen, P.F.M., Gommeren, W. and Leysen, J.E.: In vitro and in vivo binding characteristics of a new long-acting histamine H_1-antagonist, astemizole. In: Astemizole: a new, non-sedative, long-acting H_1-antagonist. The Medicine Publishing Foundation, Symposium Series 11, Oxford, pp. 11-23, 1983

Lippa, A.S., Pelham, R.W., Beer, B., Critchett, D.J., Dean, R.L. and Bartus, R.T.: Brain cholinergic dysfunction and memory in aged rats. Neurobiol. Aging, 1, 13-19 (1980)

Maeyama, K., Watanabe, T., Yamatodani, A., Taguchi, Y., Kambe, H. and Wada, H.: Effect of α-fluoromethylhistidine on the histamine content of the brain of W/W^V mice devoid of mast cells: turnover of brain histamine. J. Neurochem., 41, 128-134 (1983)

Martin, U. and Römer, D.: The pharmacological properties of a new, orally active antianaphylactic compound: ketotifen, a benzocycloheptathiophene. Arzneim.-Forsch., 28, 770-782 (1978)

McNew, J.J. and Thompson, R.: Role of the limbic system in active and passive avoidance conditioning in the rat. J. Comp. Physiol. Psychol., 61, 173-180 (1966)

Monti, J.M., D'Angelo, L., Jantos, H. and Pazos, S.: Effects of α-fluoromethylhistidine on sleep and wakefulness in the rat. J. Neural. Transm., 72, 141-145 (1988)

Muñoz, C. and Grossman, S.P.: Some behavioral effects of selective neuronal depletion by kainic acid in the dorsal hippocampus of rats. Physiol. Behav., 25, 581-587 (1980)

Nakagawa, Y., Yamazaki, A., Ishima, T., Tamura, M., Ogasawara, T., Ukai, Y. and Kimura, K.: Age-related changes in learning abilities in Fischer 344 rats. Japan. J. Pharmacol., 49 (suppl), 267P (1989)

Nakahiro, M., Fujita, N., Fukuchi, I., Saito, K., Nishimura, T. and Yoshida, H.: Pantoyl-γ-aminobutyric acid facilitates cholinergic function in the central nervous system. J. Pharmacol. Exp. Ther., 232, 501-506 (1985)

Ohmori, K., Ishii, H., Shuto, K. and Nakamizo, N.: Pharmacological studies on oxatomide: (5) Effect on the central and peripheral nervous systems. Folia pharmacol. japon., 81, 245-266 (1983a)

Ohmori, K., Ishii, H., Nito, M., Shuto, K. and Nakamizo, N.: Pharmacological studies on oxatomide (KW-4354). (7) Antagonistic effects on chemical mediators. Folia pharmacol. japon., 81, 399-409 (1983b)

Ohmori, K., Ishii, H. and Ishii, A.: Pharmacological studies on oxatomide (10) Antiallergic effects of oxatomide and other antiallergic drugs. Clin. Report, 21, 4045-4052 (1987)

Oishi, R., Itoh, Y., Nishibori, M. and Saeki, K.: Effects of the histamine H_3-agonist (R)-α-methylhistamine and the antagonist thioperamide on histamine metabolism in the mouse and rat brain. J. Neurochem., 52, 1388-1392 (1989)

Oka, M. and Shimizu, M.: A simple avoidance procedure for testing psychotropic drugs in mice. Japan. J. Pharmacol., 25, 121-127 (1975)

Onodera, K., Yamatodani, A. and Watanabe, T.: Effects of α-fluoromethylhistidine on locomotor activity, brain histamine and catecholamine contents in rats. Meth. Find. Exp. Clin. Pharmacol., 14, 97-105 (1992)

Panula, P., Yang, H.-Y.T. and Costa, E.: Histamine-containing neurons in the rat hypothalamus. Proc. Natl. Acad. Sci. U.S.A., 81, 2572-2576 (1984)

Papsdorf, J.D. and Woodruff, M.: Effects of bilateral hippocampectomy on the rabbit's acquisition of shuttle-box and passive-avoidance responses. J. Comp. Physiol. Psychol., 73, 486-489 (1970)

Paxinos, G. and Watson, C.: The rat brain in stereotaxic coordinates. Academic Press. San Diego, 1986

Perry, E.K., Blessed, K.G., Tomlinson, B.E., Perry, R.H., Crow, T.J., Cross, A.J., Dockray, G.T., Dimaline, R. and Arregui, A.: Neurochemical activities in human temporal lobe related to aging and Alzheimer's type changes. Neurobiol. Aging, 2, 251-256 (1981)

Reuse, J.J.: Comparisons of various histamine antagonists. Br. J. Pharmacol., 3, 174-180 (1948)

Rich, I. and Thompson, R.: Role of the hippocampo-septal system, thalamus, and hypothalamus in avoidance conditioning. J. Comp. Physiol. Psychol., 59, 66-72 (1965)

Rigter, H., van Riezen, H. and de Wied, D.: The effects of ACTH- and vasopressin-analogues on CO_2-induced retrograde amnesia in rats. Physiol. Behav., 13, 381-388 (1974)

Rüthrich, H.-L., Wetzel, W. and Matthies, H.: Memory retention in old rats: Improvement by orotic acid. In: Marsan, C.A. and Matthies, H. (ed) Neuronal Plasticity and Memory

Formation. Raven Press, New York, pp. 227-230, 1982.

Sakai, N., Onodera, K., Maeyama, K., Yanai, K. and Watanabe, T.: Effects of (S)-α-fluoromethylhistidine and metoprine on locomotor activity and brain histamine content in mice. Life Sci., 51, 397-405 (1992)

Sakurai, E., Niwa, H., Yamasaki, S., Maeyama, K. and Watanabe, T.: The disposition of a histidine decarboxylase inhibitor (S)-α-fluoromethylhistidine in rats. J. Pharm. Pharmacol., 42, 857-860 (1990)

Sakurai, T., Ojima, H., Yamasaki, T., Kojima, H. and Akashi, A.: Effects of N-(2,6-dimethylphenyl)-2-(2-oxo-1-pyrrolidinyl)acetamide (DM-9384) on learning and memory in rats. Japan. J. Pharmacol., 50, 47-53 (1989)

Schild, H.O.: pA, a new scale for the measurement of drug antagonism. Br. J. Pharmacol., 2, 189-206 (1947)

Schwartz, J.C., Lampart, C. and Rose, C.: Histamine formation in rat brain in vivo: Effects of histidine loads. J. Neurochem., 19, 801-810 (1972)

Shibata, K., Hirano, Y., Fujino, A., Shimazu, M., Nagakawa, N., Inoue, K. and Uemura, I.: Metabolic fate of ^{14}C-oxatomide in rats. Jpn. Pharmacol. Ther., 12, 3887-3903 (1984)

Soda, Y., Mori, I., Yokoyama, N., Horisaka, K., Shichino, F., Shimada, A., Murai, K. and Ito, N.: Metabolic fate of mequitazine (LM-209) (1) Absorption, distribution and excretion of LM-209 in rats. Iyakuhin Kenkyu, 12, 462-480 (1981)

Sunami, A. and Tasaka, K.: Two aspects of the excitatory influence of histamine on hippocampal neurons in guinea pigs. Meth. Find. Exp. Clin. Pharmacol., 13, 85-91 (1991)

Tagami, H., Sunami, A., Akagi, M. and Tasaka, K.: Effect of histamine on the hippocampal neurons in guinea-pigs. Agents Actions, 14, 538-542 (1984)

Tanaka, S., Shirasaki, Y., Yamada, F., Endo, W. and Ashida, S.: Impairment effect of DM-9384, a new cognition enhancer, on learning deficits in cerebral embolized rats. Japan. J. Pharmacol., 52 (suppl), 74P (1990)

Tasaka, K., Kamei, C., Akahori, H. and Kitazumi, K.: The effects of histamine and some related compounds on conditioned avoidance response in rats. Life Sci., 37, 2005-2014 (1985)

Tasaka, K., Kamei, C., Katayama, S., Kitazumi, K., Akahori, H. and Hokonohara, T.: Comparative study of various H_1-blockers on neuropharmacological and behavioral effects including 1-(2-ethoxyethyl)-2-(4-methyl-1-homopiperazinyl)benzimidazole difumarate (KB-2413), a new antiallergic agent. Arch. int. Pharmacodyn., 280, 275-291 (1986)

Tasaka, K.: Anti-allergic drugs. Drugs of Today, 22, 101-133 (1986)

Tasaka, K., Kamei, C., Chung, Y.H. and Nakano, S.: Pharmacological effects of oxitropium bromide on the central nervous system. Pharmacometrics, 36, 425-432 (1988)

Tasaka, K., Chung, Y.H., Sawada, K. and Mio, M.: Excitatory effect of histamine on the arousal system and its inhibition by H_1 blockers. Brain Res. Bull., 22, 271-275 (1989a)

Tasaka, K., Kamei, C., Nakano, S., Tsujimoto, S. and Chung, Y.H.: Effect of epinastine, a new antiallergic agent, on the central nervous system. Pharmacometrics, 38, 53-62 (1989b)

Tasaka, K., Kamei, C., Tsujimoto, S., Yoshida, T. and Aoki, I.: Central effect of the potent long-acting H_1-antihistamine levocabastine. Arzneim.-Forsch., 40, 1295-1299 (1990)

Tatsumi, K., Ou, T., Yamada, H. and Yoshimura, H.: Studies on metabolic fate of a new antiallergic agent, azelastine (4-(p-chlorobenzyl)-2-[N-methylperhydroazepinyl-(4)]-1-(2H)-phthalazinone hydrochloride). Japan. J. Pharmacol., 30, 37-48 (1980)

Uzan, A., Le Fur, G. and Malgouris, C.: Are antihistamines sedative via a blockade of brain H_1 receptors? J. Pharm. Pharmacol., 31, 701-702 (1979)

Van den Brink, F.G. and Lien, E.J.: Competitive and noncompetitive antagonism. In: Rocha e Silva, M. (ed) Histamine II and Anti-Histaminics, Springer-Verlag, Berlin, pp. 333-367, 1978

Wallace, J.E., Krauter, E.E. and Campbell, B.A.: Motor and reflexive behavior in the aging rat. J. Gerontol., 35, 364-370 (1980)

Wauquier, A., Van den Broeck, W.A.E., Awouters, F. and Janssen, P.A.J.: A comparison between astemizole and other antihistamines on sleep-wakefulness cycles in dogs. Neuropharmacology, 20, 853-859 (1981)

Whitehouse, P.J., Price, D.L., Struble, R.G., Clark, A.W., Coyle, J.T. and DeLong, M.R.: Alzheimer's disease and senile dementia: Loss of neurons in the basal forebrain. Science, 215, 1237-1239 (1982)

Winter, C.A. and Flataker, L.: The effect of antihistaminic drugs upon the performance of trained rats. J. Pharmacol. Exp. Ther., 101, 156-162 (1951)

Wyngaarden, J.B. and Seevers, M.H.: The toxic effects of antihistaminic drugs. J. Amer. Med. Ass., 145, 277-282 (1951)

Yamatodani, A. and Watanabe, T.: Studies with α-fluoromethylhistidine as a probe. In: Watanabe, T. and Wada, H. (ed) Histaminergic Neurons: Morphology and Function. CRC Press, Boca Raton, pp. 231-240, 1991

Yamamoto, M., Shimizu, M., Sakamoto, N., Ohtomo, H. and Kogure, K.: Protective effects of indeloxazine hydrochloride on cerebral ischemia in animals. Arch. int. Pharmacodyn., 290, 16-24 (1987)

Yamamoto, M. and Shimizu, M.: Cerebral activating properties of indeloxazine hydrochloride. Neuropharmacology, 26, 761-770 (1987)

Yamazaki, N., Take, Y., Nagaoka, A. and Nagawa, Y.: Beneficial effect of idebenone (CV-2619) on cerebral ischemia-induced amnesia in rats. Japan. J. Pharmacol., 36, 349-356 (1984)

Chapter 3
Role of brain histamine on corticosteroid release

1. Introduction

There is a considerable amount of literature that described how parenteral injection of histamine stimulates adrenocortical secretion in dogs and rats (Katsuki et al., 1967; Hirose et al., 1976; Cowan, 1975; Reilly, 1984), but few experiments have been done to clarify the role of brain histamine in releasing corticosteroids. Parenteral administration of histamine induces some side effects such as hypotensive effect and algesia, and these responses undoubtedly exert some influence on the pituitary-adrenocortical axis and adrenal steroid release. Since histamine does not enter into the brain through the blood-brain barrier, intracerebroventricular (i.c.v.) injection has been commonly employed to study the histamine effect on the central nervous system (CNS) (Feldberg and Sherwood, 1954; Tasaka et al., 1989). Rudolph et al. (1979) reported that activation of central H_1 receptors increases ACTH secretion in dogs, and that activation of central H_2 receptors decreases ACTH secretion. On the contrary, in rats, Bugajski and Gadek (1983) reported that i.c.v. injection of histamine increased plasma corticosterone concentrations and this effect was mediated by both H_1 and H_2 receptors. From these findings, it seems likely that species difference exists between dogs and rats in histamine-induced corticosteroid release especially in the participation of H_2 receptors. Recently, the presence of the H_3 receptors, which is responsible for the feedback control of histamine synthesis and release, has also been demonstrated (Arrang et al., 1987). However, it is not known whether or not the H_3 receptor is involved in histamine-induced steroidogenesis.

On the other hand, α-fluoromethylhistidine (α-FMH), a selective histamine depletor, has been widely used to study the role of histamine in the CNS (Kollonitsch et al, 1978). Therefore, it seems likely that this drug is available for investigation of the role of central histamine on the corticosteroidogenesis. In addition, it is well known that the hypothalamus contains a large amount of histamine and the cell bodies of histaminergic projections are present in the tuberal region of the posterior hypothalamus (Panula et al., 1984). Therefore, it seems likely that electric stimulation or electrocoagulation of the posterior hypothalamus may give some clues to elucidate the mechanism of the histamine-induced corticosteroidogenesis.

In this chapter, the effects of i.c.v. injection of histamine and its related compounds on plasma ACTH and corticosteroid concentrations in both dogs and rats will be mentioned in association with the mechanism connecting central histamine receptors and the glucocorticoid release.

2. Effect of parenteral administration of histamine on plasma ACTH and corticosteroid concentrations

2.1. Dogs

It is known that proteolytic enzymes are effective as anti-inflammatory agents (Cohen et al., 1955; Martin et al., 1954; Miechowski and Ercoli, 1956). However, no working hypothesis on the mechanism of anti-inflammatory action was settled in the 1970s. When Prozime-10, a proteolytic enzyme extracted from the cultured broth of *Aspergillus melleus*, was injected i.v. into rats, it was found that Prozime-10 was very effective in alleviating carrageenin edema, adjuvant arthritis and carrageenin-induced granuloma pouch (Tasaka et al., 1980). While investigating active substances to inhibit inflammatory responses, it was found that the plasma corticosterone level markedly increased after i.v. injection of Prozime-10. The amount of released corticosterone was large enough to permit expectations of an anti-inflammatory effect. Since i.v. injection of Prozime-10 often induces a hypotensive reaction, the effect of Prozime-10 on changes in plasma histamine and cortisol levels was investigated by i.v. perfusion in anesthetized dogs with the continuous monitering of blood pressure. Consequently, it became evident that the plasma histamine level increased immediately after Prozime-10 injection, though no

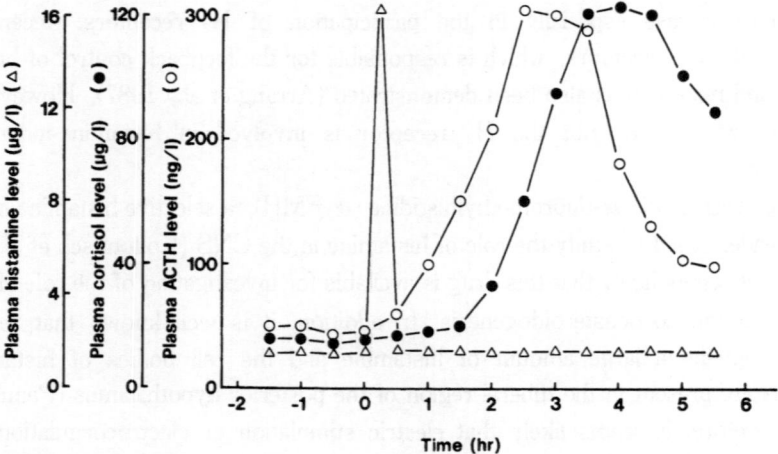

Fig. 1. Sequential changes in ACTH and cortisol concentrations of dog plasma after Prozime-10 adminis-tration (5 mg/kg, i.v.). The results were obtained from 7 dogs which exhibited sequential appearance of ACTH and cortisol. From Maki, K., Takahashi, H., Kakimoto, M., Nakajima, M. and Tasaka, K.: Pharmacology, 23, 230-236 (1981), with permission.

decrease in blood pressure was observed. Plasma ACTH increased with a latency of 30 min which was followed by remarkable elevation of plasma cortisol in many instances (Maki et al., 1981) (Fig. 1). However, in some cases, changes in plasma cortisol occurred simultaneously with ACTH after Prozime-10. This probably means that the second mediator(s) other than ACTH releases cortisol directly from the adrenal gland. Since plasma histamine significantly increased prior to ACTH elevation, such a rapid elevation of cortisol can be explained partly, by the direct stimulation of the adrenal cortex by histamine. Actually, when histamine was added to a medium containing dispersed canine adrenocortical cells, cortisol was released in a dose-dependent fashion. In this case, 85-95% of isolated adrenocortical cells were viable when the trypan blue dye exclusion test was performed (Maki et al., 1981) (Fig. 2).

Based on these findings, it was assumed that increased plasma histamine after Prozime-10 injection may cross the blood-brain barrier around the hypophyseal gland and then histamine stimulates the anterior pituitary cells so as to release ACTH. It is known that blood-brain barrier is not tight around the hypophyseal gland. After released histamine elicits an increase of ACTH concentration in the plasma, ACTH enhanced the cortisol output from the adrenal gland.

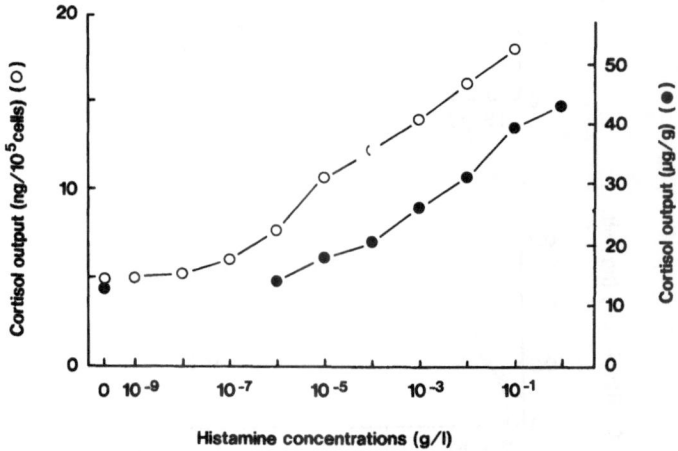

Fig. 2. Cortisol release from the dog adrenal slices and the dispersed cortical cells. ○ = cortisol output from dispersed cortical cells (10^5 cells). ● = cortisol output from 100 mg of wet adrenal slices. From Maki, K., Takahashi, H., Kakimoto, M., Nakajima, M. and Tasaka, K.: Pharmacology, 23, 230-236 (1981), with permission.

2.2. Rats

The time course of serum ACTH and corticosterone levels after i.p. histamine injection of 0.5 mg/100 g or 1 mg/100 g is shown in Fig. 3. ACTH concentration increased dose-dependently 2.5 min after the injection. The peak of ACTH concentration was achieved 10 min after the injection in both cases. Despite the difference of ACTH concentration in the serum, the time courses of corticosterone concentration were similar until 10 min after histamine injection. The corticosterone concentration reached the maximum 60 min after the injection of 1 mg/100 g of histamine.

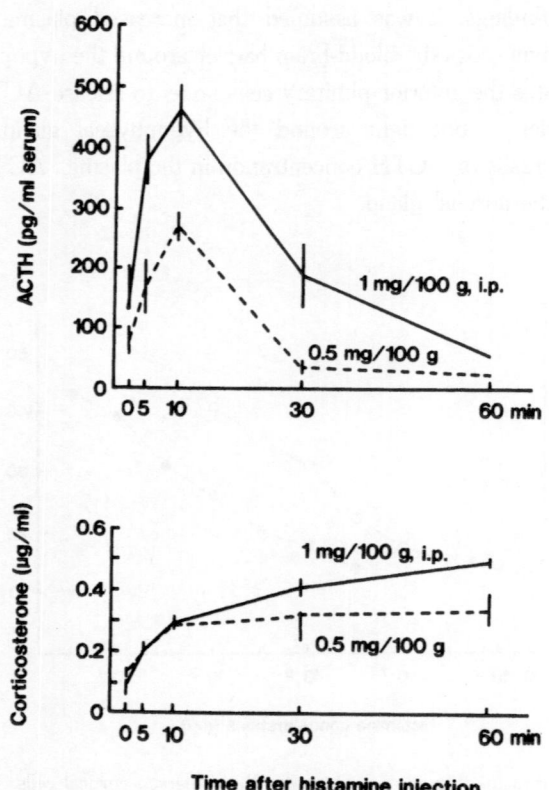

Fig. 3. Time-courses of ACTH and corticosterone concentrations in the rat serum after a single injection of histamine (i.p.). Three rats were used for each point. Vertical lines represents standard errors of the mean. From Morita, Y. and Koyama, K.: Japan. J. Pharmacol., 29, 59-65 (1979), with permission.

3. Effect of intracerebroventricular injection of histamine on plasma ACTH and corticosteroid concentrations

3.1. Dogs

As mentioned previously, when the plasma histamine concentration was increased, histamine could cross the blood-brain barrier and enter into the brain around the hypophyseal gland. However, in such cases it was impossible to estimate how much histamine was incorporated into the brain. To elucidate the histamine action on the basis of dose-response relation, only histamine applied by i.c.v. application was selected in the following studies. I.c.v. injection of histamine caused a dose-related increase in plasma ACTH concentrations ranging from 1 to 10 μg/kg. After injection of histamine at doses of 1 and 2 μg/kg, significant increase was not induced, while at doses of 5 and 10 μg/ kg, plasma ACTH concentrations significantly increased from 15 to 60 min after histamine injection (Fig. 4). Similarly, histamine at doses of 1-10 μg/kg produced a dose-dependent increase in plasma cortisol concentrations. At 2 μg/kg, a significant elevation of the plasma cortisol concentration was observed 30 and 60 min after injection. At doses of 5 and 10 μg/kg, plasma cortisol concentration increased more rapidly reaching the significant level at 15 min and lasting for 60 min. However, after i.c.v.

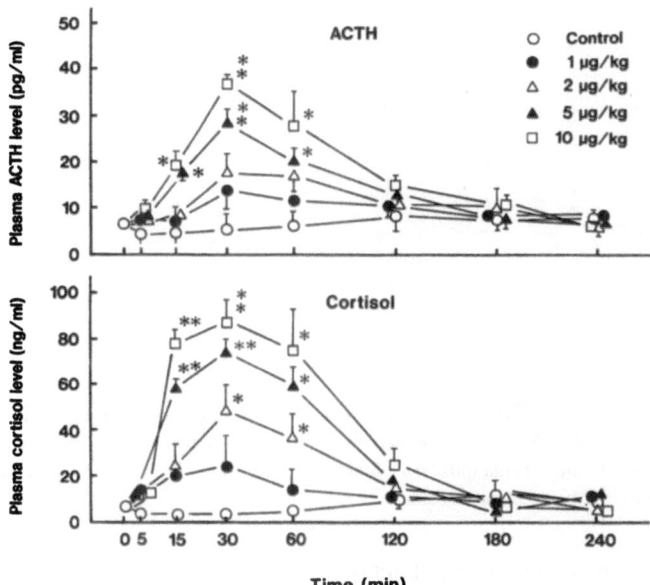

Fig. 4. Changes in plasma ACTH and cortisol concentrations induced by intracerebroventricular injection of histamine in dogs. *, **: Significantly different from the control group with P < 0.05 and P < 0.01, respectively. From Tsujimoto, S., Kamei, C., Yoshida, T. and Tasaka, K.: Pharmacology, 47, 73-83 (1993a), with permission.

Table 1. Comparison of the secretory rates of plasma ACTH and cortisol after intracerebroventricular injection of histamine in dogs

Drugs	Dose (μg/kg)	ACTH (min)	Cortisol (min)
Histamine	1	21.7 ± 5.9	16.6 ± 1.5
	2	21.9 ± 2.9	17.8 ± 1.0
	5	16.1 ± 1.5	9.2 ± 1.2 *
	10	17.6 ± 3.7	9.1 ± 2.7 *

Secretory rate was indicated by the time (min) required to reach half the maximum of ACTH or cortisol secretion. *: Significantly different from ACTH groups with P < 0.05. Reproduced from Tsujimoto, S., Kamei, C., Yoshida, T. and Tasaka, K.: Pharmacology, 47, 73-83 (1993a), with permission.

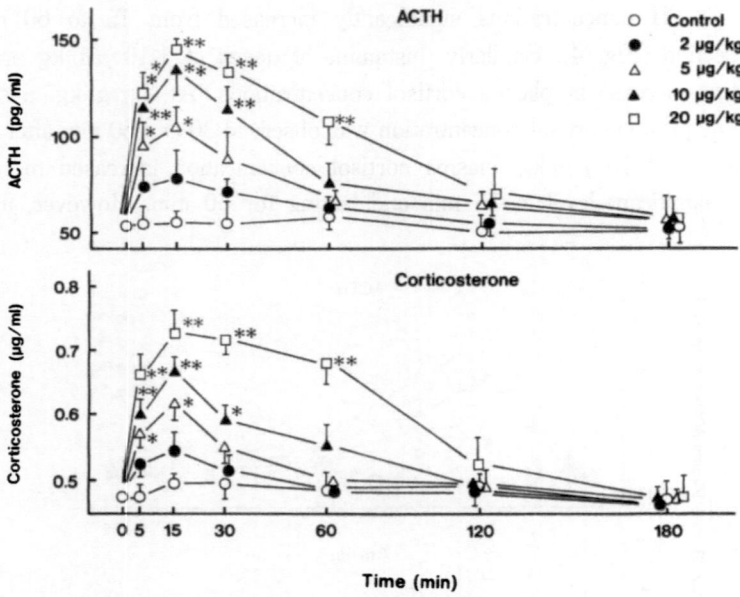

Fig. 5. Changes in plasma ACTH and corticosterone concentrations induced by intracerebroventricular injection of histamine in rats. *, **: Significantly different from the control group with P < 0.05 and P < 0.01, respectively. From Tsujimoto, S., Okumura, Y., Kamei, C. and Tasaka, K.: Br. J. Pharmacol. 109, 807-813 (1993b), with permission.

injection of histamine, no change in blood pressure was induced even at a dose of 10 μg/kg. Table 1 shows the secretory rate of plasma ACTH and cortisol release induced by histamine. Secretory rate was determined as the time (min) required to reach half the maximal blood concentrations in each case, and the value was calculated using the blood concentrations determined at 5 points (0, 5, 10, 15 and 30 min) by means of the least-squares method. Secretory rate of plasma cortisol concentrations induced by histamine (5 and 10 μg/kg) was significantly higher than those of ACTH.

It was found that the elevation of the cortisol concentrations in plasma caused by i.c.v. application of histamine appeared more rapidly than that of ACTH, and cortisol output was more sensitive to histamine than that seen in ACTH.

3.2. Rats

In rats, almost the same results as those seen in dogs were obtained, that is, the i.c.v. injection of histamine at doses of 5-20 μg/kg, histamine induced a significant elevation of plasma ACTH concentration in a dose-dependent manner. At a dose of 20 μg/kg, a significant increase was observed from 5 min which persisted even after 60 min. On the other hand, a moderate increase in plasma corticosterone occurred at a dose of 2 μg/kg histamine and the increase became more marked as the doses increased. At a dose of 20 μg/kg, plasma corticosterone increased very promptly as in the case of ACTH, and sequential changes in plasma concentrations of both ACTH and corticosterone were quite similar at all dosages of histamine injected (Fig. 5).

In rat study, i.c.v. injection of histamine induced almost simultaneous increase in plasma ACTH and glucocorticoid concentrations, though the changes in plasma concentrations were less marked than that seen in dogs.

4. Effects of H_1 and H_2 agonists on plasma ACTH and corticosteroid concentrations

4.1. Dogs

2-Methylhistamine, a representative H_1 agonist, at doses of 25 and 50 μg/kg caused a significant and dose-related increase in plasma ACTH concentrations, whereas 4-methylhistamine, an H_2 agonist induced no such elevation even at a dose of 50 μg/kg. Similar results were obtained after administration of 2-methylhistamine when plasma cortisol concentrations were measured, though the extent was less remarkable than that seen after histamine. On the contrary, no such change in plasma cortisol was observed after i.c.v. injection of 4-methylhistamine even at a dose of 50 μg/kg (Fig. 6).

H_1 agonist, 2-methylhistamine caused an elevation of plasma ACTH and cortisol concentrations, whereas H_2 agonist, 4-methylhistamine had no effect, suggesting that the corticosteroidogenesis induced by histamine is probably mediated via H_1 receptors in dogs. On the contrary, Rudolph et al. (1979) have reported that 4-methylhistamine at a dose of 25 μg/kg elicited a significant decrease in plasma ACTH concentrations 20-30 min after the start of the infusion. From this result, Rudolph et al. (1979) proposed that histamine may be having an inhibitory effect on ACTH secretion via H_2 receptors. With respect to this, it has been demonstrated that clonidine inhibits ACTH secretion in dogs

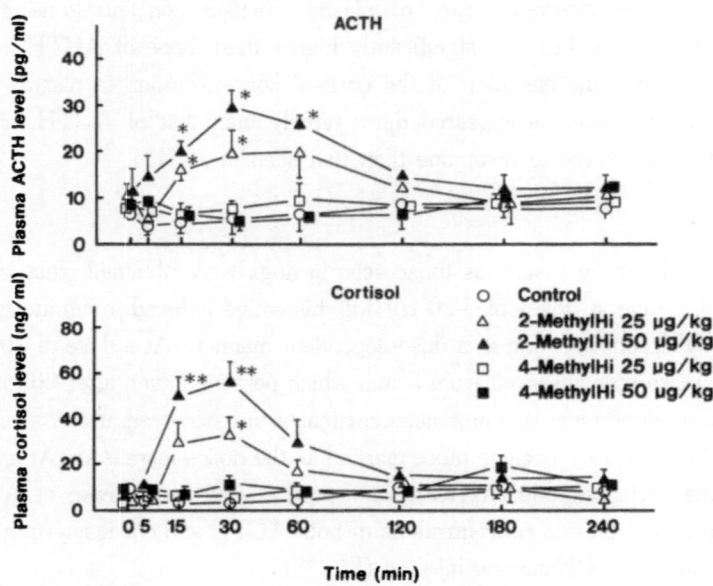

Fig. 6. Changes in plasma ACTH and cortisol concentrations induced by intracerebroventricular injection of 2-methylhistamine and 4-methylhistamine in dogs. 2-MethylHi: 2-Methylhistamine, 4-MethylHi: 4-methylhistamine. *, **: Significantly different from the control group with $P < 0.05$ and $P < 0.01$, respectively. From Tsujimoto, S., Kamei, C., Yoshida, T. and Tasaka, K.: Pharmacology, 47, 73-83 (1993a), with permission.

(Ganong et al., 1976), and it has been claimed that clonidine stimulates H_2 receptors (Karppanen et al., 1976).

4.2. Rats

When 2-methylhistamine was injected i.c.v. at doses of 10-50 μg/kg, both plasma ACTH and corticosterone concentrations increased. A significant increase was observed at doses of 20 and 50 μg/kg. When 4-methylhistamine and impromidine, H_2 agonists, were similarly injected, an increase in both plasma ACTH and corticosterone concentrations was elicited, but in both cases the potencies were slightly less than those induced by 2-methylhistamine. A significant increase was observed at doses higher than 20 μg/ kg in both cases (Fig. 7).

Differing from the findings observed in dogs, H_2 agonists, 4-methylhistamine and impromidine also caused an effect similar to that seen after histamine application, though the effects of these drugs were weaker than those of histamine or 2-methylhistamine. When histamine agonists were used for the purpose of receptor analysis, it was assumed that not only H_1 but also H_2 receptors are implicated in histamine-induced hormone secretion in rats.

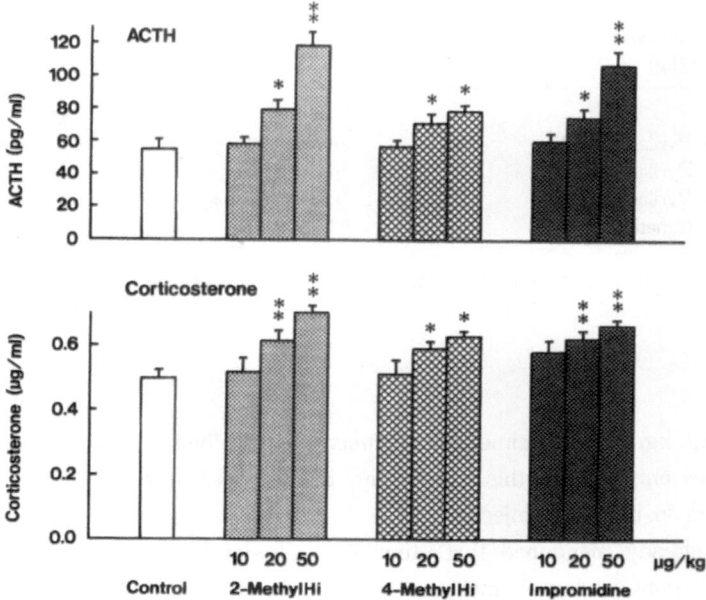

Fig. 7. Changes in plasma ACTH and corticosterone concentrations induced by intracerebroventricular injection of 2-methylhistamine, 4-methylhistamine and impromidine in rats. Drugs were injected 15 min before the blood sampling. *, ** : Significantly different from control group with $P < 0.05$ and $P < 0.01$, respectively. From Tsujimoto, S., Okumura, Y., Kamei, C. and Tasaka, K.: Br. J. Pharmacol., 109, 807-813 (1993b), with permission.

5. Effects of H_1 and H_2 antagonists on the elevation of plasma ACTH and corticosteroid concentrations induced by histamine or 4-methylhistamine

5.1. Dogs

To clarify which receptor (H_1 or H_2) is actually involved in the production of corticosteroid and ACTH release eliciting by i.c.v. histamine application. For this purpose, some H_1 and H_2 antagonists were employed. As shown in Table 2, i.c.v. injection of pyrilamine at doses of 1.3 μg/kg (a dose equivalent to 1/10 of histamine dose in molarity) or 13.0 μg/kg (a dose equivalent to histamine dose in molarity) in combination with histamine (5 μg/kg) caused a significant inhibition of histamine-induced increase in ACTH concentrations compared with that elicited by a single histamine injection. By contrast, cimetidine caused no such inhibition even at a dose of 11.0 μg/kg (a dose equivalent to histamine dose in molarity). With respect to plasma cortisol concentrations, simultaneous administration of pyrilamine also caused a significant inhibition at doses of

Table 2. Effects of H_1 and H_2 antagonists on the elevation of plasma ACTH and cortisol concentrations induced by intracerebroventricular histamine in dogs

Drugs and dose (μg/kg)	ACTH (pg/ml)	Cortisol (ng/ml)
Histamine 5	28.4 ± 3.15	75.2 ± 5.22
Histamine 5 + Pyrilamine 0.13	26.8 ± 8.20	68.8 ± 14.1
Histamine 5 + Pyrilamine 1.30	$16.7 \pm 4.65*$	$60.8 \pm 5.90*$
Histamine 5 + Pyrilamine 13.0	$9.55 \pm 2.26**$	$53.1 \pm 1.91*$
Histamine 5 + Cimetidine 11.0	24.1 ± 2.13	81.0 ± 12.0

*, ** : Significantly different from the histamine treated group with $P < 0.05$ and $P < 0.01$, respectively. From Tsujimoto, S., Kamei, C., Yoshida, T. and Tasaka, K.: Pharmacology, 47, 73-83 (1993a), with permission.

1.3 and 13.0 μg/kg, whereas cimetidine administered with histamine had no suppressive effect. At doses employed in this experiment, neither pyrilamine nor cimetidine had a significant effect in individual injection.

We have already mentioned that stimulation of steroidogenesis induced by i.c.v. injection of histamine in dogs is mediated through the H_1 receptor without participation of the H_2 receptor. This assumption seems to hold the truth. On the contrary, Rudolph et al. (1979) have reported that 4-methylhistamine at a dose of 25 μg/kg elicited a significant decrease in plasma ACTH concentrations 20-30 min after injection of the drug. Consequently, they proposed that activation of central H_2 receptors caused a decrease in ACTH secretion. In case ACTH concentration was decreased via H_2 receptor activation, it may be elevated by the administration of H_2 receptor antagonists alone. But it is not the case. Furthermore, an increase in ACTH induced by histamine should be accelerated by H_2 antagonist administration. However, no such tendency was found.

5.2. Rats

When pyrilamine, an H_1 antagonist, was injected i.c.v. before histamine (10 μg/kg) at a dose of 26 μg/kg (the molar equivalent to 10 μg/kg of histamine), it significantly inhibited histamine-induced ACTH and corticosterone elevations (Table 3). On the other hand, when cimetidine and ranitidine, at doses of 22 μg/kg and 28 μg/kg (each dose equivalent to 10 μg/kg of histamine) were given as pretreatment by i.c.v. injections, no visible changes were elicited in histamine-induced (10 μg/kg) ACTH and corticosterone release. At the doses employed in the experiment, pyrilamine, cimetidine or ranitidine did not induce an effect (Table 3). When cimetidine at a dose of 44 μg/kg (equivalent to a dose of 20 μg/kg of 4-methylhistamine) or ranitidine at a dose of 56 μg/kg (equivalent to a dose of 20 μg/kg of 4-methylhistamine) was injected i.c.v. in combination with 4-methylhistamine (20 μg/kg), the elevation of plasma ACTH and corticosterone concentrations induced by 4-methylhistamine was significantly inhibited (Table 4). Cimetidine or ranitidine alone caused no appreciable effect even at doses of 44 μg/kg and 56 μg/kg, respectively.

In the study of H_2 agonist in rats, it was demonstrated that 4-methylhistamine and impromidine caused an effect similar to that seen after histamine injection. However, both H_2 receptor antagonists caused no inhibition of plasma ACTH and corticosterone increase induced by histamine. On the other hand, both antagonists were effective in inhibiting the 4-methylhistamine-induced increase in plasma ACTH and corticosterone concentrations. Bugajski and Gadek (1983) reported that cimetidine and metiamide

Table 3. Effects of H_1 and H_2 antagonists on the elevation of plasma ACTH and corticosterone concentrations induced by histamine in rats

Drugs and dose (μg/kg)	ACTH (pg/ml)	Corticosterone (μg/ml)
Control	55.6 ± 4.4	0.50 ± 0.02
Histamine 10	134.8 ± 14.0	0.66 ± 0.02
Histamine 10 + Pyrilamine 26	68.6 ± 12.0*	0.54 ± 0.04*
Histamine 10 + Cimetidine 22	140.4 ± 19.3	0.66 ± 0.04
Histamine 10 + Ranitidine 28	133.3 ± 12.2	0.65 ± 0.08
Pyrilamine 26	57.3 ± 8.4	0.52 ± 0.05
Cimetidine 22	52.3 ± 3.9	0.50 ± 0.07
Ranitidine 28	55.7 ± 2.9	0.49 ± 0.07

Pyrilamine, cimetidine and ranitidine were injected 15 min before injection of histamine. Fifteen min after histamine injection, blood was collected. *: Significantly different from the histamine treated group with P < 0.05. From Tsujimoto, S., Okumura, Y., Kamei, C. and Tasaka, K.: Br. J. Pharmacol., 109, 807-813 (1993b), with permission.

Table 4. Effects of H_2 antagonists on the elevation of plasma ACTH and corticosterone concentrations induced by 4-methylhistamine in rats

Drugs and dose (μg/kg)	ACTH (pg/ml)	Corticosterone (μg/ml)
Control	55.6 ± 4.4	0.50 ± 0.02
4-MethylHi 20	70.4 ± 4.2	0.59 ± 0.02
4-MethylHi 20 + Cimetidine 44	56.0 ± 3.5*	0.51 ± 0.03*
4-MethylHi 20 + Ranitidine 56	55.0 ± 2.3*	0.48 ± 0.03*
Cimetidine 44	54.8 ± 1.6	0.51 ± 0.04
Ranitidine 56	54.3 ± 2.6	0.50 ± 0.02

4-MethylHi: 4-Methylhistamine. Cimetidine and ranitidine were injected 15 min before injection of 4-methylhistamine. Fifteen min after 4-methylhistamine injection, blood was collected. *: Significantly different from the 4-methylhistamine treated group with P < 0.05. From Tsujimoto, S., Okumura, Y., Kamei, C. and Tasaka, K.: Br. J. Pharmacol., 109, 807-813 (1993b), with permission.

inhibited histamine-induced corticosterone increase in rats. From these findings, it seems likely that histamine-induced hormone secretion is mediated mainly through H_1 receptor, and to a lesser extent through the H_2 receptor. Consequently, even if H_2 antagonists blocked corticosteroidogenesis due to histamine via the H_2 receptor, the effect through the H_1 receptor may appear. This postulation seems to coincide with the fact that both cimetidine and ranitidine inhibited the 4-methylhistamine-induced response but not the histamine effect.

6. Effects of H_3 agonist and antagonist on plasma ACTH and corticosterone concentrations in rats

Neither (R)-α-methylhistamine nor thioperamide, an H_3 agonist and an antagonist, respectively, caused any changes in plasma ACTH concentrations at doses of 20 and 50 μg/kg. Similarly, no changes were observed in plasma corticosterone concentrations.

Arrang et al. (1987) reported that (R)-α-methylhistamine and thioperamide are highly potent and selective agonist and antagonist, respectively, for H_3 receptors. Therefore, it is likely that the H_3 receptor may not be involved in histamine-induced ACTH and corticosterone secretions. Since it is known that thioperamide acts as an effective histamine releaser in the brain (Arrang et al., 1987), it is possible to assume that thioperamide may increase the plasma ACTH and corticosterone concentrations as that seen in histamine application. However, this is not the case and that remains to be proved.

7. Effect of CRF on plasma ACTH and corticosteroid concentrations

7.1. Dogs

As stated before, the elevation of the cortisol concentration in plasma caused by histamine application appeared more rapidly than that of ACTH. However, it is generally recognized that an increase in plasma ACTH concentration is followed by an increase in cortisol secretion. To identify the sequential changes in plasma ACTH and cortisol concentrations in dogs, the effect of CRF was studied. Plasma ACTH level rapidly increased to the peak level 15 min after i.v. injection of CRF (1 μg/kg). In contrast, plasma cortisol concentration increased gradually and reached the maximum 60 min after injection.

7.2. Rats

Almost the same results were obtained in rats. When CRF (10 μg/kg) was injected i.v., plasma ACTH concentration rapidly increased and reached a peak 5 min after injection. By contrast, plasma corticosterone concentration increased gradually and reached a peak 30-60 min after CRF injection (Fig. 8). In this study, we have found that

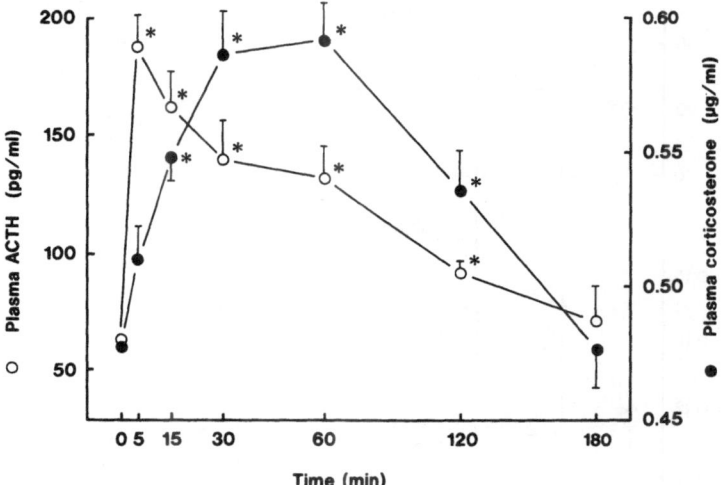

Fig. 8. Changes in plasma ACTH and corticosterone concentrations induced by intravenous injection of CRF in rats. *∗*: Significantly different from predrug value with P < 0.05. From Tsujimoto, S., Okumura, Y., Kamei, C. and Tasaka, K.: Br. J. Pharmacol., 109, 807-813 (1993b), with permission.

an increase in ACTH concentration elicited by CRF was followed by an increase in corticosteroid concentration, similar to that reported by others (Schürmeyer et al., 1985; Kemppainen et al., 1986).

8. Effect of histamine on plasma ACTH and corticosteroid concentrations in hypophysectomized animals

8.1. Dogs

To investigate whether or not corticosteroid release is elicited by i.c.v. histamine without any participation of ACTH, the effect of histamine was investigated using hypophysectomized animals. In the saline-injected control group, plasma cortisol level decreased gradually. On the other hand, when 5 μg/kg of histamine was injected i.c.v., plasma cortisol concentration increased, and a significant effect compared with that of the control group was observed from 10 to 60 min after drug application, suggesting histamine is still effective in hypophysectomized dogs, despite the fact that the histamine effect was much less than that seen in normal dogs. This finding strongly suggests that an increase in plasma cortisol concentration induced by histamine may occur partly independent of plasma ACTH.

8.2. Rats

In hypophysectomized rats, almost the same results were observed as those seen in dogs.

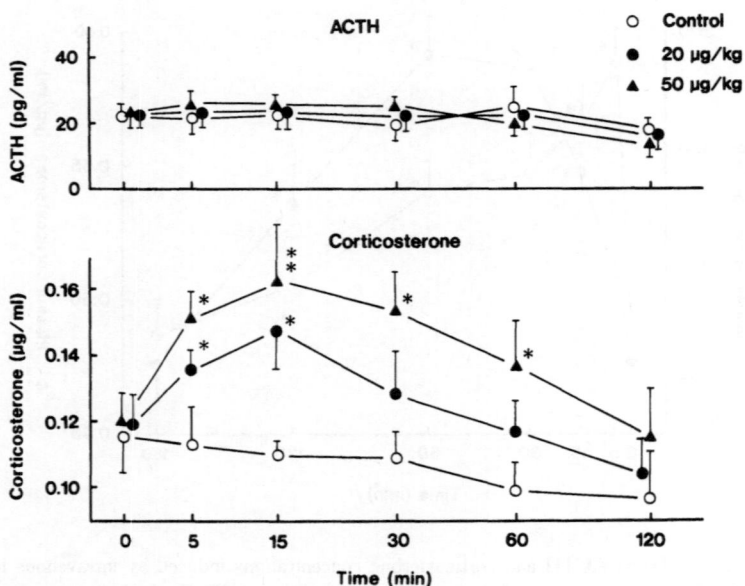

Fig. 9. Changes in plasma ACTH and corticosterone concentrations induced by intracerebroventricular injection of histamine in hypophysectomized rats. *, **: Significantly different from the control group with P < 0.05 and P < 0.01, respectively. From Tsujimoto, S., Okumura, Y., Kamei, C. and Tasaka, K.: Br. J. Pharmacol., 109, 807-813 (1993b), with permission.

When histamine was injected 1 hr after the operation at a dose of 20 μg/kg, a significant increase in corticosterone concentration was observed at 5 and 15 min, while at a dose of 50 μg/kg, a remarkable increase in plasma corticosterone concentration was noticed from 5 to 60 min. In this case, no change in plasma ACTH was induced by histamine even at a dose of 50 μg/kg (Fig. 9). Furthermore, a significant increase in plasma corticosterone concentration was induced by i.c.v. injection of histamine even in hypophysectomized rats (Fig. 9). It may be assumed, therefore, that part of the corticosterone release elicited by histamine was produced without participation of ACTH secretion. Consequently, it was postulated that electrical signals passing through the neurons connecting from the brain to the adrenal cortex may be essential to bring about histamine-induced corticosteroid release.

9. Histamine depletion in rat brain and its influence on corticosterone release

9.1. Changes in histamine content of the hypothalamus, plasma ACTH and corticosterone concentrations

It is known that a maximal decrease in histamine content of the hypothalamus occurred

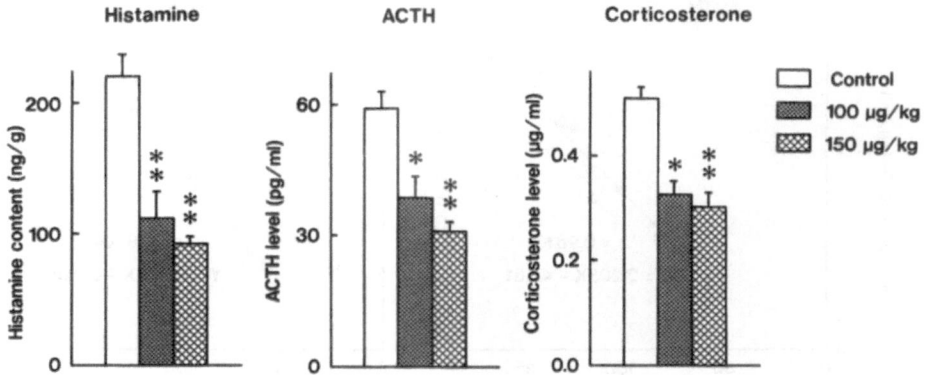

Fig. 10. Changes in histamine content of the hypothalamus, plasma ACTH and corticosterone concentrations after intracerebroventricular injection of α-fluoromethylhistidine. *, ** : Significantly different from the control group with $P < 0.05$ and $P < 0.01$, respectively. From Kamei, C., Okumura, Y., Tsujimoto, S. and Tasaka, K.: Arch. int. Pharmacodyn., 325, 35-50 (1993), with permission.

2 hr after i.c.v. injection of α-fluoromethylhistidine (α-FMH). Therefore, not only histamine content of the hypothalamus, but also plasma ACTH and corticosterone concentrations were measured 2 hr after i.c.v. injection of α-FMH. At doses of 100 and 150 μg/kg, α-FMH significantly decreased histamine contents in the hypothalamus compared with those of the control group. At the same time, plasma ACTH concentrations decreased significantly at doses of 100 and 150 μg/kg. Equivalent decrease in plasma corticosterone was also observed (Fig. 10). These findings strongly suggest that endogenous histamine plays an important role in steroidogenesis.

Seltzer et al. (1986) demonstrated that α-FMH at a dose of 20 mg/kg, i.p. elicited a decrease in plasma corticosterone concentration, however, they ascribed that this effect was induced by a direct action to the adrenal gland. Furthermore, Itowi et al. (1989) reported that plasma corticosterone concentration was not altered by i.p. injection of α-FMH at a dose of 100 mg/kg in stressed rats. This discrepancy can be ascribed due to the differences in the administration route of α-FMH.

9.2. Relationship between histamine content of the hypothalamus and plasma ACTH or corticosterone concentration

When the relationship between the decrease in histamine content of the hypothalamus and the decrease in plasma ACTH or corticosterone concentration induced by i.c.v. injection of α-FMH was investigated, high correlation was observed between the decrease in histamine content and the decrease in plasma ACTH concentration. The more remarkable relationship existed between the decrease in histamine content of the hypothalamus and the decrease in plasma corticosterone concentration (Fig. 11). The results indicate that histamine content in the hypothalamus is intimately related to plasma corticosterone

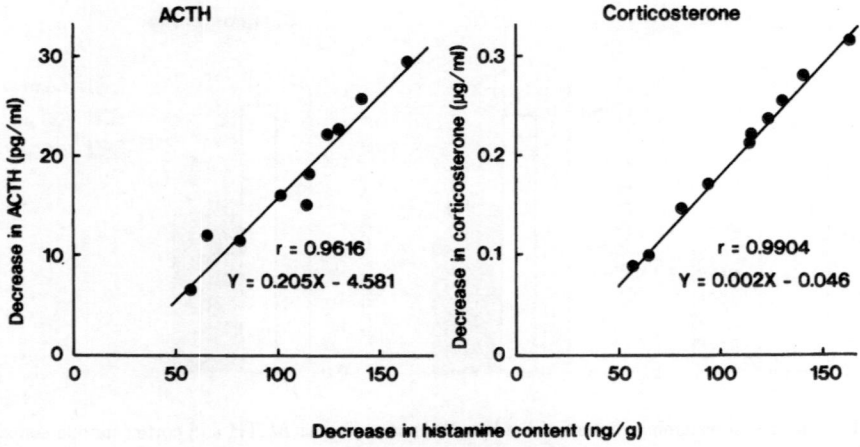

Fig. 11. Relationship between the decrease in histamine content of the hypothalamus and that in plasma ACTH or corticosterone concentration after intracerebroventricular injection of α-fluoromethylhistidine. From Kamei, C., Okumura, Y., Tsujimoto, S. and Tasaka, K.: Arch. int. Pharmacodyn., 325, 35-50 (1993), with permission.

concentration.

Mazurkiewicz-Kwilecki (1983) reported that histamine content of the hypothalamus was significantly elevated by i.p. injection of corticosterone and that plasma corticosterone concentration increased after i.p. injection of histidine which may increase the histamine contents of the brain. Based on these findings, they discussed that corticosterone induces an alteration in histamine synthesis or its metabolism.

10. Influence of lesions or stimulation to the posterior hypothalamus

10.1. Changes in histamine content of the hypothalamus, plasma ACTH and corticosterone concentrations in hypothalamic-lesioned rats

Table 5. Changes in histamine content of the hypothalamus, plasma ACTH and corticosterone concentrations after lesion of the posterior hypothalamus

	Histamine content (ng/g)	ACTH (pg/ml)	Corticosterone (μg/ml)
Control	226.2 ± 14.2	54.4 ± 2.5	0.45 ± 0.02
Lesion	135.5 ± 23.3 **	36.8 ± 6.6 *	0.33 ± 0.05 *

Seven days after lesion of the posterior hypothalamus, histamine content of the hypothalamus and plasma ACTH and corticosterone concentrations were measured. *, **: Significantly different from control group with P < 0.05 and P < 0.01, respectively. From Kamei, C., Okumura, Y., Tsujimoto, S. and Tasaka, K.: Arch. int. Pharmacodyn., 325, 35-50 (1993), with permission.

When the posterior hypothalamus was electrically lesioned, a large portion of the posterior hypothalamic nucleus, and a part of the premammillary dorsal nucleus and the dorsomedial hypothalamic nucleus were lesioned (Paxinos and Watson, 1986). When histamine content in the remaining areas, including the ventromedial hypothalamic nucleus and the premammillary ventral nucleus, were measured 7 days after electrocoagulation,

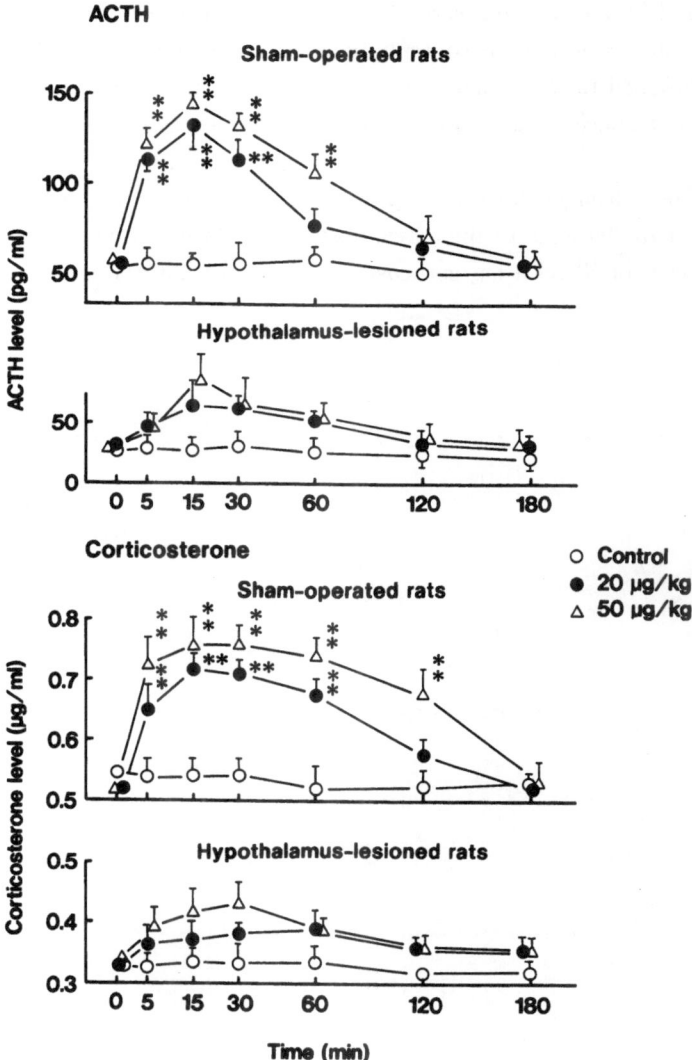

Fig. 12. Changes in plasma ACTH and corticosterone concentrations induced by intracerebroventricular injection of histamine in hypothalamic-lesioned rats. **∗∗** : Significantly different from the control group with P < 0.01. From Kamei, C., Okumura, Y., Tsujimoto, S. and Tasaka, K.: Arch. int. Pharmacodyn., 325, 35-50 (1993), with permission.

histamine content markedly decreased compared with that of the control group. At the same time, both plasma ACTH and corticosterone concentrations decreased significantly compared with those of sham-operated controls (Table 5).

10.2. Effect of histamine on plasma ACTH and corticosterone concentrations in hypothalamic-lesioned rats

I.c.v. injection of histamine at doses of 20 and 50 μg/kg caused a significant increase in plasma ACTH and corticosterone concentrations in sham-operated rats. However, in the hypothalamic-lesioned rats, no significant increase was observed both in plasma ACTH and corticosterone concentrations after histamine application even at a dose of 50 μg/kg (Fig. 12).

These findings strongly support our proposal that 1) a direct neuronal pathway exists connecting the hypothalamus to the adrenal cortex 2) which can be stimulated by i.c.v. injected histamine for 3) releasing corticosterone from the adrenal cortex.

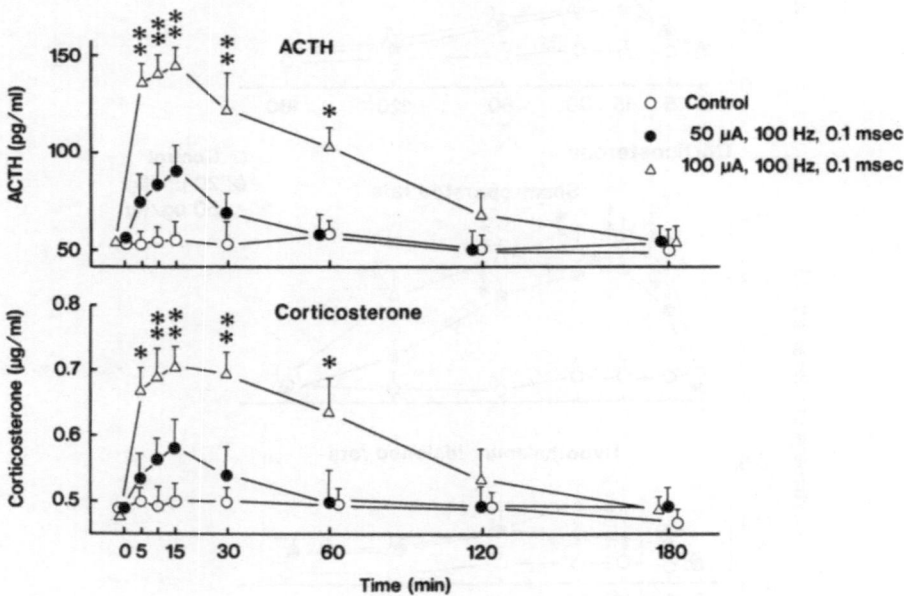

Fig. 13. Changes in plasma ACTH and corticosterone concentrations after electric stimulation to the posterior hypothalamus. The posterior hypothalamus was stimulated electrically (10 sec on, 5 sec off, for 60 sec). *, **: Significantly different from the control group with P < 0.05 and P < 0.01, respectively. From Kamei, C., Okumura, Y., Tsujimoto, S. and Tasaka, K.: Arch. int. Pharmacodyn., 325, 35-50 (1993), with permission.

10.3. Changes in plasma ACTH and corticosterone concentrations after electric stimulation to the posterior hypothalamus

It is known that the posterior hypothalamus contains a large amount of histamine (Panula et al., 1984). When electric stimulation to the posterior hypothalamus was given through chronically implanted electrodes (100 Hz, 0.1 msec, 50 μA, 10 sec on, 5 sec off, for 60 sec), a moderate but not significant increase in plasma ACTH and corticosterone concentrations was observed. However, as the electric current was increased to 100 μA, a significant elevation of plasma ACTH and corticosterone concentrations were observed from 5 to 60 min after stimulation (Fig. 13). These results seem to indicate that electrical stimulation may cause an activation of the synthesis and release of histamine in the hypothalamus and this may induce an increase in the plasma ACTH and corticosterone concentrations.

10.4. Effect of electric stimulation to the posterior hypothalamus on plasma corticosterone concentrations after hypophysectomy

Electric stimulation to the posterior hypothalamus caused a significant increase in plasma corticosterone concentration, even after hypophysectomy. Significant effects were observed after electric stimulation (Table 6). However, plasma ACTH concentration was not elevated by electric stimulation to the posterior hypothalamus after hypophysectomy.

Redgate and Fahringer (1973) also reported that electric stimulation to the tuberal and mammillary nucleus of the posterior hypothalamus resulted in an increase in plasma corticosterone concentrations in rats. From this finding, they postulated that electric stimulation to the posterior hypothalamus may cause an excitation of the CRF-ACTH-corticosterone system. However, as shown in Table 6, it is evident that plasma corticosterone release may partly be brought about without participation of CRF-ACTH system.

Table 6. Changes in plasma corticosterone concentration induced by electric stimulation to the posterior hypothalamus in hypophysectomized rats

Time	Corticosterone concentration (μg/ml)	
(min)	Control	Stimulation
Before	0.128 ± 0.01	0.131 ± 0.01
5	0.111 ± 0.01	0.136 ± 0.01
10	0.098 ± 0.01	0.140 ± 0.01 *
15	0.091 ± 0.01	0.149 ± 0.01 **
30	0.091 ± 0.01	0.140 ± 0.01 *
60	0.090 ± 0.01	0.126 ± 0.02

The posterior hypothalamus was stimulated electrically (100 Hz, 0.1 msec, 100 μA, 10 sec on, 5 sec off, for 60 sec). *, **: Significantly different from control group with $P < 0.05$ and $P < 0.01$, respectively. From Kamei, C., Okumura, Y., Tsujimoto, S. and Tasaka, K.: Arch. int. Pharmacodyn., 325, 35-50 (1993), with permission.

10.5. Effects of histamine receptor agonist and antagonists on the increase in plasma ACTH and corticosterone concentrations after electric stimulation to the posterior hypothalamus

As shown in Table 7, when pyrilamine, at doses of 10 and 20 μg/kg, and cimetidine, at a dose of 20 μg/kg, were applied i.c.v., these drugs effectively antagonized the increase in plasma ACTH and corticosterone concentrations induced by electric stimulation to the posterior hypothalamus. α-Methylhistamine also caused a significant inhibitory effect at doses of 10 and 20 μg/kg, i.c.v. However, the effect of α-methylhistamine (10 μg/kg) was significantly inhibited by simultaneous administration of thioperamide (10 μg/kg).

Not only H_1 and H_2 but also H_3 receptors seem to participate in the process leading to the increase of plasma ACTH and corticosterone concentrations after stimulation to the posterior hypothalamus. We have shown that both H_1 and H_2 receptors are involved in an increase in steroidogenesis induced by i.c.v. injection of histamine. In addition, we mentioned that H_3 receptors may not be involved in histamine-induced ACTH and corticosterone secretions (Tsujimoto et al., 1993b). However, in this study, α-methylhistamine showed an inhibitory effect on the increase in plasma ACTH and corticosterone concentrations induced by hypothalamic stimulation. Arrang et al. (1987) reported that H_3 receptors control not only the synthesis but also the release of histamine at the level of nerve endings. It seems, therefore, that the inhibitory effect of α-

Table 7. Effects of histamine receptor agonist and antagonists on the increase in plasma ACTH and corticosterone concentrations after electric stimulation to the posterior hypothalamus

Treatment	Dose (μg/kg)	ACTH (pg/ml)	Corticosterone (μg/ml)
Before stimulation		54.4 ± 2.0	0.453 ± 0.02
After stimulation		133.6 ± 8.8	0.703 ± 0.03
Pyrilamine	10	93.7 ± 6.6**	0.534 ± 0.03**
	20	83.4 ± 7.1**	0.512 ± 0.04**
Cimetidine	10	117.2 ± 10.8	0.623 ± 0.03
	20	97.5 ± 10.4*	0.551 ± 0.02*
α-MethylHi	10	94.5 ± 4.3**	0.536 ± 0.05**
	20	85.1 ± 5.4**	0.510 ± 0.03**
α-MethylHi (10) +	1	112.6 ± 10.8	0.625 ± 0.01
Thioperamide	10	124.9 ± 8.2#	0.673 ± 0.03#

α-MethylHi: (R)-α-Methylhistamine. The posterior hypothalamus was stimulated electrically (100 Hz, 0.1 msec, 10 sec on, 5 sec off, for 60 sec). *, **: Significantly different from "After stimulation" group with P < 0.05 and P < 0.01, respectively. #: Significantly different from α-methylhistamine treated group with P < 0.05. From Kamei, C., Okumura, Y., Tsujimoto, S. and Tasaka, K.: Arch. int. Pharmacodyn., 325, 35-50 (1993), with permission.

methylhistamine may be related in some way to the inhibition of histamine release due to posterior hypothalamic stimulation.

11. Interaction between the posterior hypothalamus and the adrenal gland

11.1. Evoked potential recorded from the adrenal nerve after electric stimulation to the posterior hypothalamus

To investigate whether or not the neuronal pathway from the hypothalamus to the adrenal gland is involved in histamine-induced steroidogenesis, evoked potentials in response to electric stimulation to the posterior hypothalamus or cervical cord were recorded from the left adrenal nerve. When electric stimulation was applied to the posterior hypothalamus (1 Hz, 0.01 msec, 500 μA, 50 times), evoked potentials were recorded consisting of 2 peak components (Fig. 14). Negative-positive deflections were observed alternatively following the artifact, and these were designated as N_1-P_1 and N_2-P_2, respectively. Latency for the N_1 was 5.0 ± 0.4 msec. However, when the intensity of electric stimulation was lowered (200 μA), the second component disappeared. The sites of the electrode tip eliciting the positive response in evoked potentials were located in the posterior hypothalamic nucleus. The amplitudes of both the first and second components

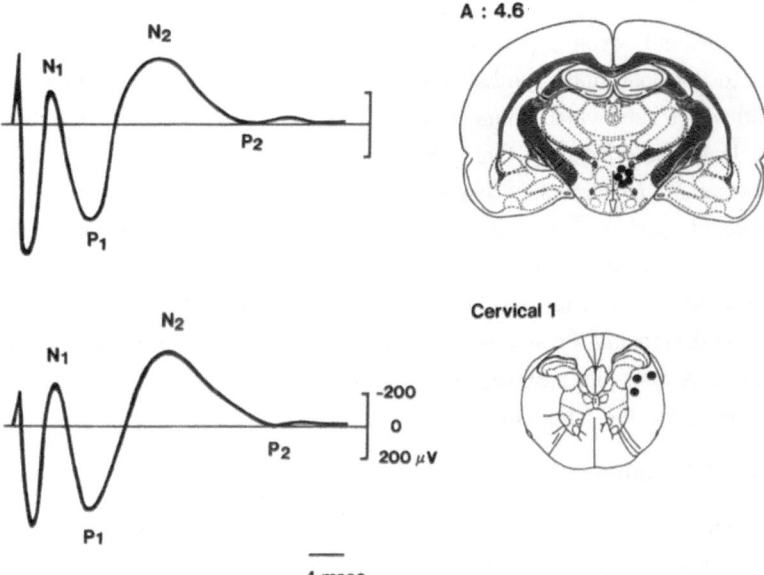

Fig. 14. Evoked potential recorded from the left adrenal nerve induced by electric stimulation to the right posterior hypothalamus or the lateral funiculus of the cervical cord and localization of the electrode tips. The posterior hypothalamus or the lateral funiculus of the cervical cord was stimulated electrically (1 Hz, 0.01 msec, 500 μA, 50 times). From Kamei, C., Okumura, Y., Tsujimoto, S. and Tasaka, K.: Arch. int. Pharmacodyn., 325, 35-50 (1993), with permission.

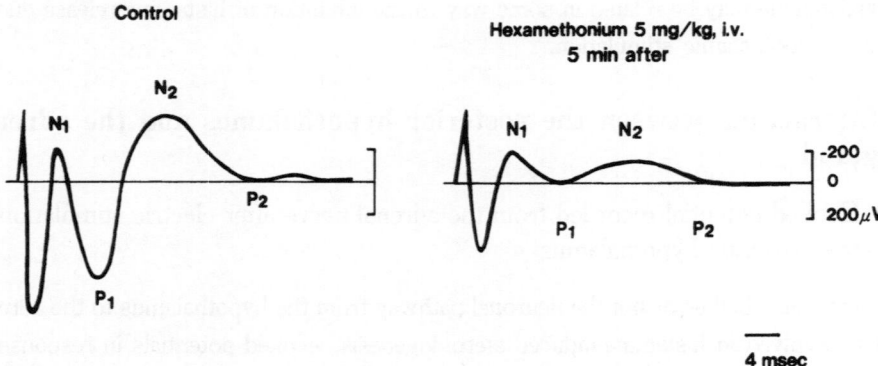

Fig. 15. Effect of hexamethonium on evoked potential recorded from the left adrenal nerve induced by electric stimulation to the posterior hypothalamus. Hexamethonium was injected intravenously. From Kamei, C., Okumura, Y., Tsujimoto, S. and Tasaka, K.: Arch. int. Pharmacodyn., 325, 35-50 (1993), with permission.

remarkably decreased by intravenous injection of 5 mg/kg of hexamethonium (Fig. 15). Interestingly, almost the same evoked potential was also recorded by electric stimulation to the cervical cord (1 Hz, 0.01 msec, 500 μA, 50 times). Localization of the electrode tip showing positive response was located in the lateral funiculus of the cervical cord (Fig. 14).

It became evident, evoked potential was really recorded from the adrenal nerve after stimulations to the posterior hypothalamus as well as the cervical cord, indicating that a neuronal pathway exists which connects hypothalamus to the adrenal organ through the spinal cord. Halász and Szentágothai (1959) reported that an afferent nervous connection may exist between the adrenal cortex and the opposite site of the ventromedial hypothalamus in histological studies. Saper et al. (1976) found that there are direct projections from the hypothalamus to the spinal cord by means of the horse-radish peroxidase (HRP) technique. As shown in Fig. 15, hexamethonium, a representative ganglion blocker, was effective in decreasing the amplitude of the potential. These results suggest that there exists at least one neuronal pathway that originates in the hypothalamus and reaches to the adrenal nerve via the ganglion(s).

11.2. Effect of histamine on adrenal nerve activity

To confirm whether or not i.c.v. injection of histamine is effective in stimulating the adrenal gland through the neuronal pathway, electrical signals were recorded on the adrenal nerve. Discharge rates and the amplitudes of adrenal nerve activity were 104.0 ± 2.5 Hz and 18.2 ± 1.0 μV, respectively. I.c.v. injection of histamine at doses of 10 and 20 μg/kg immediately increased the amplitude of electrical signals. The increase became significant 5 min after histamine application and lasted for 30 min. However, histamine caused no significant effect on the rate of discharges (Table 8). When C_1 transection was performed, the adrenal nerve activity disappeared almost instantaneously in all cases. The amplitude of the adrenal nerve activity increased after i.c.v. injection of histamine. Therefore, it seems likely that i.c.v. injected histamine stimulates the adrenal gland via a neuronal pathway. Unsicker (1971) reported that nerve fibers are regularly found in the fasciculata zone of the rat adrenal cortex, and nerve fiber containing synaptic vesicles and two types of dense-cored vesicles in contact with endocrine cells have been observed by electron microscopy.

11.3. Changes in plasma ACTH and corticosterone concentrations after stimulation to the adrenal nerves

Table 8. Effect of intracerebroventricular injection of histamine on the adrenal nerve activity in rats

Drug	Dose	Discharge rates (Hz)					
	(μg/kg)	Before	5	15	30	60	120 min
Control	—	101.0 ± 2.5	102.0 ± 4.1	105.0 ± 3.4	104.0 ± 3.3	104.0 ± 3.3	102.0 ± 3.2
Histamine	10	105.0 ± 4.7	108.0 ± 3.8	111.0 ± 3.5	109.5 ± 4.0	108.0 ± 3.8	106.5 ± 4.0
	20	106.5 ± 4.6	106.5 ± 4.6	109.5 ± 3.4	112.5 ± 2.6	111.0 ± 2.7	106.5 ± 4.6

Drug	Dose	Amplitudes (μV)					
	(μg/kg)	Before	5	15	30	60	120 min
Control	—	18.2 ± 1.0	18.6 ± 1.0	18.3 ± 1.4	18.1 ± 1.2	18.5 ± 1.9	18.2 ± 1.4
Histamine	10	18.4 ± 1.0	$23.2 \pm 1.2*$	$30.2 \pm 2.0**$	$28.5 \pm 1.4**$	23.6 ± 0.9	18.7 ± 0.9
	20	18.2 ± 1.3	$30.2 \pm 1.8**$	$36.6 \pm 0.9**$	$29.1 \pm 1.3**$	26.4 ± 1.5	19.7 ± 1.2

*, ** : Significantly different from the control group with $P < 0.05$ and $P < 0.01$, respectively. From Tsujimoto, S., Okumura, Y., Kamei, C. and Tasaka, K.: Br. J. Pharmacol., 109, 807-813 (1993b), with permission.

Table 9. Effect of stimulation of the adrenal nerves on plasma ACTH and corticosterone concentrations

Treatment	ACTH concentrations (pg/ml)					
	Before	5	10	15	30	60 min
Control	61.5 ± 2.8	63.2 ± 3.6	61.2 ± 2.6	62.5 ± 3.8	62.4 ± 2.9	62.3 ± 4.7
ES	61.7 ± 3.7	64.9 ± 4.3	59.4 ± 4.8	63.9 ± 4.2	64.7 ± 4.5	60.0 ± 6.6

Treatment	Corticosterone concentrations (μg/ml)					
	Before	5	10	15	30	60 min
Control	0.54 ± 0.02	0.55 ± 0.02	0.55 ± 0.01	0.54 ± 0.02	0.53 ± 0.02	0.55 ± 0.02
ES	0.55 ± 0.02	0.60 ± 0.02	$0.63 \pm 0.01**$	$0.65 \pm 0.02**$	$0.64 \pm 0.02**$	0.61 ± 0.02

ES: Electric stimulation. ** : Significantly different from the control group with $P < 0.01$. From Tsujimoto, S., Okumura, Y., Kamei, C. and Tasaka, K.: Br. J. Pharmacol., 109, 807-813 (1993b), with permission.

Stimulation of the adrenal nerves were performed as follows: The left adrenal nerve was retroperitoneally dissected under urethane anesthesia, 6 hr after the cannulation of the left carotid artery had been performed. After 2 hr of rest, the adrenal nerve was stimulated electrically through bipolar platinum wire electrodes in which covered with warm liquid paraffin (100 Hz, 1 msec, 500 μA, for 60 sec). Electrical stimulation to the adrenal nerve resulted in an increase in plasma corticosterone concentration. A significant effect was observed from 10 to 30 min after stimulation to the adrenal nerve. However, plasma ACTH concentration was not affected (Table 9). This clearly shows that corticosterone can be released from the adrenal gland without involvement of the hypophysis-adrenocortical system. In addition, an increase in plasma corticosterone concentration may support the assumption that the adrenal cortex is also innervated by the adrenal nerve. Actually it has been recognized that the adrenal cortex contains a complex autonomic innervation, one part of which innervates the cortex independency of the splanchnic nerve, while the other part appears to originate in the adrenal medulla to be regulated by splanchnic nerve activity (Holzwarth et al., 1987). Morphological evidence for adrenal cortical nerves indicates that several peptidergic neuronal systems exists in the adrenal cortex such as vasoactive intestinal peptide (VIP), neuropeptide Y and substance P (Holzwarth et al., 1987). Infusion of VIP to the capsule-glomerulosa system caused a dose-dependent output of corticosterone (Holzwarth et al., 1987). Furthermore, when primary cultured bovine adrenocortical cells were treated with substance P, cortisol output increased in a dose-dependent fashion (Yoshida et al. 1992). As a result of these findings, it can be postulated that i.c.v. injection of histamine stimulates the adrenal nerves, and this may, in turn, result in the release of substance P or VIP in the adrenal cortex which results in corticosteroid release. As shown in Table 10, plasma ACTH concentration was not elevated after electrical stimulation to the adrenal nerves, while corticosterone release was induced. This may clearly indicate that corticosterone can be released from the adrenal gland without involvement of the hypophysis-adrenocortical system.

12. Effects of some neurotransmitters on steroidogenesis in dog adrenocortical cells

It has been demonstrated that when adrenal cells were incubated with histamine, a slight but significant increase was observed in the production of cortisol in dogs (Hirose et al., 1978). In our study, histamine elicited a significant increase in cortisol output in dog adrenocortical cells (Table 10). In addition, it has been demonstrated that not only ACTH but also epinephrine, norepinephrine and acetylcholine caused a concentration-dependent increase in cortisol output, though the potency of the latter compounds was much less than that of ACTH (Table 10). Bugajski and Gadek (1984) reported that cholinergic and muscarinic receptors are not involved in central histaminergic stimulation

Table 10. Effects of some neurotransmitters on steroidogenesis in dog adrenocortical cells

Drugs	Concentrations (M)	Cortisol output (ng/10^5 cells)
ACTH	—	33.0 ± 1.7
	10^{-12}	130.0 ± 7.9**
	10^{-11}	176.4 ± 6.4**
	10^{-10}	245.6 ± 11.9**
	10^{-9}	339.8 ± 12.2**
Histamine	—	28.9 ± 2.5
	10^{-7}	32.7 ± 3.4
	10^{-6}	46.2 ± 0.6**
	10^{-5}	69.5 ± 0.6**
	10^{-4}	119.0 ± 2.1**
Epinephrine	—	26.5 ± 0.5
	10^{-8}	62.0 ± 1.6**
	10^{-7}	76.2 ± 1.8**
	10^{-6}	125.9 ± 1.9**
	10^{-5}	135.6 ± 3.6**
Norepinephrine	—	31.4 ± 0.7
	10^{-8}	48.8 ± 2.5**
	10^{-7}	69.1 ± 2.0**
	10^{-6}	120.5 ± 3.2**
	10^{-5}	137.4 ± 4.3**
Acetylcholine	—	27.9 ± 2.2
	10^{-7}	29.4 ± 1.1
	10^{-6}	85.5 ± 2.6**
	10^{-5}	153.9 ± 4.6**
	10^{-4}	348.3 ± 17.6**

** : Significantly different from the control group with $P < 0.01$. Reproduced From Tsujimoto, S., Kamei, C., Yoshida, T. and Tasaka, K.: Pharmacology, 47, 73-83 (1993a), with permission.

eliciting the increased pituitary-adrenocortical response in rats. As mentioned above, substance P caused a marked increase in cortisol secretion from bovine adrenocortical cells.

13. Conclusion

I.c.v. injection of histamine caused a significant and dose-related increase in plasma ACTH and corticosteroid in both dogs and rats. In addition, an increase in plasma cortisol was brought about much faster than that of plasma ACTH, and the extent of cortisol release induced by histamine was more remarkable than that of ACTH release in dogs. Almost the same results were observed in rats. Histamine-induced increase in ACTH and corticosteroid secretion were mediated only via H_1 receptors in dogs. However, in rats, not only H_1 but also H_2 receptors are involved in histamine-induced hormone release. An increase in corticosteroid release induced by histamine was also occurred in hypophysectomized animals, suggesting that an increase in plasma cortico-steroid concentrations may occur partly independently of plasma ACTH. Histamine contents in the hypothalamus as well as plasma ACTH and corticosterone concentrations in rats were decreased by i.c.v. injection of α-FMH. Almost the same result was obtained when the posterior hypothalamus was lesioned electrically. These results suggest that endogenous hypothalamic histamine plays an important role in corticoste-roidogenesis. The increase in plasma ACTH and corticosterone concentrations induced by i.c.v. injection of histamine in sham-operated rats was abolished after hypothalamic lesions. When electric stimulation were applied to the posterior hypothalamus, plasma ACTH and corticosterone concentrations increased significantly. Increased plasma ACTH and corticosterone concentrations were inhibited not only H_1 and H_2 antagonists but also H_3 agonist.

When the posterior hypothalamus was stimulated electrically, evoked potentials were recorded on the adrenal nerve. In addition, histamine elicited an increase in the amplitude of adrenal nerve activity, and electrical stimulation to the adrenal nerves resulted in an increase in plasma corticosterone concentrations. From these findings, it was concluded that a neuronal pathway from the hypothalamus to the adrenal cortex may be essential for the rapid increase in the plasma corticosterone concentration induced by i.c.v. injection of histamine.

References

Arrang, J.-M., Garbarg, M. and Schwartz, J.-C.: Autoinhibition of histamine synthesis mediated by presynaptic H_3-receptors. Neuroscience, 23, 149-157 (1987)

Bugajski, J. and Gadek, A.: Central H_1- and H_2-histaminergic stimulation of pituitary-adrenocortical response under stress in rats. Neuroendocrinology, 36, 424-430 (1983)

Bugajski, J. and Gadek, A.: The effect of adrenergic and cholinergic antagonists on central histaminergic stimulation of pituitary-adrenocortical response under stress in rats. Neuroen-docrinology, 38, 447-452 (1984)

Cohen, H., Graff, M. and Kleinberg, W.: Inhibition of dextran edema by proteolytic enzymes. Proc. Soc. exp. Biol. Med., 88, 517-519 (1955)

Cowan, J.S.: Adrenocorticotropin secretion rates following histamine injection in adult and newborn dogs. Can. J. Physiol. Pharmacol., 53, 592-602 (1975)

Feldberg, W. and Sherwood, S.L.: Injections of drugs into the lateral ventricle of the cat. J. Physiol., 123, 148-167 (1954)

Ganong, W.F., Kramer, N., Salmon, J., Reid, I.A., Lovinger, R., Scapagnini, U., Boryc-zka, A.T. and Shackelford, R.: Pharmacological evidence for inhibition of ACTH secre-tion by a central adrenergic system in the dog. Neuroscience, 1, 167-174 (1976)

Halász, B. and Szentágothai, J.: Histologischer Beweis einer nervösen signalübermittlung von der Nebennierenrinde zum Hypothalamus. Z. Zellforsch., 50, 297-306 (1959)

Hirose, T., Matsumoto, I. and Suzuki, T.: Adrenal cortical secretory responses to histamine and cyanide in dogs with hypothalamic lesions. Neuroendocrinology, 21, 304-311 (1976)

Hirose, T., Matsumoto, I. and Aikawa, T.: Direct effect of histamine on cortisol and corticosterone production by isolated dog adrenal cells. J. Endocrinol., 76, 369-370 (1978)

Holzwarth, M.A., Cunningham, L.A. and Kleitman, N.: The role of adrenal nerves in the regulation of adrenocortical functions. Ann. N.Y. Acad. Sci. U.S.A., 512, 449-464 (1987)

Itowi, N., Yamatodani, A., Cacabelos, R., Goto, M. and Wada, H.: Effect of histamine depletion on circadian variations of corticotropin and corticosterone in rats. Neuroendocrinology, 50, 187-192 (1989)

Kamei, C., Okumura, Y., Tsujimoto, S. and Tasaka, K.: Role of hypothalamic histamine in stimulating the corticosterone release in rats. Arch. int. Pharmacodyn., 325, 35-50 (1993)

Karppanen, H., Paakkari, I., Paakkari, P., Huotari, R. and Orma, A.-L.: Possible involvement of central histamine H_2-receptors in the hypotensive effect of clonidine. Nature, 259, 587-588 (1976)

Katsuki, S., Ito, M., Watanabe, A., Iino, K., Yuji, S. and Kondo, S.: Effect of hypothalamic lesions on pituitary-adrenocortical responses to histamine and methopyrapone. Endocrinology, 81, 941-945 (1967)

Kemppainen, R.J., Filer, D.V., Sartin, J.L. and Reed, R.B.: Ovine corticotrophin-releasing factor in dogs: Dose-response relationships and effects of dexamethasone. Acta Endocrinol., 112, 12-19 (1986)

Kollonitsch, J., Patchett, A.A., Marburg, S., Maycock, A.L., Perkins, L.M., Doldouras, G.A., Duggan, D.E. and Aster, S.D.: Selective inhibitors of biosynthesis of aminergic neurotransmitters. Nature, 274, 906-908 (1978)

Maki, K., Takahashi, H., Kakimoto, M., Nakajima, M. and Tasaka, K.: Anti-inflammatory mechanism of Prozime-10, a proteolytic enzyme. Pharmacology, 23, 230-236 (1981)

Martin, G.J., Brendel, R. and Beiler, J.M.: Inhibition of egg-white edema by proteolytic enzymes. Proc. Soc. exp. Biol. Med., 86, 636-638 (1954)

Mazurkiewicz-Kwilecki, I.M.: Brain histamine-plasma corticosterone interactions. Life Sci., 32, 1099-1106 (1983)

Miechowski, W.L. and Ercoli, N.: Studies on proteolytic enzymes. II. Trypsin and chymotrypsin in relation to inflammatory processes. J. Pharmacol. Exp. Ther., 116, 43-44 (1956)

Morita, Y. and Koyama, K.: Histamine-induced ACTH secretion and inhibitory effect of antihistaminic drugs. Japan. J. Pharmacol., 29, 59-65 (1979)

Panula, P., Yang, H.-Y.T. and Costa, E.: Histamine-containing neurons in the rat hypothalamus. Proc. Natl. Acad. Sci. U.S.A., 81, 2572-2576 (1984)

Paxinos, G. and Watson, C.: The rat brain in stereotaxic coordinates. Academic Press, San Diego, 1986

Redgate, E.S. and Fahringer, E.E.: A comparison of the pituitary adrenal activity elicited by electrical stimulation of preoptic, amygdaloid and hypothalamic sites in the rat brain. Neuroendocrinology, 12, 334-343 (1973)

Reilly, M.A.: Biogenic amine participation in histamine stimulation of ACTH release. Agents Actions, 14, 630-632 (1984)

Rudolph, C., Richards, G.E., Kaplan, S. and Ganong, W.F.: Effect of intraventricular histamine on hormone secretion in dogs. Neuroendocrinology, 29, 169-177 (1979)

Seltzer, A.M., Donoso, A.O. and Podestá, E.: Restraint stress stimulation of prolactin and ACTH secretion: Role of brain histamine. Physiol. Behav., 36, 251-255 (1986)

Saper, C.B., Loewy, A.D., Swanson, L.W. and Cowan, W.M.: Direct hypothalamo-autonomic connections. Brain Res., 117, 305-312 (1976)

Schürmeyer, T.H., Gold, P.W., Gallucci, W.T., Tomai, T.P., Cutler, G.B. Jr., Loriaux, D.L. and Chrousos, G.P.: Effects and pharmacokinetic properties of the rat/human corticotropin-releasing factor in rhesus monkeys. Endocrinology, 117, 300-306 (1985)

Tasaka, K., Meshi, T., Akagi, M., Kakimoto, M., Saito, R., Okada, I. and Maki, K.: Anti-inflammatory activity of a proteolytic enzyme, Prozime-10. Pharmacology, 21, 43-52 (1980)

Tasaka, K., Chung, Y.H., Sawada, K. and Mio, M.: Excitatory effect of histamine on the arousal system and its inhibition by H_1 blockers. Brain Res. Bull., 22, 271-275 (1989)

Tsujimoto, S., Kamei, C., Yoshida, T. and Tasaka, K.: Changes in plasma adrenocorticotropic hormone and cortisol levels induced by intracerebroventricular injection of histamine and its related compounds in dogs. Pharmacology, 47, 73-83 (1993a)

Tsujimoto, S., Okumura, Y., Kamei, C. and Tasaka, K.: Effects of intracerebroventricular injection of histamine and related compounds on corticosterone release in rats. Br. J. Pharmacol., 109, 807-813 (1993b)

Unsicker, K.: On the innervation of the rat and pig adrenal cortex. Z. Zellforsch., 116, 151-156 (1971)

Yoshida, T., Mio, M. and Tasaka, K.: Cortisol secretion induced by substance P from bovine adrenocortical cells and its inhibition by calmodulin inhibitors. Biochem. Pharmacol., 43, 513-517 (1992)

Chapter 4

Role of Ca^{2+} and cAMP in histamine release from mast cells

Introduction

It is well known that an increase in intracellular Ca^{2+} concentrations and subsequent activation of Ca^{2+}-dependent pathways, such as calmodulin, protein kinase C and cytoskeletons, are prerequisite for the histamine release from mast cells. On the other hand, an increase in intracellular concentrations of cAMP is effective in inhibiting the histamine release from mast cells. Many antiallergic drugs have been developed based on this consequence either to inhibit the increase in intracellular Ca^{2+} level or to increase intracellular cAMP concentrations of mast cells. In this chapter, the roles of Ca^{2+} and cAMP in the histamine release from mast cells are reviewed.

1. Intracellular Ca^{2+} release induced by histamine releasers and its inhibition by some antiallergic drugs

1.1. Introduction

It has been shown that some compounds release histamine from mast cells even in a Ca-free medium. Compound 48/80 and substance P are typical agents causing histamine release in such conditions (Fewtrell et al., 1982; Ennis et al., 1980a). It has been assumed that the process leading to histamine release may be mediated by Ca^{2+} released from an intracellular store (Ennis et al., 1980b), but direct evidence has not yet been presented. On the other hand, it has been shown that quin 2/AM incorporated into lymphocytes will be catalyzed to quin 2, thereby providing high affinity for free Ca^{2+}. Quin 2 chelated with Ca^{2+} emits fluorescence when it has been excited by a 340 nm beam (Tsein et al., 1982). Also, it has been suggested that the quin 2 signal may be useful in investigating intracellular Ca^{2+} dynamics in the histamine release from mast cells (White et al., 1984).

So far, it is known that many antiallergic agents are useful in inhibiting histamine release from mast cells. However, the inhibitory mechanism was not elucidated in most of the compounds. Since compound 48/80 and substance P seem to utilize intracellular Ca^{2+} (Fewtrell et al., 1982; Ennis et al., 1982), a quin 2 signal can be used to clarify whether or not these releasers actually release Ca^{2+} from the Ca store in the process leading to histamine release. Furthermore, it is known that an increase of cAMP content in the mast cell is effective to prevent histamine release (Tasaka, 1986). In relation to this, it is necessary to clarify whether or not an increase of the cAMP level in mast cells is effective in preventing the mobilization of Ca^{2+} from an intracellular store. If this is the case, at least one of the mechanisms of histamine release inhibition due to cAMP can be definitely elucidated.

1.2. Changes in fluorescence intensity in an individual mast cell before and after exposure to histamine releasers

Rat peritoneal mast cells harvested from the abdominal cavity of male Wistar rats were incubated with 5 μM quin 2/AM dissolved in a Ca-free physiological buffered solution (PBS) at 37°C for 30 min (Tasaka et al., 1986a, b). Fluorescence intensity derived from quin 2 chelated with intracellular free Ca^{2+} was measured in an individual cell (n = 30, at least) using a fluorescence microscope connected to a photon counter. In some cases, a video-intensified microscopy (VIM) system (Hamamatsu, C-1966-20) was employed in combination with an invert-type fluorescence microscope. The system consisted of an image processor including video frame memories and an ultra-high sensitive TV camera equipped with 2-dimensional microchannel plates. Fluorescence images of mast cells were recorded with a VIM system coupled to a Sony 3/4 inch casette video tape recorder. Photographs of fluorescent images of mast cell were analyzed by an online image processing system (Okayama University Computer Center, NEC ACOS-1000, image system for online processing). Using this system, photographs were read into frame memories, contrasts were enhanced and converted into pseudo-color, from red to violet by a rainbow color system according to the intensities of the pixels. A profile of the pixel values displayed as a three dimensional projection, and a histogram displaying the distribution of pixel values was performed.

When a mast cell loaded with quin 2 was exposed to compound 48/80 (0.1 μg/ml) for 30 sec at 37°C, fluorescence intensity of mast cell was increased significantly, exhibiting the value of 148.2 ± 2.1 (n = 1890, p < 0.01) when the mean intensity of the control cells was taken as 100. In mast cells exposed to substance P at 2 mM for one min at 37°C, fluorescence increased to 130.4 ± 1.4 (n = 330, p < 0.01). When db-cAMP was pretreated at concentrations ranging from 0.01 mM to 1 mM for 5 min, an increase of fluorescence intensity due to compound 48/80 and substance P was inhibited in a concentration-dependent fashion (Fig. 1). Similar inhibitory effects were also seen after pretreatment with theophylline, disodium cromoglycate, ketotifen, mequitazine, terfenadine and metabolite I of terfenadine as shown in Table 1. It has been indicated that terfenadine is converted into metabolite I and II after administration in vivo (Garteiz et al., 1982). In this experiment, it was shown that metabolite I is active in inhibiting both the fluorescence increase and histamine release but no such action was observed in metabolite II (Table 2). These results indicate that either compound 48/80 or substance

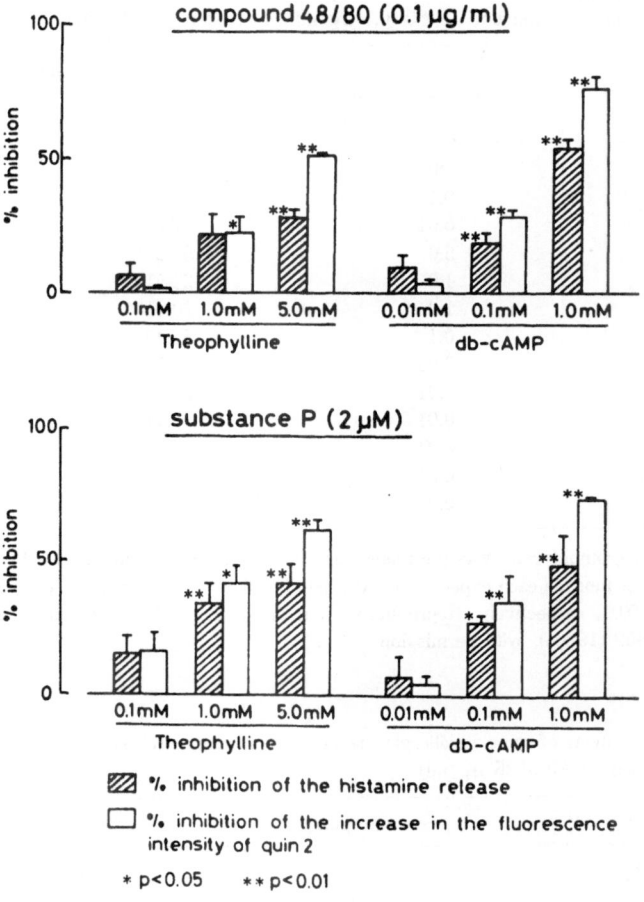

Fig. 1. Inhibitory effects of theophylline and db-cAMP on histamine release and fluorescence intensity. Mast cells were exposed to either compound 48/80 (0.1 μg/ml) or substance P (2 μM) at 37 °C for 30 sec. Reproduced from Tasaka, K., Mio, M. and Okamoto, M.: Ann. Allergy, **56**, 464-469 (1986a), with permission.

P induces release of Ca^{2+} from an intracellular Ca store; this can be the cause for histamine release (Ennis et al., 1980b; White et al., 1984) since the extents of histamine release inhibition were in accordance with those of fluorescent intensities in each experiment. Also, a compound not effective in preventing histamine release, such as metabolite II of terfenadine, was not efficient in inhibiting the increase of fluorescent intensity due to compound 48/80.

When mast cells were treated with quin 2, histamine release due to compound 48/ 80 (0.1 μg/ml) was investigated, since quin 2 has a high affinity for Ca^{2+} and could be the reason for inhibition of histamine release. Discernible inhibition, however, was not

Table 1. Inhibitory effects of some antiallergic compounds on the increase of fluorescence intensity of quin 2-loaded mast cells due to compound 48/80 (0.1 μg/ml) or substance P (2 μM)

Drugs	concentration(mM)	% inhibition	
		compound 48/80	substance P
db-cAMP	1.0	76.40**	74.74**
	0.1	28.98**	35.39**
	0.01	7.20	4.07
theophylline	5.0	51.27**	62.12**
	1.0	22.51	42.19**
DSCG†	1.0	70.91**	70.09**
	0.1	52.49**	68.92**
	0.02	32.58**	38.36**
	0.01	26.38*	30.36*
terfenadine	0.01	63.41**	96.45**
metabolite I ††	0.01	25.78*	45.38**
metabolite II ††	0.01	7.85	8.53
mequitazine	0.015	31.09**	65.63**

† DSCG (disodium cromoglycate) was pretreated for 30 sec. ††: Metabolite I and II are metabolites of terfenadine. n = 3 or more in each experiment. At least 25 cells were determined in each experiment. * p < 0.05 and ** p < 0.01, respectively. Reproduced from Tasaka, K., Mio, M. and Okamoto, M.: Ann. Allergy, 56, 464-469 (1986a), with permission.

Table 2. Inhibitory effects of some antiallergic agents on the histamine release from rat peritoneal mast cells induced by compound 48/80 (0.35 μg/ml)

Drugs	concentration(μM)	% inhibition
terfenadine	2	8.7 ± 0.6
	10	77.2 ± 6.1**
	20	99.0 ± 5.1**
metabolite I	2	29.0 ± 1.2
	10	43.0 ± 9.1**
	20	67.8 ± 1.1**
	100	97.3 ± 2.9**
metabolite II	2	−6.0 ± 1.2
	10	−3.0 ± 7.1
	20	−1.4 ± 4.5
	100	2.6 ± 1.0
ketotifen	2	19.7 ± 6.3
	10	25.0 ± 3.2*
	20	27.7 ± 1.4*
mequitazine	2	6.2 ± 1.1
	10	34.7 ± 5.7**
	20	88.2 ± 3.6**

In each experiment, at least n = 3 were performed. * p < 0.05 and ** p < 0.01, respectively. Reproduced from Tasaka, K., Mio, M. and Okamoto, M.: Ann. Allergy, 56, 464-469 (1986a), with permission.

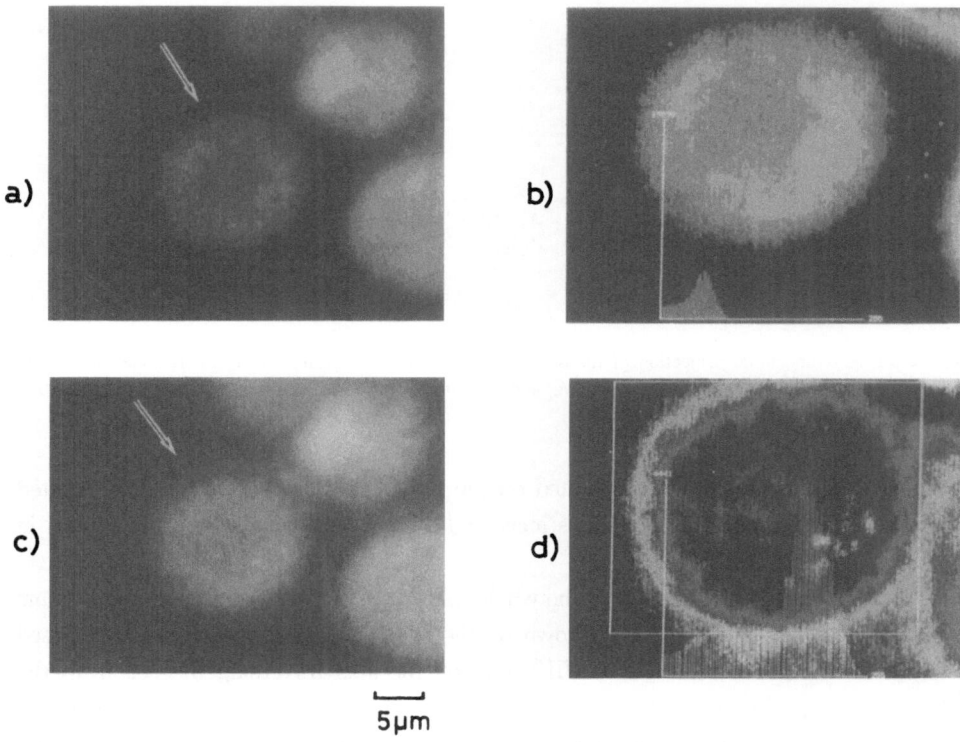

5μm

Fig. 2. Fluorescent image of rat mast cell loaded with quin 2. (a) Fluorescent image before exposure to compound 48/80 (the target cell is marked with the arrow). (b) Image processing appearance of the target cell. Pseudo-color conversion was made and histogram of the pixel values was displayed. (c) Fluorescent image of the target cell after exposure to compound 48/80 (0.1 μg/ml). (d) Image processing appearance of the target cell after exposure to compound 48/80 (0.1 μg/ml). Reproduced from Tasaka, K., Mio, M. and Okamoto, M.: Ann. Allergy, 56, 464-469 (1986a), with permission.

induced at concentration lower than 5 μM but was affected at concentrations higher than 10 μM.

Fig. 2a shows a fluorescent image of a normal mast cell loaded with quin 2 and the mast cell marked with the arrow in Fig. 2a is shown in a pseudo-color image in Fig. 2b. Distribution of the height of pixel values in the window is displayed at the bottom. Low values are located on the left and higher values are shown in the right. All the pixel values were gathered by scanning the axis traversing the cell image. When the mast cells were exposed to compound 48/80, fluorescent intensity increased markedly as shown in Fig. 2c. In these two figures (Fig. 2a, c), the fluorescent images were taken from the same cell, allowing a comparison of fluorescent intensity before and after exposure to compound 48/80. In the pseudo-color display shown in Fig. 2d, a remarkable increase of fluorescent intensity was easily assessed by appearance of the red color covering almost

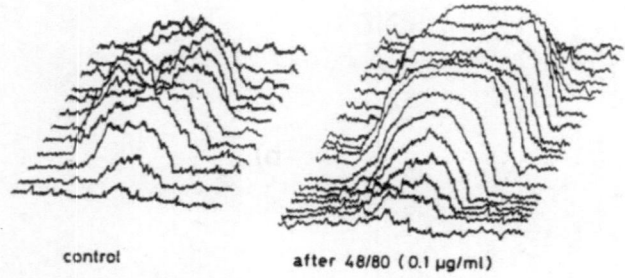

control   after 48/80 (0.1 µg/ml)

Fig. 3. Three dimensional projection of fluorescent intensity derived from quin 2 incorporated into mast cells. Reproduced from Tasaka, K., Mio, M. and Okamoto, M.: Ann. Allergy, 56, 464-469 (1986a), with permission.

the whole cell. High pixel values located on the right in the histogram were augmented tremendously, although the low values increased a little comparing with those shown in Fig. 2b.

The profile of the pixel values of the whole cell image is shown in a three-dimensional projection (Fig. 3). The control is shown on the left hand side. Each line is composed of the pixel values determined from 512 dots on the axis traversing the cell from the bottom to the top. In the middle of the figure, certain lines show two peaks and the valley surrounded by those lines corresponds to the nucleus. After it was exposed to compound 48/80 (0.1 µg/ml), the pixel value valley has almost disappeared.

1.3. Discussion

It has been reported that either compound 48/80 or substance P releases histamine from mast cells even in a Ca-free medium and that such histamine release was inhibited by some antiallergic drugs (Ennis et al., 1980b; Fewtrell et al., 1982). There is support for Ca^{2+} release from the intracellular Ca store into the cytoplasm in the case of histamine release taking place in a Ca-free medium. If histamine release in the Ca-free medium is inhibited by pretreatment with antiallergic agents, it might be interpreted that antiallergic agents elicit that effect by inhibiting Ca^{2+} release from the intracellular Ca store. Ennis et al. (1980a) reported that antiallergic compounds such as disodium cromoglycate, theophylline and db-cAMP inhibit histamine release from mast cells induced by compound 48/80 in a Ca-free medium. But they have not yet presented direct evidence for antiallergic drugs actually inhibiting Ca^{2+} release from the intracellular Ca store. Recently, quin 2 has been employed as a highly sensitive indicator for intracellular Ca^{2+} (Tsein et al., 1982; Beaven et al., 1984). It has been shown that a rapid release of Ca^{2+} from an intracellular Ca store was brought about through an antigen-antibody reaction by means of quin 2 (White et al., 1984; Beaven et al., 1984).

Fluorescent intensity of quin 2 incorporated into mast cells was apparently increased by compound 48/80 as well as substance P. Theophylline, db-cAMP, and certain antiallergic drugs (except metabolite II) inhibited an increase of the fluorescence intensity of quin 2 induced by these two stimuli. It thus became clear that one of the mechanisms of these antiallergic compounds to inhibit histamine release from mast cells may be due to the suppression of Ca^{2+} release from the intracellular Ca store. The fact that cAMP elevating drugs such as theophylline and db-cAMP inhibited Ca^{2+} release from intracellular Ca store suggests the regulatory role of cAMP in histamine release.

Time course of the increase of the quin 2 signal after exposure to these stimuli has been reported by White et al. (1984) and Beaven et al. (1984), but the detail of the movement of Ca^{2+} in a single cell has not yet been observed. As previously shown, we analyzed the distribution of the quin 2 signal using VIM system and image processing, and it was revealed that free Ca^{2+} was distributed uniformly in the cytoplasm except for around the region occupied by the nucleus. After the addition of compound 48/80, distribution of the quin 2 signal became wider indicating swelling of the mast cell, and the height of the intensity of the quin 2 signal increased markedly. Corresponding observation was seen in the analysis of quin 2 in a three dimensional projection. This was in accordance with a pseudo-color image (Tasaka et al., 1986a).

2. Inhibition of intracellular Ca mobilization, Ca uptake and histamine release induced by some antiallergic drugs

2.1. Introduction

It has been reported that some calcium antagonists effectively inhibit antigen-induced bronchospasm and exercise-induced asthma in humans (Ritchie et al., 1984; Patel, 1981). However, the calcium antagonist concentrations necessary to exert antiallergic reactions are much higher than those required to evoke cardiovascular action (Ritchie et al., 1984). In connection with this, it has been reported that oxatomide, which structurally resembles calcium antagonists of diphenylpiperazines, is effective not only in inhibiting experimental asthma in rats and guinea pigs (Ohmori et al., 1982a), but also in preventing histamine release from human lung samples and basophils induced by antigen or by anti-IgE (Church and Gradidge, 1980). However, the mechanism which prevents histamine release is not understood in detail. In order to analyze its antiallergic properties, Ca uptake and intracellular Ca mobilization in mast cells treated with oxatomide were studied in comparison with other antiallergic drugs and calcium antagonists (Tasaka et al., 1987a).

2.2. Anaphylactic histamine release from sensitized guinea pig lung samples

When the chopped lungs of actively sensitized guinea pigs were exposed to antigen,

Table 3. Effects of oxatomide and ketotifen on anaphylactic histamine release from actively sensitized guinea pig lungs

compounds	concentration (μM)	% histamine release − ovalbumin	% histamine release + ovalbumin	% inhibition
control		4.9 ± 0.5	27.6 ± 3.6	—
oxatomide	0.01	5.8 ± 1.1	23.6 ± 1.4	15.3
	0.1	4.9 ± 0.4	20.2 ± 3.5*	27.7
	1	4.7 ± 0.4	18.3 ± 0.6**	40.1
	10	4.2 ± 0.5	15.9 ± 0.8**	48.5
ketotifen	0.1	3.3 ± 0.3	24.4 ± 3.5	7.0
	1	3.8 ± 0.4	20.2 ± 0.7*	27.8
	10	4.7 ± 0.7	18.8 ± 1.0**	37.9
	100	7.5 ± 0.7	22.9 ± 2.0*	32.2

Each value represents the mean ± SEM (n = 5).
Significantly different from the control at * p < 0.05, ** p < 0.01.
Reproduced from Tasaka, K., Akagi, M., Mio, M., Miyoshi, K. and Nakaya, N.: Int. Arch. Allergy Appl. Immunol., 83, 348-353 (1987a), with permission.

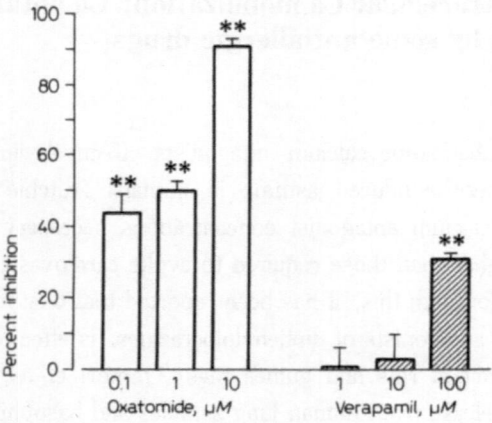

Fig. 4. Inhibitory effects of oxatomide and verapamil on the ^{45}Ca uptake into rat peritoneal mast cells induced by compound 48/80 (0.5 μg/ml). Each column represents the mean ± SEM of 5 experiments. Significantly different from the control at * p < 0.05 and ** p < 0.01. Reproduced from Tasaka, K., Akagi, M., Mio, M., Miyoshi, K. and Nakaya, N.: Int. Arch. Allergy Appl. Immunol., 83, 348-353 (1987a), with permission.

histamine release was induced to 27.6 ± 3.6 %. As indicated in Table 3, oxatomide at concentrations higher than 0.1 μM effectively inhibited histamine release in a dose-dependent fashion. Ketotifen caused less marked inhibition than did oxatomide.

2.3. ^{45}Ca uptake into rat peritoneal mast cells induced by compound 48/80

The effect of oxatomide on the ^{45}Ca uptake into mast cells induced by compound 48/80 was investigated in comparison with that of verapamil. The Ca uptake in cells treated with compound 48/80 (0.5 μg/ml) was about 3 times higher than that found in non-treated control cells; an uptake of 378.6 $\pm$ 16.9 dpm/10^6 cells in resting cells was augmented to 1,163.5 $\pm$ 58.8 dpm/10^6 cells. As shown in Fig. 4, oxatomide treatment at concentrations higher than 0.1 μM significantly inhibited ^{45}Ca uptake in a dose-dependent fashion and the effective concentration range was similar to that observed in the case of histamine release inhibition (Church and Gradidge, 1980). Oxatomide alone did not affect the spontaneous influx of ^{45}Ca into normal mast cell. Verapamil was much less effective in preventing ^{45}Ca uptake; significant inhibition was elicited at 100 μM.

2.4. Intracellular mobilization of Ca^{2+} in rat peritoneal mast cells

2.4.1. Fluorescence measurement by photon counting

When mast cells loaded with quin 2 were exposed to compound 48/80 (0.2 μg/ml) for 1 min at 37 °C, their fluorescence intensity increased significantly, exhibiting a value of 149.6 $\pm$ 8.7 (n $=$ 250, p $<$ 0.01), as compared with the control value of 100 $\pm$ 6.8 (n $=$

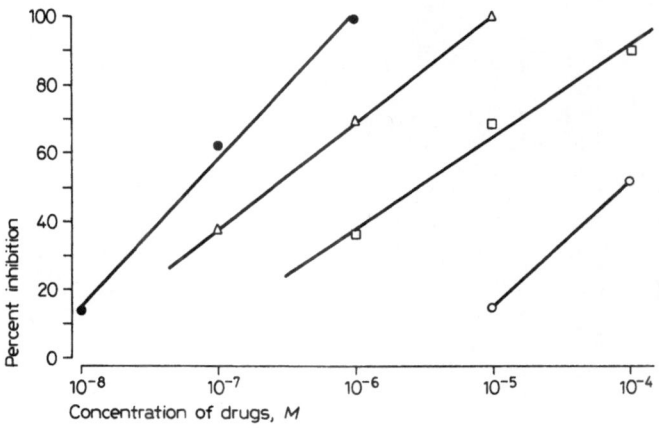

Fig. 5. Inhibitory effects of some antiallergic compounds and calcium antagonists on the fluorescence intensity increase in quin 2-loaded mast cells induced by compound 48/80 (0.2 μg/ml). Oxatomide (●), verapamil (△), D-600 (□), ketotifen (○). IC50 values (95 % confidence limits) were as follows: oxatomide 5.0 $\times$ 10^{-8}M (3.2 $\times$ 10^{-8} $-$ 9.5 $\times$ 10^{-8}), verapamil 2.7 $\times$ 10^{-7}M (1.6 $\times$ 10^{-7} $-$ 5.1 $\times$ 10^{-7}) and D-600 3.7 $\times$ 10^{-6}M (1.9 $\times$ 10^{-6} $-$ 7.2 $\times$ 10^{-6}). Reproduced from Tasaka, K., Akagi, M., Mio, M., Miyoshi, K. and Nakaya, N.: Int. Arch. Allergy Appl. Immunol., 83, 348-353 (1987a), with permission.

106

260). Since the experiment was carried out in a Ca-free medium, an increase in fluorescence intensity reflects the amount of Ca^{2+} released from the intracellular Ca store. The inhibitory effects of oxatomide, verapamil, D-600 and ketotifen on the intracellular Ca^{2+} release induced by compound 48/80 are shown in Fig. 5. The inhibitory effect of oxatomide was more potent than those of verapamil and D-600. Ketotifen exerted only a weak inhibition.

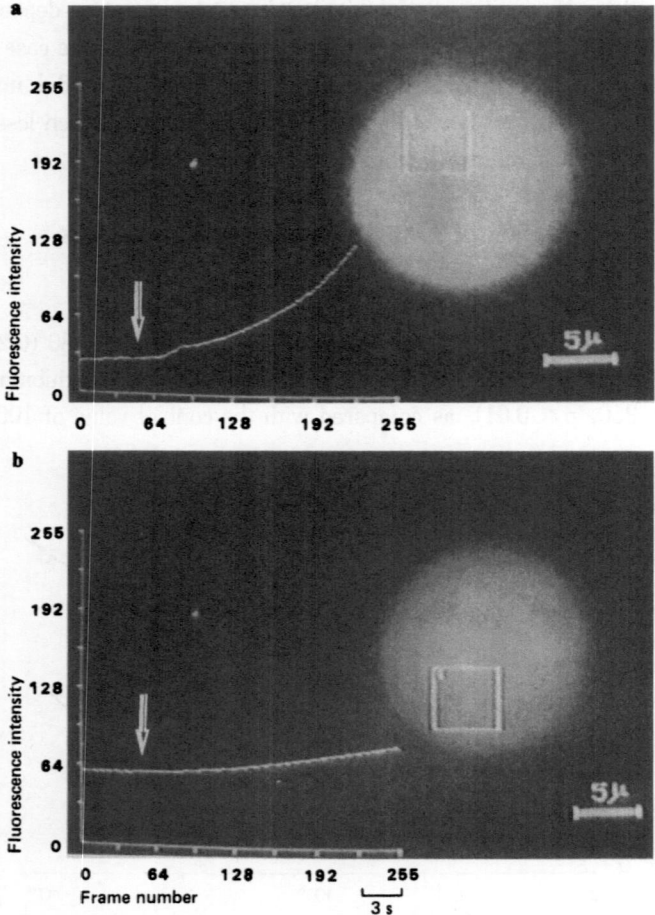

Fig. 6. Real time measurements of the fluorescence intensity of single, quin 2-loaded mast cells stimulated with compound 48/80 (0.2 μg/ml). a: Normal mast cell exposed to compound 48/80. b: Mast cell pretreated with oxatomide (1 μM). The target area used for the fluorescence measurement is shown in a square. An arrow indicates the addition of compound 48/80. Reproduced from Tasaka, K., Akagi, M., Mio, M., Miyoshi, K. and Nakaya, N.: Int. Arch. Allergy Appl. Immunol., 83, 348-353 (1987a), with permission.

2.4.2. Real time measurement of the fluorescence intensity in single cell

By means of a VIM system, the effect of compound 48/80 on fluorescence intensity was measured in real time. The target area for fluorescence intensity measurement was indicated as a square on the cell (Fig. 6). The fluorescence intensities of all pixels in this area were integrated in real time and expressed as a relative value in eight bits (from 0 to 255) on the Y-axis. Measurement was performed every 10 frames, i.e. every 1/3 sec. The X-axis represents the number of video frames measured; the film runs 30 frame/

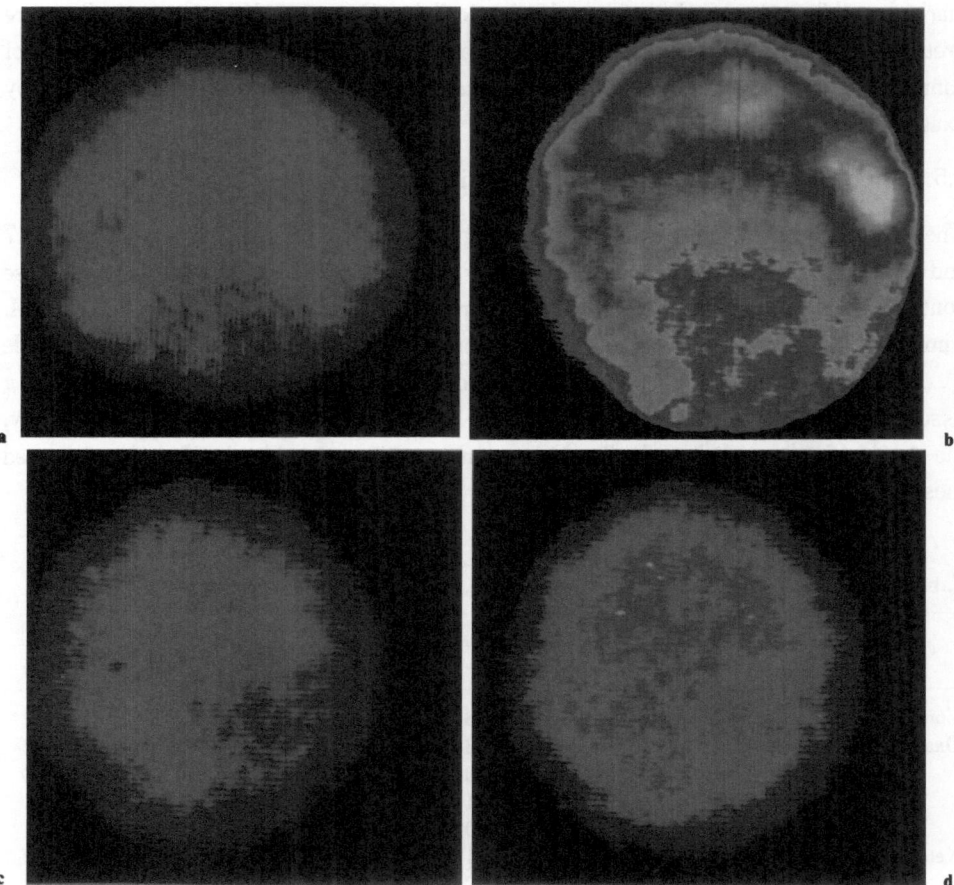

Fig. 7. Pseudo-color conversion of the fluorescence images of quin 2-loaded mast cells. a: Normal mast cell. b: 1 min after addition of compound 48/80 (0.2 μg/ml) to a. c: Mast cell treated with oxatomide (1 μM). d: 1 min after addition of compound 48/80 (0.2 μg/ml) to c. Reproduced from Tasaka, K., Akagi, M., Mio, M., Miyoshi, K. and Nakaya, N.: Int. Arch. Allergy Appl. Immunol., 83, 348-353 (1987a), with permission.

sec. As seen in Fig. 6a, in a Ca-free medium, compound 48/80 (0.2 μg/ml) induced rapid and significant increase in fluorescence intensity. When the mast cell was pretreated with 1 μM of oxatomide, no such elevation was observed (Fig. 6b). The inhibitory effect of oxatomide may be associated with the inhibition of Ca^{2+} release from the intracellular Ca store.

2.4.3. Processing of the fluorescence images of quin 2-loaded mast cells

In normal mast cells, fluorescence intensity was weak and the pseudo-color image of whole cell area was almost blue (Fig. 7a). After addition of compound 48/80, fluorescence intensity increased tremendously over the entire cell surface (Fig. 7b), indicating marked mobilization of Ca^{2+} from the intracellular Ca store. When mast cells were pretreated with oxatomide, the resulting color images were quite similar to those of control cells (Fig. 7c, d), indicating that intracellular Ca^{2+} release was clearly inhibited by oxatomide pretreatment.

2.5. Effects on cyclic nucleotide contents in guinea pig lung samples

The cAMP and cGMP contents of guinea pig lungs in control animals were 348.8 ± 3.7 and 31.7 ± 0.4 pmole/g tissue, respectively. Oxatomide alone did not affect these contents or the A/G ratios (cAMP/cGMP ratio) at concentrations lower than 1 μM. Similarly, in the case of ketotifen, no alteration in these parameters was observed (Table 4). After exposure to antigen, the cAMP content decreased to 223.9 ± 1.4 pmole/g tissue, while the cGMP content increased to 63.8 ± 3.4 pmole/g tissue; consequently, the A/G ratio decreased markedly. Pretreatment with oxatomide significantly prevented these changes (Table 5).

Table 4. Effects of oxatomide and ketotifen on cyclic nucleotide content in sensitized guinea pig lungs.

Compounds	concentration (μM)	cAMP pmole/g tissue	cGMP pmole/g tissue	A/G ratio
Control (PBS)		348.8 ± 3.7	31.7 ± 0.4	11.0
Oxatomide	0.01	341.3 ± 1.3	30.5 ± 0.4	11.2
	0.1	342.0 ± 1.7	30.1 ± 1.8	11.4
	1	343.1 ± 3.0	33.1 ± 3.6	10.4
	10	325.0 ± 4.8**	39.9 ± 1.1**	8.1
Ketotifen	0.1	346.1 ± 1.5	33.8 ± 0.1	10.2
	1	342.0 ± 1.4	33.2 ± 1.9	10.3
	10	341.3 ± 2.7	31.6 ± 2.3	10.8
	100	349.9 ± 2.1	32.0 ± 3.4	10.9

Each value represents the mean $\pm$ SEM (n = 5). A/G ratio represents the ratio of cAMP to cGMP. ** Significantly different from the control (p < 0.01). Reproduced from Tasaka, K., Akagi, M., Mio, M., Miyoshi, K. and Nakaya, N.: Int. Arch. Allergy Appl. Immunol., 83, 348-353 (1987a), with permission.

Table 5. Effects of oxatomide and ketotifen on cyclic nucleotide content in sensitized guinea pig lung samples after exposure to antigen

Compounds	concentration (μM)	cAMP pmole/g tissue	cGMP pmole/g tissue	A/G ratio
Control (+ ovalbumin)		223.9 ± 1.4	63.8 ± 3.4	3.5
Oxatomide	0.01	256.5 ± 3.5**	68.7 ± 5.3	3.7
	0.1	331.5 ± 1.3**	50.8 ± 4.3*	6.5
	1	333.5 ± 4.2**	44.9 ± 3.5*	7.4
	10	357.0 ± 1.2**	60.0 ± 6.4**	6.0
Ketotifen	0.1	243.0 ± 4.4**	58.4 ± 1.8	4.2
	1	323.3 ± 0.5**	57.1 ± 1.6	5.7
	10	352.1 ± 2.5**	50.3 ± 4.4*	7.0
	100	309.0 ± 1.7**	40.5 ± 1.6**	7.6

Antigen (ovalbumin) was added to a final concentration of 50 μg/ml. Each value represents the mean ± SEM of 5 samples; each assay was carried out in duplicate. Significantly different from the control at * p < 0.05 and ** p < 0.01. Reproduced from Tasaka, K., Akagi, M., Mio, M., Miyoshi, K. and Nakaya, N.: Int. Arch. Allergy Appl. Immunol., 83, 348-353 (1987a), with permission.

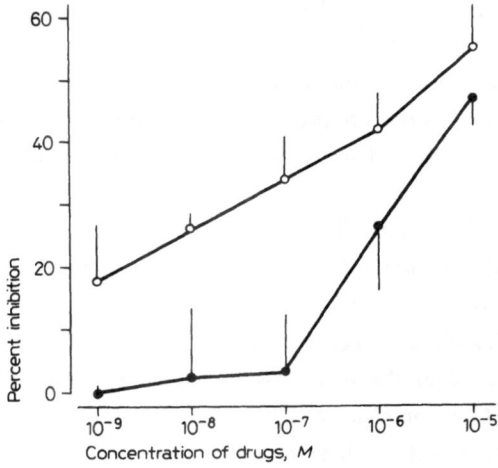

Fig. 8. Inhibitory effect of oxatomide (●) and verapamil (○) on ^{45}Ca uptake into cultured neonatal rat heart cells. Each plot represents the mean ± SEM of 4 experiments. Reproduced from Tasaka, K., Akagi, M., Mio, M., Miyoshi, K. and Nakaya, N.: Int. Arch. Allergy Appl. Immunol., 83, 348-353 (1987a), with permission.

2.6. ^{45}Ca uptake into cultured neonatal rat heart cells

When ^{45}Ca uptake into cultured rat heart cells was examined, beating neonatal heart cells exhibited active uptake of ^{45}Ca (444.8 ± 29.2 dpm/10^6 cells, n = 4). Verapamil at concentrations higher than 10^{-9} M strongly inhibited this uptake in a dose-dependent fashion (Fig. 8). Although oxatomide exhibited similar inhibition at concentrations higher than 10^{-6} M, it was about 100 times less potent than verapamil.

2.7. Discussion

It has been reported that oxatomide inhibited the histamine release from rat peritoneal mast cells induced by compound 48/80, antigen, anti-IgE and A23187 and that its potency was higher than those of clemastine, promethazine and ketotifen (De Clerk et al., 1981; Ohmori et al., 1982b). In the present study, the concentration range of oxatomide required for the inhibition of histamine release from sensitized guinea pig samples was similar to that reported in relation to rat mast cells (De Clerk et al., 1981; Ohmori et al., 1982b). Furthermore, it has been reported that histamine release inhibition produced by oxatomide involves part of a Ca^{2+}-required activation step in mast cells such as antigen exposure, or compound 48/80 or A23187 induction and oxatomide is equally active against all three inducers (De Clerk et al., 1981). The structural resemblance of oxatomide to calcium antagonists of diphenylpiperazines also suggests that this compound may act on a Ca^{2+}-mediated stage of cell activation in the cell membrane. Such an assumption is supported by the fact that oxatomide effectively inhibited the compound 48/80-induced uptake of ^{45}Ca. The inhibitory effect of oxatomide on calcium uptake into mast cells was much more potent than that of verapamil. Interestingly enough, oxatomide was much less potent in preventing ^{45}Ca uptake into neonatal heart cells. These findings strongly suggest that oxatomide is a calcium antagonist highly selective toward mast cells.

When the quin 2-loaded mast cells were pretreated with oxatomide, the fluorescence increase was remarkably suppressed. This indicates that one of the inhibitory mechanisms of oxatomide involved in histamine release is the inhibition of Ca^{2+} release from intracellular store. Although we reported that the storage site of Ca^{2+} in rat peritoneal mast cells may be located in the endoplasmic reticulum (ER) (Tasaka et al., 1987b; Tasaka et al., 1988), it is not clear whether oxatomide acts directly on the ER. As shown in Fig. 5, Ca antagonists such as verapamil and D-600 inhibited Ca^{2+} release from the intracellular store, though the extent of inhibition was approximately 1/10th to 1/100th that of oxatomide.

It has been reported that a reduction in cAMP content in rat peritoneal mast cells was associated with the histamine release reaction, and that pretreatment with certain antiallergic agents such as theophylline and dibutyryl cAMP, which are capable of increasing cAMP content, resulted in histamine release inhibition (Tasaka, 1986; Tasaka et al., 1986a). Moreover, in bovine chromaffin cells and rat liver cells, an increase in the intracellular Ca^{2+} level may be effected by either muscarinic stimulation or calcium ionophore treatment, resulting in an elevation of cGMP content. One interpretation of this finding is that an increase in guanylate cyclase activity takes place after increased uptake of Ca^{2+} (Ohsako and Deguchi, 1983; Pointer et al., 1976). Therefore, a Ca^{2+}-mediated cell activating process may be caused by an increase in cGMP content and the A/G ratio seems to be an indication of the histamine release response. Indeed, antigen-antibody reaction decreased cAMP content and increased cGMP content in chopped guinea pig lung samples, with a consequent decrease in the A/G ratio. Although neither oxatomide nor ketotifen affected contents of these nucleotides in normal chopped lungs, pretreatment with either drug effectively prevented both the decrease in cAMP and increase in cGMP elicited by antigen challenge. Since both compounds exhibit an inhibitory effect on the increase in intracellular Ca^{2+} level, the decrease in cGMP content may be considered the effect which prevents intracellular Ca^{2+} increase. In connection with changes in cAMP content, it has been shown that cAMP effectively prevents Ca^{2+} release from the intracellular Ca store of rat mast cells (Tasaka et al., 1986). Because of this, inhibiting cAMP decrease may effectively prevent mast cell activation by suppressing the Ca-mediated processes.

3. Sequential analysis of histamine release and intracellular Ca^{2+} release from murine mast cells

3.1. Introduction

It has been shown that the rate of phosphatidylinositol (PI) turnover increases after cellular stimulation in connection with the activation of phospholipase C, resulting in the release of inositol-1, 4, 5-trisphosphate (IP_3); IP_3 is capable of releasing Ca^{2+} from the endoplasmic reticulum (ER) which is one of the important Ca stores involved in cellular activation (Mitchell, 1975; Berridge and Irvine, 1984). In rat mast cells, it is also reported that PI turnover increases in association with histamine release (Imai et al., 1984a) and that IP_3 releases Ca^{2+} from the ER (Yoshii et al., 1988). However, the rapid onset of histamine release from the rat mast cell makes it difficult to clarify the sequential relationships between the increase in PI turnover and the histamine release induced by histamine releasers such as compound 48/80 or by antigen-antibody reaction. That histamine release in murine mast cells is slower than that seen in rat mast cells has been previously observed (Fox et al., 1982; Barrett and Pearce, 1983). This suggests that murine mast cells may be more suitable for such sequential analysis. In this section, the changes in inositol phosphate production in murine mast cells were reviewed in association with the changes in intracellular Ca^{2+} concentration and histamine release (Tasaka et al., 1990).

3.2. Sequential changes in histamine release, inositol phosphate contents and intracellular Ca^{2+} concentration in murine mast cells after exposure to compound 48/80

The onset of the histamine release from murine mast cells induced by compound 48/80 (1 μg/ml) was at 5 sec, and reached a plateau at about 60 sec (Fig. 9). As shown in this figure, both IP_2 and IP_3 increased to three to four times the control level at 5 sec. The IP_3 content decreased rapidly and returned to the control level within 60 sec. The rate of decrease in IP_2 content was slower than that of IP_3. IP_1 increased gradually and reached a maximum (1.6 times of the control) 30 sec after stimulation. Thereafter, the IP_1 content gradually decreased to the control level. The changes in the intracellular Ca^{2+} concentrations determined with quin 2 are indicated in Fig. 10. The fluorescence intensity derived from Ca^{2+}-quin 2 complex increased immediately after addition of compound 48/80 and reached a maximum at about 7 sec. Thereafter, the fluorescence intensity gradually decreased to the control level.

3.3. Discussion

Histamine can be released from mast cells even in a Ca-free medium, suggesting that Ca^{2+} release from the intracellular Ca store after exposure to histamine-releasing stimuli (Ennis et al., 1980b). In such cases, it is assumed that an increase in PI turnover may be the event leading to cellular activation, since Ca^{2+} can be released from the ER by IP_3 (Yoshii et al., 1988; Abdel-Latif, 1986). Actually, it is reported that both an increase in IP_3 content after exposure to histamine releasing stimuli (Imai et al., 1984a) and IP_3 at micromolar range release Ca^{2+} from the isolated ER of rat mast cells (Yoshii et al., 1988). However, in these cells the histamine release induced by compound 48/80

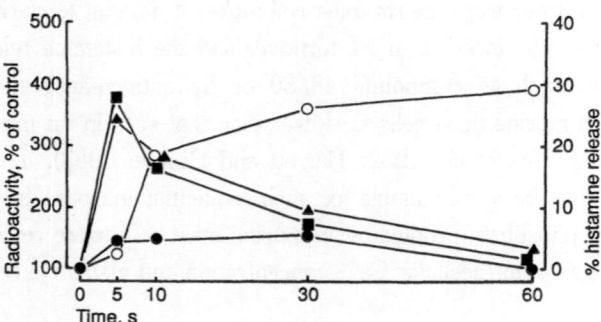

Fig. 9. Sequential changes in histamine release and inositol phosphate contents in murine peritoneal mast cells induced by compound 48/80 (1 μg/ml). ○: Histamine release; ■: IP_3; ▲: IP_2; ●: IP_1. Each point represents the mean ± SEM (n = 5). Reproduced from Tasaka, K., Sugimoto, Y. and Mio, M.: Int. Arch. Allergy Appl. Immunol., 91, 211-213 (1990), with permission.

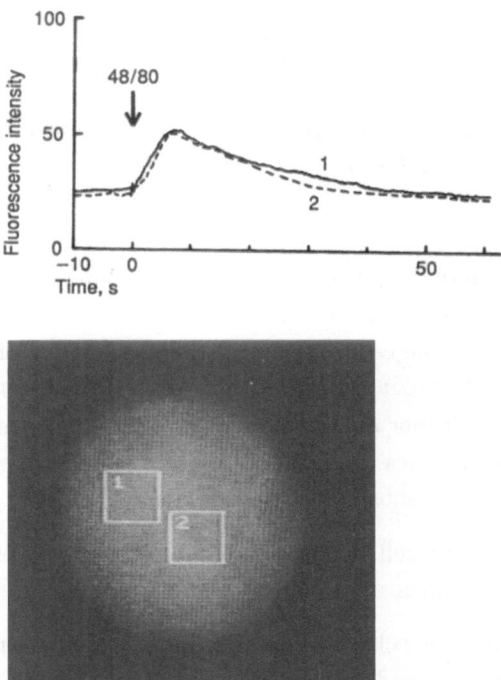

Fig. 10. Changes in the fluorescence intensity of quin 2-loaded murine peritoneal mast cells induced by compound 48/80 (1 μg/ml) in a Ca-free medium. The squares 1 and 2 represent the target areas for fluorescence intensity measurement. Reproduced from Tasaka, K., Sugimoto, Y. and Mio, M.: Int. Arch. Allergy Appl. Immunol., 91, 211-213 (1990), with permission.

reaches a maximum within a few seconds after stimulation, so that accurate analysis of the time sequences between IP_3 formation, Ca^{2+} increase, and histamine release is not possible. It was revealed in this section that histamine release from murine mast cells takes place more slowly than that from rat mast cells. Similar observations have been reported by several authors (Fox et al., 1982; Barrett and Pearce, 1983). Both IP_3 and IP_2 contents reached their maximum at approximately the onset of histamine release. Since it is reported that IP_2 is formed by a dephosphorylation of IP_3 (Abdel-Latif, 1986), the slower decrease in IP_2 seems to correspond with this observation. The fact that the decrease in IP_1 content was much slower than those of the other two compounds may suggest that IP_1 formation is derived from a dephosphorylation of IP_2. The maximum increase in intracellular Ca^{2+} was slightly slower than that of IP_3, but much faster than that of histamine release. Among the inositol phosphates, since only IP_3 is active in releasing Ca^{2+} from the ER (Yoshii et al., 1988b), it is reasonable to assume that a rapid increase in Ca^{2+} concentration depends on the Ca^{2+} release from the ER by IP_3. When mast cells are stimulated with compound 48/80, IP_3 is liberated initially from PI-4, 5-bisphosphate by activated phospholipase C and as a consequence of this, IP_3 releases Ca^{2+} from the ER. Finally, histamine release takes place.

4. Ca uptake and Ca releasing properties of the endoplasmic reticulum in rat peritoneal mast cells

4.1. Introduction

When mast cells are stimulated with compound 48/80, or after antigen-antibody reaction, there is an increase in PI turnover (Kennerly et al., 1979; Imai et al., 1984a). Subsequently, IP_3 liberated from the cell membrane acts on the intracellular Ca store so as to release Ca^{2+} (Abdel-Latif, 1986; Tasaka et al., 1990). However, neither the existence of the Ca store in mast cells nor the effect of IP_3 on such a store has been reported. In order to study the characteristics of the intracellular Ca store of mast cells, fractionation of the organelles and identification of Ca stores were performed.

4.2. Fractionation of mast cell organelles and determination of ^{45}Ca binding and marker enzyme activities of endoplasmic reticulum and mitochondria

Peritoneal rat mast cells were isolated from the abdominal cavity of male Wistar rats and purified by Percoll density gradient centrifugation to more than 95 % homogeneity (Tasaka et al., 1987b). The mast cells were washed with sucrose/HEPES buffer (pH 7.2) (buffer A) and suspended in buffer A containing chymostatin, p-amidinophenylmethylethanesulfonylfluoride (p-APMSF) and bovine serum albumin (BSA). The cells were disrupted by ultrasonication and centrifuged to remove the nuclei and cell debris. The supernatant was layered onto Percoll solution and centrifuged at 4 °C. After that, the sediment was divided into 18 fractions. Each fraction (500 μl) was mixed with the same volume of buffer B (sucrose, $MgCl_2$, HEPES; pH 7.2) containing $^{45}CaCl_2$ and incubated. Thereafter, the reaction mixture was passed through a glass-fiber filter. The filter was washed with buffer A and the radioactivity of the filter was determined by means of a liquid scintillation counter (Yoshii et al., 1988). Glucose-6-phosphatase (G-6-Pase) was chosen as a marker enzyme for the ER, and its activity was determined according to the method of de Duve et al. (1955). As a marker enzyme for the mitochondria, the activity of succinate dehydrogenase was determined according to the method of Amende and Donlon (1985).

Fig. 11A shows profiles of ^{45}Ca binding to mast cell fractions and the density of each fraction. Significant binding of ^{45}Ca was observed in fractions 6-8, with densities of 1.068-1.072. Histamine contents of the fractions are shown in Fig. 11B. High histamine contents were detected in both the initial (1-2) and terminal (16-17) fractions, while very low levels of histamine were observed in between. Fractions 1 probably corresponds the

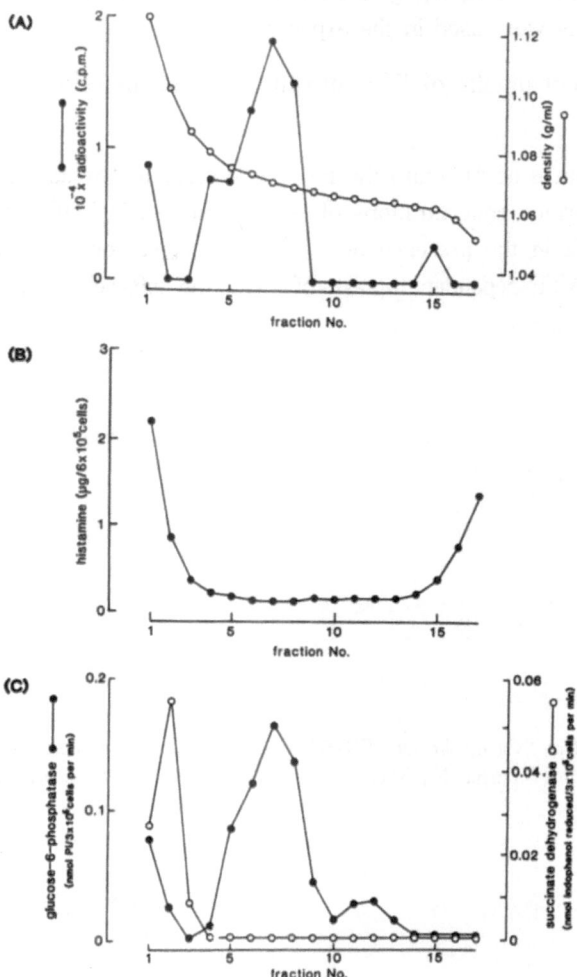

Fig. 11. Characteristics of the fractionated mast cell organelles. (A) Profiles of density and ^{45}Ca binding capacity. ●, ^{45}Ca binding; ○, density of each fraction. (B) Profile of histamine content in each fraction. (C) Activity profiles of the marker enzymes for the endoplasmic reticulum (ER) and mitochondria. ●, glucose-6-phosphatase (marker for ER); ○, succinate dehydrogenase (marker for mitochondria). Reproduced from Yoshii, N., Mio, M. and Tasaka, K.: Immunopharmacology, 16, 107-113 (1988), with permission.

granule fraction, and the histamine liberated from disrupted granules during sonication may appear in the terminal fractions. As shown in Fig. 11C, prominent succinate dehydrogenase activity was observed in fraction 2, while activity of this enzyme was not detected in the other fractions. Noticeable activity of G-6-Pase was detected in fractions 6-8. Since high Ca binding capacity was determined in these fractions, they are thought to be ER-rich fractions having negligible amounts of contaminated granules and mitochondria. These fractions were used in the experiments.

4.3. ATP-dependent uptake of ^{45}Ca into the ER and the effect of LaCl$_3$ and Na$_3$VO$_4$

ATP-dependent uptake of ^{45}Ca into the ER was determined by adding the ER fraction of mast cells to various concentrations of ATP dissolved in buffer B and these were incubated with ^{45}Ca in the presence of NaN$_3$. To investigate the effects of ATPase inhibitors on the ATP-dependent uptake of ^{45}Ca, the ER-rich fraction was added to

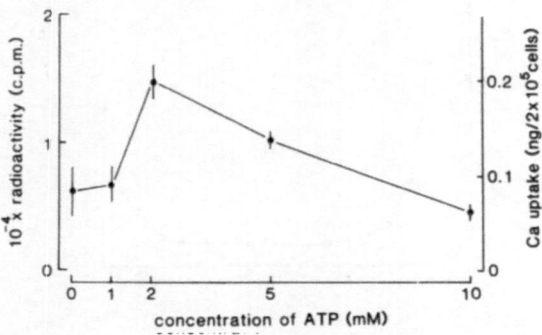

Fig. 12. ATP-dependent ^{45}Ca uptake into ER-rich fractions. Each point represents the mean $\pm$ SEM; n = 6. Reproduced from Yoshii, N., Mio, M. and Tasaka, K.: Immunopharmacology, 16, 107-113 (1988), with permission.

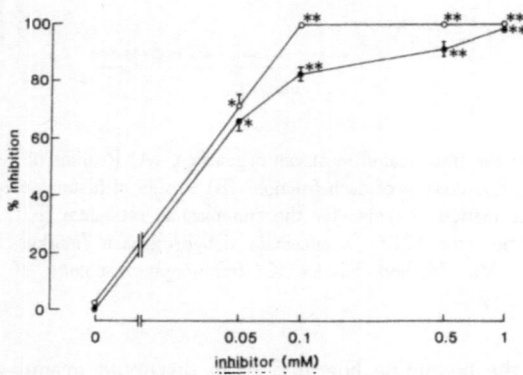

Fig. 13. Influence of ATPase inhibitors on the ATP-dependent ^{45}Ca uptake into ER-rich fractions. ●, LaCl$_3$; ○, Na$_3$VO$_4$. Each point represents the mean $\pm$ SEM; n = 6. *, **, p < 0.05 and p < 0.01, respectively. Reproduced from Yoshii, N., Mio, M. and Tasaka, K.: Immunopharmacology, 16, 107-113 (1988), with permission.

buffer A containing ATP, NaN_3 and $^{45}CaCl_2$ in the presence or in the absence of various concentrations of Na_3VO_4 or $LaCl_3$. The radioactivity of ^{45}Ca incorporated into the ER was determined (Yoshii et al., 1988).

Fig. 12 represents a profile of ATP-dependent ^{45}Ca uptake into the ER-rich fraction. Maximum uptake was observed in the presence of 2 mM ATP. NaN_3 was added into the incubation medium to prevent uptake of ^{45}Ca into the mitochondria, although the amount of contaminating mitochondria in these fractions were negligible. When the time course of ^{45}Ca uptake into the ER-rich fraction was studied in the presence of 2 mM ATP, maximum binding was reached 10 min after addition of Ca into the medium.

As indicated in Fig. 13, ATP-dependent ^{45}Ca uptake into the ER-rich fractions was inhibited, markedly and dose-dependently, by Na_3VO_4 and $LaCl_3$; both are known as efficient ATPase inhibitors preventing active Ca uptake into the sarcoplasmic reticulum (Wang et al., 1979; Weiss, 1974). Complete inhibition was achieved by 0.1 mM Na_3VO_4 and 1 mM $LaCl_3$, respectively.

4.4. Ca release from ER fraction induced by inositol phosphates

After incubation of ER-rich fraction with 2 mM ATP and 1 μCi ^{45}Ca for 10 min, ^{45}Ca uptake was terminated by the addition of 2 mM $LaCl_3$. Thereafter, various concentrations of IP_1, IP_2 or IP_3 were added and incubated for another 2 min at 37 °C. The residual contents of ^{45}Ca in ER fractions were measured (Yoshii et al., 1988).

Fig. 14 shows the effect of IP_3 as an intracellular messenger for Ca^{2+} release (Abdel-Latif, 1986). Nearly maximum and half-maximum releases of ^{45}Ca from the ER-rich fractions were obtained at 2 μM and 0.8 μM, respectively; the results correspond closely to those reported by Hirata et al. (1984). In contrast, IP_1 and IP_2 did not

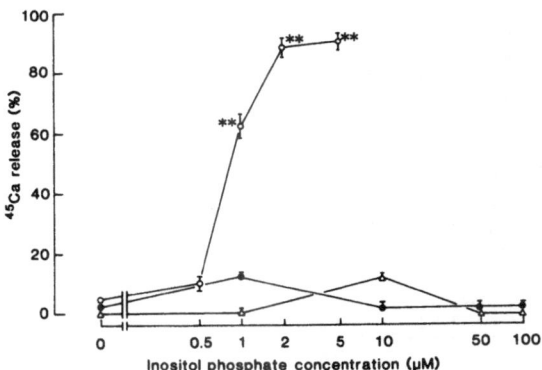

Fig. 14. Dose-response curves of inositol phosphate-stimulated release of ^{45}Ca from ER-rich fractions. ●, IP_1; △, IP_2; ○, IP_3. Each point represents the mean $\pm$ SEM; n = 6. ** $p < 0.01$. Reproduced from Yoshii, N., Mio, M. and Tasaka, K.: Immunopharmacology, 16, 107-113 (1988), with permission.

release ⁴⁵Ca from the ER at all. Although IP₃ effectively releases Ca from the ER, negligible amounts of ⁴⁵Ca were released from the mitochondrial fraction, as shown in Fig. 15. This clearly shows that IP₃-sensitive Ca stores in rat mast cells are in the ER.

4.5. ³H-IP₃ binding to ER

ER-rich fractions were incubated with various concentrations of ³H-IP₃ dissolved in the same volume of buffer A at 4°C for 2 min. Non-specific binding was determined in the presence of 10 μM non-labeled IP₃. Thereafter, the reaction mixture was passed

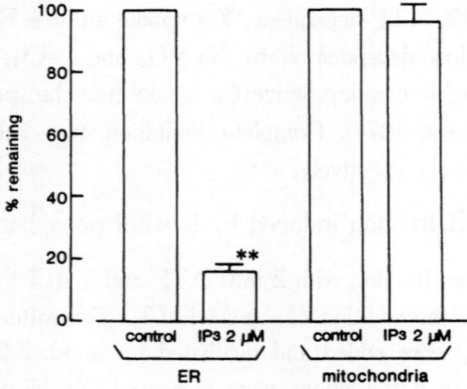

Fig. 15. Effects of IP₃ on the ⁴⁵Ca contents in the ER and mitochondrial fractions. Each column represents the mean ± SEM; n = 10., ** p < 0.01. Reproduced from Yoshii, N., Mio, M. and Tasaka, K.: Immunopharmacology, 16, 107-113 (1988), with permission.

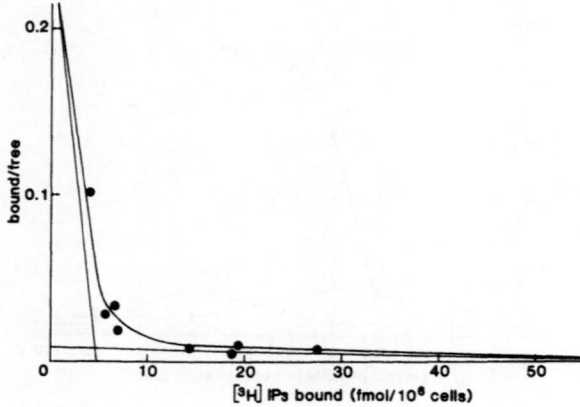

Fig. 16. Scatchard plot of ³H-IP₃-binding to the ER fraction of rat mast cells. Reproduced from Yoshii, N., Mio, M. and Tasaka, K.: Immunopharmacology, 16, 107-113 (1988), with permission.

through a glass fiber filter. The filter was promptly washed with 5 ml of buffer A and the radioactivity of the filter was determined by means of a liquid scintillation counter (Yoshii et al., 1988).

Fig. 16 shows Scatchard analysis of IP_3 binding to the ER. The binding forms a curvilinear plot, which could consist of two specific binding components. The mean value for the apparent K_D, calculated from the slope of the high-affinity component, was 5.20×10^{-11} M and B_{max} was 4.25 fmol/10^6 cells. The apparent K_D for the low-affinity site was 2.17×10^{-8} M and B_{max} was 65 fmol/10^6 cells.

4.6. Effects of cyclic nucleotides on the Ca release from the ER induced by IP_3

ER-rich fractions of rat mast cells were incubated with ^{45}Ca in the presence of 2 mM ATP and Ca uptake was terminated by the addition of 2 mM $LaCl_3$. In order to determine the effects of the following nucleotides on Ca^{2+} release from ER-rich fractions, the reaction mixture was incubated with various concentrations of cAMP, AMP or cGMP for 5 min before addition of IP_3. Thereafter, IP_3 was added to the ER fraction and the remaining radioactivity in the ER was determined (Yoshii et al,. 1988).

As shown in Fig. 17, when the ER was pretreated with cAMP at concentrations higher than 0.1 μM, IP_3-induced ^{45}Ca release was significantly inhibited in a dose-dependent fashion. However, neither AMP nor cGMP influenced the action of IP_3 on the ER-rich fraction.

4.7. Discussion

Since Ca uptake into ER was inhibited by ATPase inhibitors such as La^{3+} or vanadate,

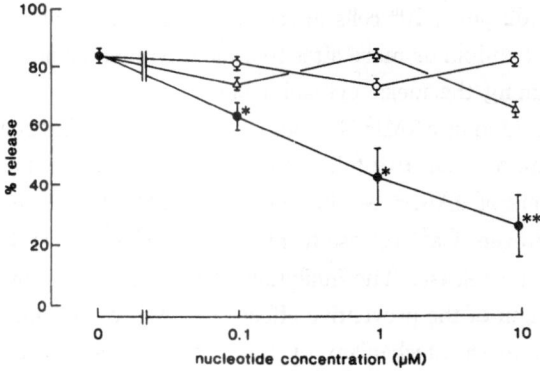

Fig. 17. Influence of nucleotides on IP_3-stimulated release of ^{45}Ca from ER-rich fractions. ●, cAMP; ○, cGMP; △, AMP. Each point represents the mean ± SEM; n = 6. *, **, p < 0.05 and p < 0.01, respectively. Reproduced from Yoshii, N., Mio, M. and Tasaka, K.: Immunopharmacology, 16, 107-113 (1988), with permission.

the process may be dependent on a Ca^{2+}-ATPase pump system (Martonosi, 1984; Hirata et al., 1985). At higher concentrations of ATP, however, ^{45}Ca uptake into the ER decreased. This result suggests that an active Ca-extrusion system may exist on the ER membrane. Furthermore, it was shown that IP_3 efficiently liberated Ca^{2+} from ER fractions, although IP_1 and IP_2 did not affect Ca^{2+} release. Moreover, when the mitochondrial fractions were exposed to IP_3, insignificant amounts of Ca^{2+} were released. This clearly indicates that an IP_3-sensitive Ca store in the mast cells is located in the ER but not in the mitochondria. Since Ca^{2+} release from ER was induced by IP_3 in the presence of an ATPase inhibitor, La^{3+}, which is capable of eliminating active Ca^{2+} incorporation, it was supposed that an La^{3+}-sensitive Ca^{2+}-ATPase system dose not participate in Ca^{2+} release from the ER. Although the mechanism of Ca^{2+} release from ER is not clear at present, a high selectivity for IP_3 suggests the existence of IP_3-receptive sites. Actually, results of the ^3H-IP_3 binding assay showed that there exist two kinds of IP_3 binding site in the ER fractions of rat mast cells; one has a high affinity and a low capacity for IP_3 binding, while the other has a low affinity accompanied by a large capacity. The results were in good agreement with the values reported for rat liver (Spat et al,. 1986). Since the concentration of IP_3 necessary to release Ca^{2+} from the ER was higher than 0.1 μM, the low-affinity site may be the functional receptor for IP_3.

It is known that an increase in the cAMP content of mast cell effectively prevents histamine release (Sullivan et al., 1975; Hayashi et al., 1976). By means of a Ca-sensitive fluorescent probe, quin 2, and VIM systems, it was shown that pretreatment with either db-cAMP or theophylline inhibits Ca^{2+} release from the intracellular Ca store (Tasaka et al., 1986a). However, no solid evidence for the interaction between cAMP and the Ca store has yet been presented. In this section, it was clearly shown that IP_3-stimulated Ca^{2+} release from the ER was inhibited in the presence of cAMP. A basal cAMP content of 0.62 $pmol/10^6$ cells in rat peritoneal mast cells has been reported and this value increased two-fold or more after treatment with some antiallergic agents (Akagi et al., 1983). Assuming the mean cytosol volume of mast cells to be about 1 pl, the intracellular concentration of cAMP is about 0.62 μM in a resting state. Since the cAMP content increases more than two-fold, after treatment with antiallergic agents, intra-cellular concentrations of cAMP would be about 2 μM. This level of cAMP seems to be high enough to prevent Ca^{2+} release from the intracellular Ca store (Fig. 17) and block the resulting histamine release. The inhibition of Ca^{2+} release from the intracellular Ca store may be the origin of the preventive effect of cAMP on histamine release from mast cells. Although the exact mechanism of this effect is not clear, it is possible that phosphorylation of Ca channel may induce the modulation of the Ca-releasing activity of IP_3 (Curtis and Catterall, 1985).

5. Role of endoplasmic reticulum, an intracellular Ca²⁺ store, in histamine release from mast cell

5.1. Introduction

In a previous section, we have shown that the endoplasmic reticulum (ER) of rat peritoneal mast cells may be the most relevant Ca store available for histamine release, since significant Ca^{2+} uptake into the ER takes place in the presence of ATP and Ca^{2+} release takes place in response to IP_3 (Yoshii et al., 1988). Furthermore, the inhibitory effect of cAMP on Ca^{2+} release from the ER caused by IP_3 was also demonstrated (Yoshii et al., 1988). However, the mechanism of Ca^{2+} incorporation into the ER is not clearly understood. Although it has been indicated that various antiallergic drugs are effective in inhibiting Ca^{2+} release from intracellular Ca store (Tasaka et al., 1986a, 1987a), there is no documentation regarding the direct action of such compounds on the ER. In this section, the characteristics of the ER isolated from rat peritoneal mast cells

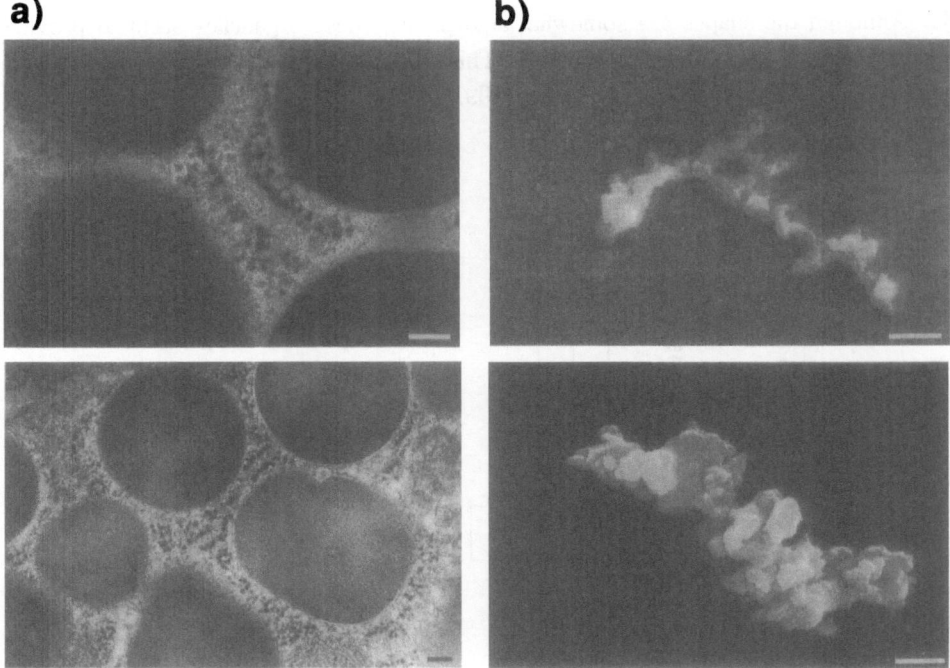

Fig. 18. Electron micrographs of the Ca^{2+} storage site of rat mast cells. (a) TEM appearance of β-escin-treated mast cell. Precipitations of Ca-antimonate complex were observed on the surface of the ER. (b) SEM appearance of ER-rich fraction obtained from rat mast cells. Bar indicates 100 nm. Reproduced from Yoshii, N., Mio, M., Akagi, M. and Tasaka, K.: Immunopharmacology, 21, 13-22 (1991), with permission.

were studied in association with histamine release (Yoshii et al., 1991).

5.2. Electron microscopic observation for the selective localization of intracellular Ca store in rat peritoneal mast cells

In order to identify the intracellular Ca store morphologically, rat mast cells were permeabilized with β-escin, and Ca^{2+} in the cells was precipitated as a calcium antimonate by reacting it with potassium antimonate. Thereafter, the specimen was thin-sectioned and observed by means of transmission electron microcopy (TEM) (Wick and Hepler, 1982; Hirata et al., 1985; Yoshii et al., 1991). In addition, the organelles of rat mast cells were fractionated according to the method described in previous section, and the ER fraction was observed by means of scanning electron microscopy (SEM) (Yoshii et al. 1991).

As shown in Fig. 18a, an assembly of grains consisting of Ca-antimonate was observed to apparently coincide with the surface membrane of the ER, suggesting that Ca^{2+} may be stored in this organelle in the resting state. The SEM appearance of the mast cell fraction showing the highest activity of the ER marker enzyme is shown in Fig. 18b. Although the shapes are somewhat amorphous, these organelles seem to provide the characteristic appearance of the ER. This observation suggests that the fraction consists mainly of the ER (Yoshii et al., 1991).

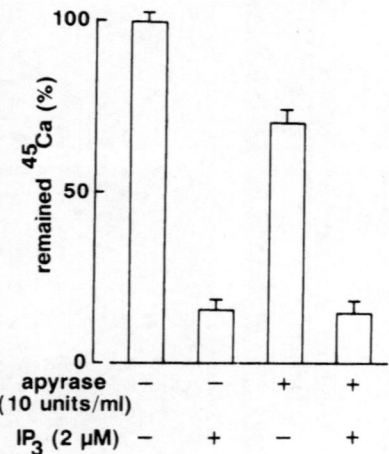

Fig. 19. Effect of ATP depletion with apyrase on the ^{45}Ca release from ER-rich fractions induced by IP_3. Apyrase treatment significantly reduced the ^{45}Ca content in the ER ($p < 0.01$; $n = 6$). Further, IP_3 caused a marked ^{45}Ca release from the ER unrelated to the presence or absence of apyrase ($p < 0.01$; $n = 6$). Reproduced from Yoshii, N., Mio, M., Akagi, M. and Tasaka, K.: Immunopharmacology, 21, 13-22 (1991), with permission.

5.3. Influence of ATP depletion on Ca^{2+} release from the ER due to IP_3

ER-rich fractions of mast cell were incubated with $^{45}CaCl_2$ in the presence of 2 mM ATP. Subsequently, apyrase was added to the reaction mixture and incubated at 37 °C; in the preliminary experiments it was confirmed that 10 units/ml of apyrase is enough to decompose 2 mM ATP within a few seconds. Thereafter, 2 μM of IP_3 was added and incubated for 2 min at 37 °C. The radioactivity in the ER was determined (Yoshii et al., 1991).

Fig. 19 indicates the effects of ATP depletion on ^{45}Ca content in the ER and ^{45}Ca release induced by IP_3. When 10 units/ml of apyrase were added to the reaction mixture after ATP dependent ^{45}Ca uptake was completed, the ^{45}Ca content in the ER was reduced significantly but not completely (28.6 ± 6.2 % of reduction), suggesting that ATP is important in retaining Ca^{2+} in the ER. IP_3 effectively liberated ^{45}Ca from the ER in both the presence and the absence of ATP.

5.4. Ca^{2+} release from the ER by GTP

After incubation of ER-rich fractions with 2 mM of ATP and 1 μCi of $^{45}CaCl_2$ for 10 min, various concentrations of GTP or GTP-γ-S were added and incubated for 2 min at 37 °C. The residual contents of ^{45}Ca in the ER fractions were measured (Yoshii et al., 1991).

As shown in Fig. 20, GTP induced ^{45}Ca release from the ER dose-dependently. Nearly maximum and half maximum release of ^{45}Ca were observed at 10 μM and 4 μM, respectively. In contrast, a non-hydrolysable analogue of GTP, GTP-γ-S, did not alter

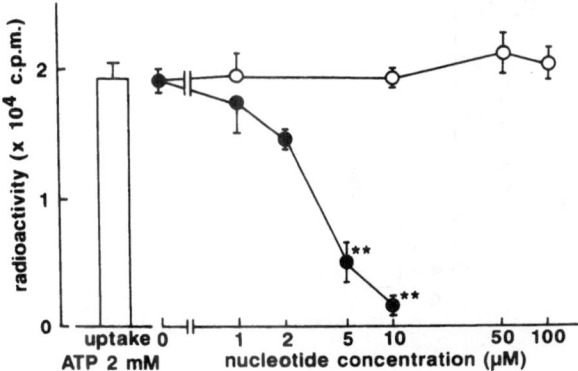

Fig. 20. Effects of GTP analogues on the ^{45}Ca contents in ER-rich fractions. ●: GTP, ○: GTP-γ-S. ** $p < 0.01$. N = 6. In the case of GTP, a significant reduction of ^{45}Ca content was detected when compared to the control value. Reproduced from Yoshii, N., Mio, M., Akagi, M. and Tasaka, K.: Immunopharmacology, 21, 13-22 (1991), with permission.

the ^{45}Ca content in the ER.

5.5. Effects of Ca blockers on the histamine release from mast cells induced by compound 48/80 and Ca^{2+} release from the ER induced by IP_3

The inhibitory effects of Ca blockers (D-600, flunarizine, diltiazem, nifedipine) and intracellular Ca blockers (TMB-8, dantrolene sodium) on the histamine release from rat peritoneal mast cells induced by compound 48/80 are shown in Fig. 21a and b, respectively. When mast cells were incubated with the Ca blockers for 1 hr, a significant inhibition was elicited, as shown in Fig. 21a. The most potent inhibition was produced by flunarizine; almost complete inhibition was achieved at 10 μM. D-600 was less potent; the efficacy was approximately 1/10th of flunarizine. Diltiazem and nifedipine caused only a weak inhibition. In Fig. 21b, the effects of intracellular Ca blockers on the histamine release after 15 min of incubation with mast cells were indicated. TMB-8

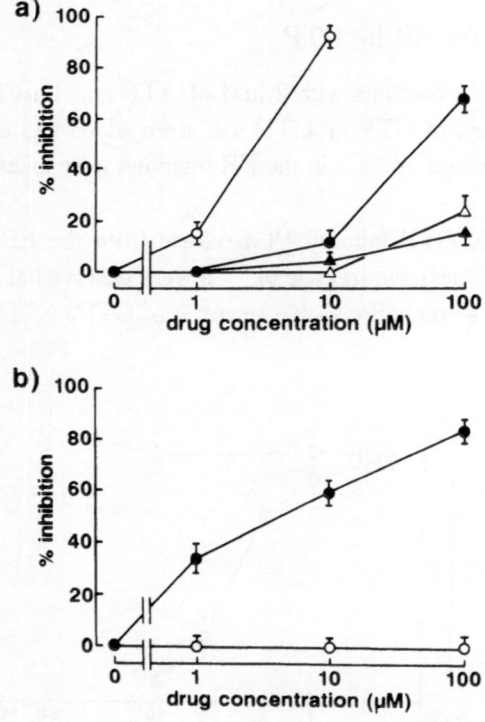

Fig. 21. Effects of Ca blockers on the histamine release from rat peritoneal mast cells induced by compound 48/80 (0.35 μg/ml). (a) Ca blockers. ●: D-600; ○: flunarizine; ▲: diltiazem; △: nifedipine. N = 6. (b) intracellular Ca blockers. ●: TMB-8; ○: dantrolene sodium. N = 6. Reproduced from Yoshii, N., Mio, M., Akagi, M. and Tasaka, K.: Immunopharmacology, 21, 13-22 (1991), with permission.

exerted a potent inhibition even at lower concentrations, but it was less effective than flunarizine. Dantrolene sodium did not affect the histamine release from mast cells even after an incubation period of 1 hr.

The inhibitory effects of Ca blockers and intracellular Ca blockers on ^{45}Ca release from the ER induced by IP$_3$ are shown in Fig. 22a and b, respectively. The most potent inhibition was brought about by flunarizine; almost complete inhibition was achieved at 10 μM. D-600 was less potent; the efficacy was approximately 1/10th of flunarizine in accordance with histamine release from mast cells. Both diltiazem and nifedipine caused only a weak inhibition. The inhibitory effects of intracellular Ca blockers, TMB-8 and dantrolene sodium, were less potent than that of flunarizine.

5.6. Effects of antiallergic drugs on the Ca^{2+} release from the ER induced by IP$_3$

As shown in Fig. 23, some antiallergic drugs (oxatomide, DSCG, ketotifen) inhibited ^{45}Ca release from the ER induced by IP$_3$ in a dose-dependent fashion. Oxatomide almost

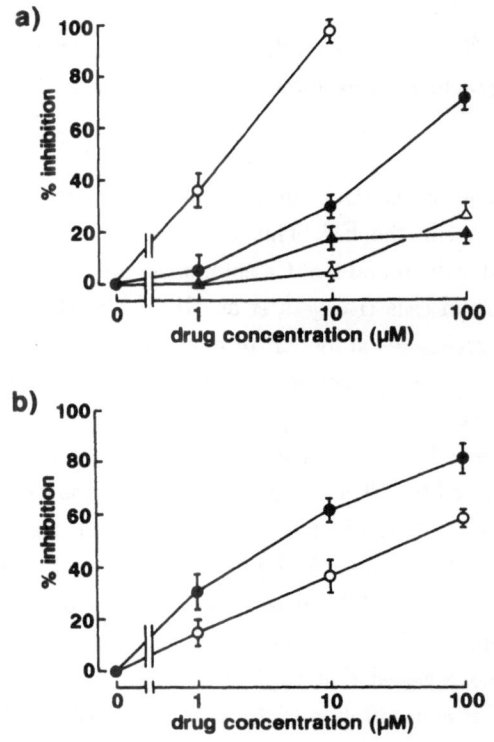

Fig. 22. Effects of Ca blockers on IP$_3$-stimulated ^{45}Ca release from ER-rich fractions. (a) Ca blockers. ●: D-600; ○: flunarizine; ▲: diltiazem; △: nifedipine. N = 6. (b) intracellular Ca blockers. ●: TMB-8; ○: dantrolene sodium. N = 6. Reproduced from Yoshii, N., Mio, M., Akagi, M. and Tasaka, K.: Immunopharmacology, 21, 13-22 (1991), with permission.

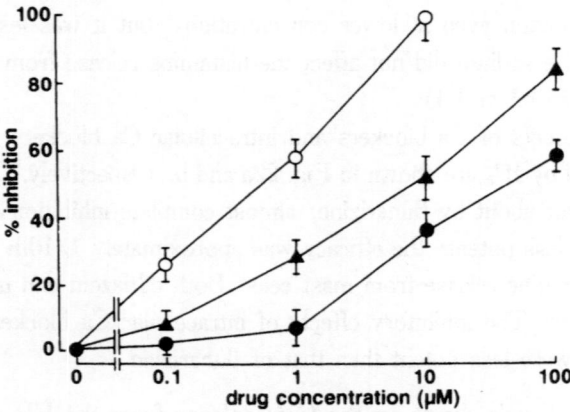

Fig. 23. Effects of antiallergic drugs on IP$_3$-stimulated ^{45}Ca release from ER-rich fraction. ○: oxatomide; ●: DSCG; ▲: ketotifen. N = 6. Reproduced from Yoshii, N., Mio, M., Akagi, M. and Tasaka, K.: Immunopharmacology, 21, 13-22 (1991), with permission.

completely inhibited ^{45}Ca release from the ER at a concentration of 10 μM. DSCG and ketotifen were less potent than oxatomide in inhibiting the ^{45}Ca release from the ER.

5.7. Discussion

In this section, using an electron microscope, many precipitates of Ca-antimonate complex were observed in the ER. This observation was similar to that reported by Borgers et al. (1983), who found that Ca-antimonate precipitates exist in Ca^{2+} accumulating vesicles in red blood cells (Borgers et al., 1983). Hirata et al. (1985) also reported that Ca-antimonate precipitates in the ER of macrophages. It was suggested that the ER may accumulate Ca^{2+} in the cytosol of mast cells at the resting states and may act as an intracellular Ca storage site.

Furthermore, the addition of apyrase into the medium containing the ^{45}Ca-incorporated ER reduced the ^{45}Ca contents in ER. This indicates that ATP may also be required in retaining Ca^{2+} inside the ER. However, at higher concentrations, ATP induced a decrease in ^{45}Ca content in the ER as shown in Fig. 12 (Yoshii et al., 1988). From the point of view of the ATP-dependent Ca^{2+} transport system of the ER, it was considered that bidirectional translocation of Ca^{2+} takes place in the ER membranes. However, it was also revealed that Ca^{2+} release induced by IP$_3$ does not necessarily depend upon the ATP concentration in the medium, since ATP depletion did not affect ^{45}Ca release from the ER as shown in Fig. 19. This observation may be in agreement with that reported by Hirata et al. (1985).

In the present section, it became evident that not only IP$_3$ but also GTP is effective in releasing Ca^{2+} from the ER at concentrations encountered in physiological conditions.

However, GTP-γ-S was not effective in inducing Ca^{2+} release from the ER. It was considered that a hydrolysis of GTP may be required to induce Ca^{2+} release from the ER, and simple binding of GTP-γ-S to the GTP binding protein may not be effective, as reported by Wolf et al. (1987). However, it was reported that when β-escin-permeabilized mast cells were exposed to GTP-γ-S, both histamine release and IP_3 production were elicited in a dose-dependent fashion (Izushi and Tasaka, 1989). Together with these findings, it can be supposed that GTP-γ-S may be effective in activating GTP-binding protein in association with phospholipase C so as to liberate IP_3, though a hydrolysis of GTP in the ER is necessary to release Ca^{2+}. Since cAMP inhibits Ca^{2+} release from the ER (Yoshii et al., 1988), it was suggested that at least these three substances (IP_3, GTP and cAMP) may participate in the regulation of Ca^{2+} release from the ER.

When mast cells were pretreated with Ca^{2+} blockers for 15 min at 37°C, the histamine release induced by compound 48/80 in a Ca-free medium was not suppressed (data not shown). However, prolonged pretreatment of the mast cells with these compounds (37°C for 1 hr) was effective in inhibiting the histamine release induced by compound 48/80. Based on the estimated IC50 values, the relative order of the inhibitory activities of these compounds was as follows; flunarizine $>>$ D-600 $>$ nifedipine $>$ diltiazem. These Ca blockers effectively inhibited ^{45}Ca release from the ER induced by IP_3. In the cases of nifedipine and diltiazem, the inhibitory effects on the histamine release were less potent than that observed in ^{45}Ca release from the ER. In some smooth muscle preparations, it is reported that nifedipine and diltiazem are not easily incorporated into the cell (bepridil $>$ nifedipine $>$ diltiazem; cat ileal smooth muscle, chick embryonic ventricular muscle and rabbit papillary muscle) (Pang and Sperelakis, 1983).

It has been proposed that TMB-8 may (1) act on the cell membrane to inhibit the influx of extracellular Ca^{2+}; (2) act on intracellular Ca^{2+} storage site so as to block the release of Ca^{2+} from the stores; (3) act on both sites to reduce the availability of Ca^{2+} (Chiou and Malagodi, 1975); (4) interfere with oxidative metabolism in mast cells (Grosman, 1986); (5) inhibit calmodulin (Spearman and Butcher, 1983). Although it is not clear whether the histamine release inhibition due to TMB-8 was exerted by intracellular Ca^{2+} blockade or by some other mechanisms, it became clear that TMB-8 actually inhibits Ca^{2+} release from the isolated ER of mast cells. Since the experiment concerning histamine release from mast cells was performed in a Ca-free medium, it can be considered that TMB-8 may act directly on the ER to prevent Ca^{2+} release from the intracellular stores. In contrast, dantrolene sodium did not alter histamine release from mast cells at all, although this compound inhibited ^{45}Ca release from ER induced by IP_3. It has been reported that dantrolene sodium acts on the excitation-contraction coupling (EC coupling) in skeletal muscle and isolated sarcoplasmic reticulum (Martonosi, 1984). It was assumed that the difference in the histamine release inhibition caused by dantrolene sodium and TMB-8 may be ascribed either to the difference in their capability to pass through the cell membrane or to the difference in the profile of pharmacological actions other than intracellular Ca^{2+} blockade.

As shown previously (Tasaka et al., 1986b), the inhibitory effects of DSCG and ketotifen were almost the same regarding the increase of fluorescence derived from quin 2-Ca complex caused by compound 48/80, while the inhibitory effect of oxatomide on Ca^{2+} mobilization from the intracellular store of rat mast cells was more potent than that of the other two drugs. A similar order of potency was found when histamine release was tested with two different combinations of these three drugs (Tasaka et al., 1987a; Ohmori 1982b). In this section, it became apparent that one of the inhibitory mechanisms of antiallergic drugs is the direct inhibition of Ca^{2+} release from the ER. As shown in Fig. 23, when the direct effect on the ER was compared, ketotifen is more potent than DSCG.

Based on these findings, it is assumed that in mast cells the ER may play important roles in the regulation of the intracellular Ca^{2+} concentration, which can be controlled by intracellular concentrations of each of ATP, cAMP, GTP and IP_3. Alteration of the balance between these substances may lead to Ca^{2+} release from the ER. Prolonged incubation of mast cells with Ca blockers or antiallergic agents may induce an incorporation of these drugs into the cells, and the Ca^{2+} release from ER can be inhibited; this may be one of the reasons for the inhibitory effects of these compounds on the histamine release from mast cells.

6. Histamine release from β-escin-permeabilized rat peritoneal mast cells and its inhibition by intracellular Ca^{2+} blockers, calmodulin inhibitors and cAMP

6.1. Introduction

It has been reported that there are certain kinds of GTP-binding proteins which play important roles in the signal transduction of the cell membrane, and that the breakdown of polyphosphoinositides (PI breakdown) by phospholipase C seems to be regulated by a G protein in the signal transduction of the secretory process (Burgoyne, 1987). In connection with this, Yoshii et al. (1988) reported that the ER is the Ca^{2+} store of rat peritoneal mast cells and that IP_3 released Ca^{2+} from the mast cell ER. To clarify the intracellular events leading to histamine release, the effects of several drugs which act on the cytoskeletal systems and cAMP were studied using permeabilized rat mast cells (Izushi and Tasaka, 1989).

6.2. Mast cell permeabilization and histamine release

Rat peritoneal mast cells were collected from the abdominal cavity of male Wistar rats.

The cells were suspended in Ca^{2+} free physiological buffered solution containing glucose $(PS(-Ca^{2+}, +glu))$ and centrifuged after 10 min of preincubation. The supernatant was decanted as thoroughly as possible. A permeabilizing buffer consisting of K-glutamate, $MgSO_4$, ATP, HEPES, EGTA and β-escin 7.5 $\mu g/ml$ (pH 6.8) was added into the test tube in the presence or absence of the test compound and incubated at 37 °C. Next, the medium was centrifuged and the supernatant was replaced with KG buffer consisting of K-glutamate, $MgSO_4$, ATP, HEPES and EGTA (pH 6.8). When Ca^{2+}-induced histamine release was examined, Ca^{2+} was added at various concentrations into the medium. At that time, submicromolar concentrations of Ca^{2+} in the medium (pCa) were calculated using apparent binding constants between $EGTA/CaCl_2$ and $EGTA/MgCl_2$ considering ATP concentration and pH value (Pershadsingh and McDonald, 1980). The reaction was terminated by the addition of ice-cold KG buffer containing 5 mM EGTA and tubes were placed in an ice-bath. The released and residual histamine in the cell pellet was measured separately. Trypan blue, which is not incorporated into the intact cells, was used to stain the β-escin-treated mast cells, and the percentage of permeabilized mast cells was counted (Izushi and Tasaka, 1989).

β-Escin treatment at 7.5 $\mu g/ml$ caused the histamine release slightly more than in the non-treated control, though about 70 % of the treated cells were stained with trypan blue. Mast cell damage due to exposure to detergents was monitored by measuring the amounts of LDH and ATP released into the medium. As shown in Table 6, β-escin treatment at 7.5 $\mu g/ml$ released a small amount of LDH, while more than half of the ATP in the

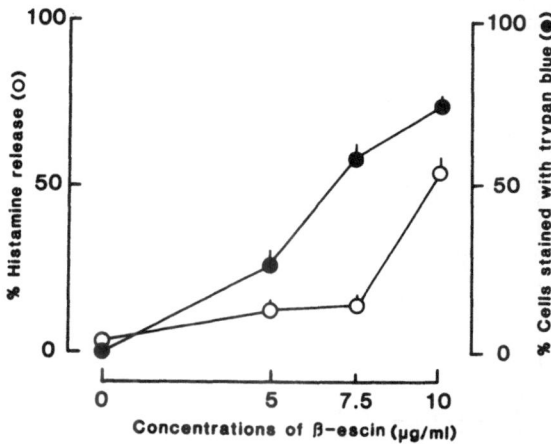

Fig. 24. Changes in percent histamine release and percentage of cells stained with trypan blue in rat peritoneal mast cells after treatment with various concentrations of β-escin. Rat mast cells were incubated for 10 min in the presence of various concentrations of β-escin. Subsequently, histamine contents (○) and cells stained with trypan blue (●) were measured. Each value indicates the mean ± SEM (n = 5). Reproduced from Izushi, K. and Tasaka, K.: Immunopharmacology, 18, 177-186 (1989), with permission.

cytosol was released into the medium. Since LDH is an indicator of the leakage of high molecular weight components and ATP that of low molecular weight substances in cytosol, the results seem to indicate that at a concentration of 7.5 $\mu g/ml$ β-escin may selectively permeabilize the cell membrane for low molecular weight substances. In connection with this, when membrane-intact granules isolated from normal mast cells were collected and exposed to 7.5 $\mu g/ml$ β-escin, histamine release was not induced and no granule were stained with ruthenium red. This clearly indicates that the permeability of the granule membrane was not affected by this treatment. As shown in Fig. 24, 7.5 $\mu g/ml$ of β-escin released histamine at a level lower than 10 %, indicating that 7.5 $\mu g/ml$ of β-escin is suitable for permeabilization of mast cells. As shown in Table 6, permeabilization of mast cells with 0.1 % Triton X-100 caused severe damage of the cell membrane and this was confirmed in the SEM images.

Table 6. LDH and ATP release from permeabilized rat peritoneal mast cells.

Treatment	LDH (%)	ATP (%)
Non-treated control	8.9 ± 0.5	13.2 ± 1.9
β-Escin (7.5μg/ml) treated cells	28.9 ± 1.3	52.4 ± 3.3
β-Escin (10 μg/ml) treated cells	75.4 ± 4.9	68.8 ± 7.1
0.1 % Triton X-100 treated cells	86.9 ± 3.0	69.8 ± 4.8

Mast cells were permeabilized by either β-escin or Triton X-100. The amounts of LDH and ATP leaked into the medium are expressed as percent of the total values determined in intact cells. Each value indicates the mean $\pm$ SEM (n = 5). Reproduced from Izushi, K. and Tasaka, K.: Immunopharmacology, 18, 177-186 (1989), with permission.

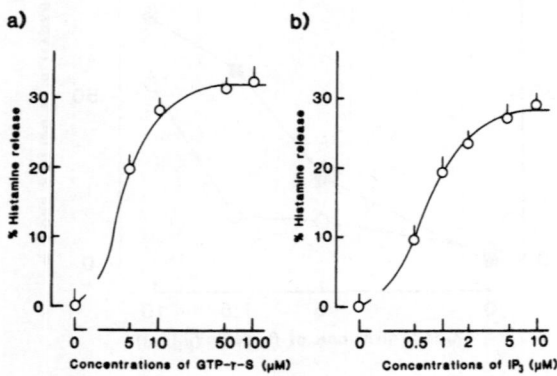

Fig. 25. Histamine release from permeabilized mast cells induced by GTP-γ-S or IP$_3$. Mast cells were permeabilized with 7.5 $\mu g/ml$ β-escin in the presence of EGTA 50 μM and then (a) GTP-γ-S or (b) IP$_3$ was added. Each value indicates the mean $\pm$ SEM (n = 5). Reproduced from Izushi, K. and Tasaka, K.: Immunopharmacology, 18, 177-186 (1989), with permission.

When IP_3 and GTP-γ-S were incubated with mast cells permeabilized with β-escin in KG buffer, histamine release was elicited dose-dependently (Fig. 25a, b), reaching a maximum at 5 μM and 50 μM, respectively. Observation with a phase-contrast microscope showed that marked degranulation was induced in almost all cells, even in a Ca^{2+}-free medium, in both cases. To investigate whether the histamine release induced by either GTP-γ-S or IP_3 is due to the direct action of these compounds on the granules, membrane-intact granules were exposed to these stimulants. However no histamine release was elicited at any of the concentrations employed: 0.1-10 μM of IP_3 and 1-100 μM of GTP-γ-S.

6.3. Effect of TMB-8 and neomycin on the histamine release from permeabilized mast cells induced by GTP-γ-S

In permeabilized mast cells pretreated with TMB-8, an intracellular Ca^{2+} blocker, histamine release due to either 2 μM IP_3 or 50 μM GTP-γ-S in KG buffer was inhibited dose-dependently (Fig. 26a). In addition, histamine release from the permeabilized cells elicited by IP_3 or GTP-γ-S was inhibited almost completely in the presence of a high concentration of EGTA (1 mM). This indicates that Ca^{2+} is a prerequisite for initiation of the process of histamine release; IP_3 effectively released Ca^{2+} from the Ca^{2+} store (Yoshii et al., 1988). It is known that chemically permeabilized cells can be easily resealed. It is possible that after pretreatment with TMB-8, mast cells are resealed, so that either IP_3 or GTP-γ-S added into the medium could not reach the cytosol. This could be a reason for the histamine release inhibition caused by TMB-8. However, when permeabilized cells were stained with trypan blue before and after exposure to the test

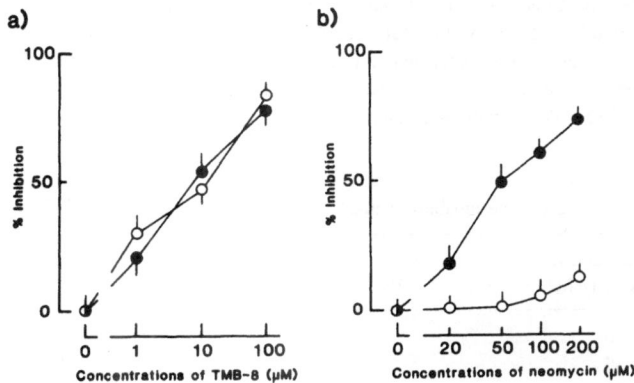

Fig. 26. Inhibitory effects of TMB-8 and neomycin on histamine release from permeabilized mast cells induced by IP3 or GTP-γ-S. Mast cells were permeabilized and treated with (a) TMB-8 or (b) neomycin, and then exposed to IP_3 (2 μM; $\bigcirc$) or GTP-γ-S (50 μM; $\bullet$) for 5 min. Each value indicates the mean $\pm$ SEM (n = 5). Reproduced from Izushi, K. and Tasaka, K.: Immunopharmacology, 18, 177-186 (1989), with permission.

compound, the cells were stained distinctly. This verified that the permeability of the cell membrane was retained, as seen in the control cells. To avoid ambiguous interpretation of the results in relation to the possibility of membrane resealing, trypan blue staining was performed in each experiment.

It is known that the activity of phospholipase C, which generates IP_3 from the parent compound (PIP_2), is regulated by G-protein. In the case of GTP-γ-S, increased IP_3 production may be elicited via G-protein activation. When permeabilized cells were pretreated with neomycin, a typical phospholipase C inhibitor, histamine release induced by GTP-γ-S was markedly suppressed but no such inhibition was found in the case of IP_3 (Fig. 26b).

6.4. Formation of inositol phosphates in permeabilized mast cells

When inositol phosphate formation in permeabilized mast cells was measured sequentially up to 5 min using an HPLC system, the formation of inositol phosphates increased, as

Table 7. Sequential changes in the formation of inositol phoshates in permeabilized mast cells.

Time (min)	Inositol phosphate formation (dpm)		
	IP_1	IP_2	IP_3
0	683.1 ± 52.1	182.3 ± 20.4	45.6 ± 5.6
1	672.9 ± 51.3	$381.6 \pm 41.4^{**}$	250.8 ± 29.8
2	724.3 ± 76.6	$563.8 \pm 61.3^{**}$	$398.4 \pm 35.5^{**}$
5	842.4 ± 84.3	$672.5 \pm 76.6^{**}$	$429.6 \pm 36.8^{**}$

[^{3}H]-Inositol-labeled mast cells were permeabilized with 7.5 μg/ml of β-escin, and subsequently the cells were exposed to 50 μM GTP-γ-S. Inositol phosphates were fractionated by means of HPLC and the radioactivity in each specimen was determined. Each value indicates the mean $\pm$ SEM (n = 5). ** Significantly different from the control (p < 0.01). Reproduced from Izushi, K. and Tasaka, K.: Immunopharmacology, 18, 177-186 (1989), with permission.

Table 8. Formation of IP_3 in permeabilized mast cells after stimulation with GTP-γ-S.

Concentration of GTP-γ-S (μM)	IP_3 formation (dpm)
0	42.6 ± 4.9
5	$253.8 \pm 30.1^{**}$
50	$432.8 \pm 40.8^{**}$
100	$452.3 \pm 43.2^{**}$

[^{3}H]-Inositol-labeled mast cells were permeabilized and exposed to various concentrations of GTP-γ-S for 2 min. Each value indicates the mean $\pm$ SEM (n = 5). ** Significantly different from the control (p < 0.01). Reproduced from Izushi, K. and Tasaka, K.: Immunopharmacology, 18, 177-186 (1989), with permission.

shown in Table 7. The amounts of IP_3 formed within 1 min increased approximately 6 times, while those of IP_1 and IP_2 were less prominent. With exposure of permeabilized cells to GTP-γ-S for 2 min at various concentrations up to 100 μM, the amounts of IP_3 formed increased dose-dependently. At 50 μM, IP_3 formation reached nearly a maximum, an increase of approximately 10 times that produced in the control group (Table 8). This probably indicates that IP_3 is produced during the process leading to histamine release through the activation of a G-protein, and that IP_3 releases Ca^{2+} from the intracellular Ca store.

6.5. Ca^{2+}-induced histamine release and degranulation from permeabilized mast cells

Fig. 27 shows the dose-response curve of Ca^{2+}-induced histamine release from permeabilized rat mast cells. At concentrations lower than 0.1 μM, Ca^{2+} did not induce histamine release. However, at concentrations higher than 0.2 μM, the ion elicited histamine release and the dose-response curve rose very promptly; histamine release achieved nearly a maximum at 1 μM free Ca^{2+}. As in the case of IP_3 or GTP-γ-S, when membrane-intact granules were incubated with Ca^{2+} at various concentrations, no histamine release was induced even at 0.1 mM, indicating that no cation-exchange takes place between histamine and Ca^{2+} in membrane-intact granules.

As shown in Fig. 28a, the SEM appearance of the permeabilized mast cell is very

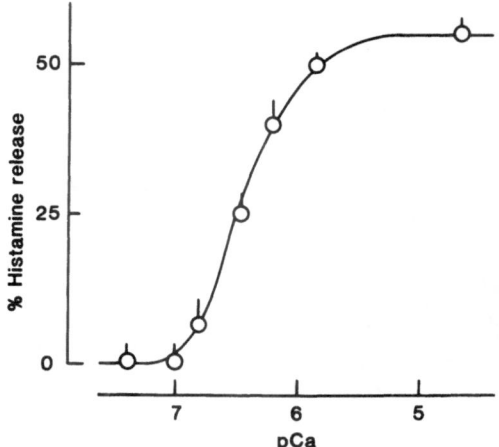

Fig. 27. Histamine release from permeabilized mast cells induced by various concentrations of Ca^{2+}. Mast cells were stimulated by various Ca^{2+} concentrations and incubated for 5 min. Free Ca^{2+} concentrations were calculated using apparent binding constants between EGTA/CaCl₂ and EGTA/MgCl₂. Each value indicates the mean ± SEM (n = 5). Reproduced from Izushi, K. and Tasaka, K.: Immunopharmacology, 18, 177-186 (1989), with permission.

134

similar to that of intact cells; no degranulation was observed. In the presence of 0.64 μM Ca^{2+} (pCa = 6.19), marked degranulation occurred (Fig. 28b).

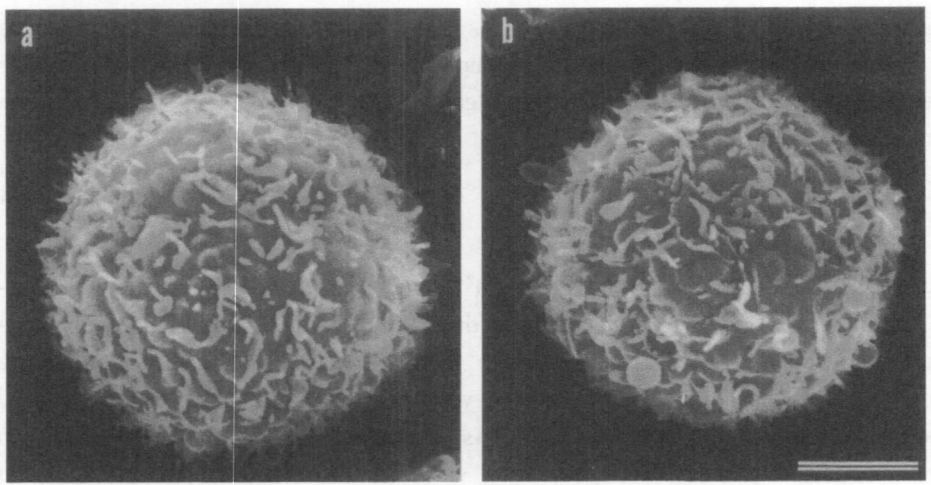

Fig. 28. SEM images of permeabilized mast cells before and after exposure to Ca^{2+}. Mast cells were permeabilized (a) with 7.5 μg/ml of β-escin and (b) exposed to Ca^{2+} (pCa = 6.19). Cells were incubated for 5 min and then fixed with 2 % glutaraldehyde. Bar indicates 5 μm. Reproduced from Izushi, K. and Tasaka, K.: Immunopharmacology, 18, 177-186 (1989), with permission.

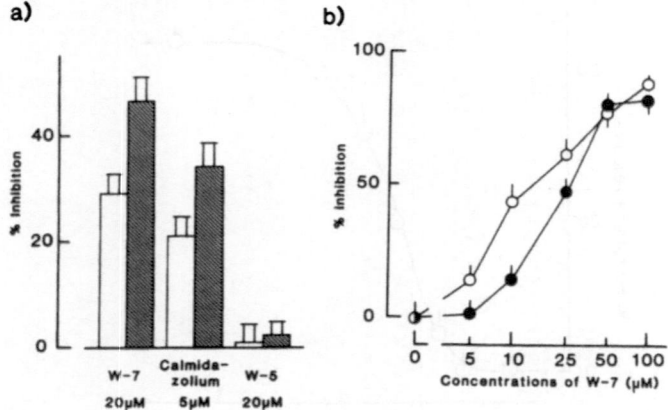

Fig. 29. Inhibitory effects of calmodulin inhibitors on histamine release from permeabilized mast cells induced by IP$_3$ or Ca^{2+}. (a) Mast cells were stimulated with IP$_3$ (2 μM; hatched columns) or Ca^{2+} (0.64 μM; open columns) in the presence of calmodulin inhibitors. (b) Dose-dependent inhibition of W-7 on IP$_3$- (○) or Ca^{2+}-induced (●) histamine release. Each value indicates the mean ± SEM (n = 5). Reproduced from Izushi, K. and Tasaka, K.: Immunopharmacology, 18, 177-186 (1989), with permission.

6.6. Effects of calmodulin inhibitors, cytoskeleton inhibitors and cAMP on the histamine release from permeabilized mast cells

Fig. 29a shows the inhibitory effect of calmodulin inhibitors on histamine release from permeabilized mast cells. When W-7 or calmidazolium was added simultaneously with

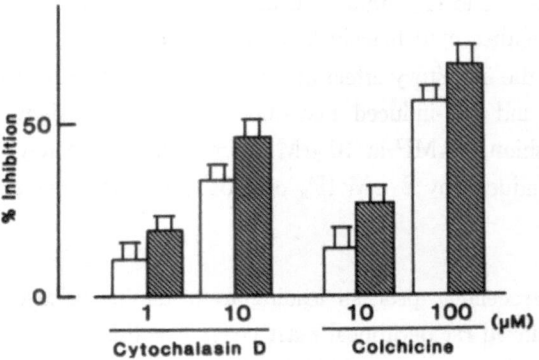

Fig. 30. Inhibitory effects of cytochalasin D and colchicine on histamine release from permeabilized mast cells. Histamine release induced by IP$_3$ (2 μM) or Ca^{2+} (0.64 μM) is represented by hatched or open columns, respectively. Each value indicates the mean $\pm$ SEM (n = 5). Reproduced from Izushi, K. and Tasaka, K.: Immunopharmacology, 18, 177-186 (1989), with permission.

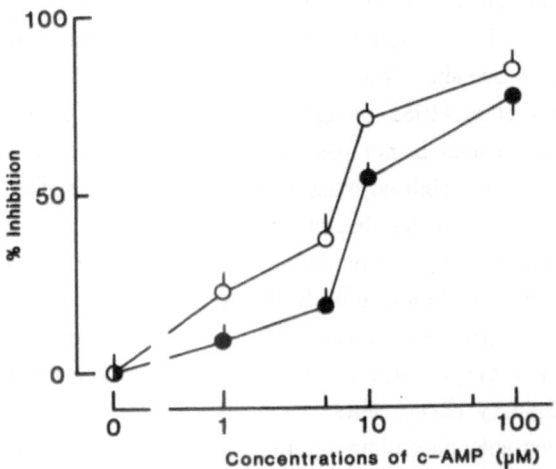

Fig. 31. Effect of cAMP on histamine release from permeabilized mast cells. Permeabilized cells were pretreated with cAMP and subsequently exposed to IP$_3$ (2 μM; $\bigcirc$) or Ca^{2+} (0.64 μM; $\bullet$) and incubated for 5 min. Each value indicates the mean $\pm$ SEM (n = 5). Reproduced from Izushi, K. and Tasaka, K.: Immunopharmacology, 18, 177-186 (1989), with permission.

Ca^{2+} or IP_3, both of the calmodulin inhibitors significantly suppressed Ca^{2+}- or IP_3-induced histamine release, while W-5 was not effective. W-7 inhibited histamine release dose-dependently; 25 μM W-7 inhibited approximately 50 % or 60 % of Ca^{2+}- or IP_3-induced histamine release, respectively.

Furthermore, cytochalasin D, an F-actin-destabilizing agent, and colchicine, a microtubule-depolymerizing agent, also effectively inhibited histamine release. As shown in Fig. 30, both Ca^{2+}- and IP_3-induced histamine release were inhibited dose-dependently in the presence of either cytochalasin D or cholchicine.

Fig. 31 shows the inhibitory effect of cAMP on histamine release from permeabilized mast cells. Ca^{2+}- and IP_3-induced histamine release were significantly inhibited in a dose-dependent fashion. cAMP at 10 μM inhibited approximately 80 % or 60 % of the histamine release induced by 2 μM IP_3 or 0.64 μM Ca^{2+}, respectively.

6.7. Discussion

To analyze the intracellular process leading to histamine release from mast cells, we employed permeabilized rat peritoneal mast cells. For this purpose, considerable efforts have been made to develop permeabilized cell preparations in which the composition of the cytosol can be controlled, yet which retain their ability to respond to the stimulus leading to histamine release. In general, it is desirable that the pore sizes generated in the cell membrane have dimensions just large enough to allow flux, in both inward and outward directions, although the sustained interconnection should be retained. If the membrane lesions are large enough to allow the complete efflux of high molecular weight substances such as LDH, accurate understanding of the sequence that has taken place in the cytoplasma may become difficult. A method suitable for this purpose has been developed by Bennet et al. (1981) using micromolar concentrations of ATP^{4-}. However, the interpretation of results is not easy in some cases, since the extent of histamine release from ATP^{4-}-permeabilized mast cells after exposure to Ca^{2+} is regulated by the concentration of ATP^{4-} but not that of Ca^{2+}.

Saponin is one of the typical compounds which have been employed to permeabilize a variety of cells, such as smooth muscle, neutrophils (Prentki et al., 1984), chromaffin cells and hepatocytes (Burgess et al., 1983); it is known that saponin interacts with cholesterol in the lipid bilayer with a 1:1 stoichiometry and forms pores (Gennis, 1989). However, in preliminary experiments, it became apparent that the permeable effect of saponin is quite variable depending on the lot number. The proper concentration of saponin for permeabilization ranged from 25 to 100 μg/ml. Since such diversity can be ascribed to chemical impurity, β-escin, a major constituent of saponins obtained from the seed of the horse chestnut tree, was employed; β-escin is a typical triterpenoid having the characteristic properties of saponin.

As shown in Fig. 24 and Table 6, at 10 μg/ml, β-escin treatment damaged the mast cells and caused a histamine release of more than 50 % and high LDH activity was detected in the medium (higher than 75 %). By contrast, at 7.5 μg/ml, β-escin caused much less histamine release, though the percentage of cells stained with trypan blue was higher than 70 %, indicating that permeability between the inside and outside of the cell membrane occurred in more than 70 % of the cells without noticeable damage. Also, in the SEM images of the permeabilized mast cells, neither degranulation nor damage was seen on the cell surface. These results indicate that β-escin at 7.5 μg/ml definitely increased membrane permeability, especially for low molecular weight compounds; the molecular weight of ATP is 507.91, while that of LDH is more than 1×10^5.

As shown in Fig. 25b, IP$_3$ caused a dose-dependent histamine release from per-meabilized mast cells. The effect of IP$_3$ is inhibited by a high concentration of EGTA and by pretreatment with TMB-8, which inhibits Ca^{2+} release from the sarcoplasmic reticulum (Chiou and Malagodi, 1975) and ER (Yoshii et al., 1991). This probably indicates that IP$_3$ releases Ca^{2+} from the ER and, in turn, the increased Ca^{2+} concentration may trigger histamine release. In relation to this, Brass and Joseph (1985) reported that IP$_3$ plays an important role in intracellular Ca^{2+} mobilization and serotonin release from saponin-treated platelets. IP$_3$ is formed when PIP$_2$ is degraded following the activation of phospholipase C (Berridge, 1984). Moreover, it has been reported that activation of phospholipase C is regulated by G-protein (Burgoyne, 1987), which requires about 1 μM Ca^{2+} to induce the secretory process (Lindau and Nusse, 1987). In this experiment, nearly maximum histamine release occurred when permeabilized mast cells were exposed to 1 μM Ca^{2+}. Since histamine release induced by GTP-γ-S was significantly inhibited by pretreatment with neomycin, it is clear that IP$_3$ formation is a prerequisite for histamine release. It was clearly demonstrated IP$_3$ formation in the mast cells increased after stimulation with GTP-γ-S (Tables 7 and 8). No resealing was found before or after exposure to GTP-γ-S.

In the present experiment, at concentrations higher than 0.2 μM, Ca^{2+} induced histamine release from permeabilized mast cells. There was a sharp rise in the dose-response curve, indicating that a minute increase in Ca^{2+} concentration caused a marked increase in histamine release. It has been reported that Ca^{2+} acts on calmodulin (Cha-kravarty and Nielsen, 1985), Ca^{2+}-dependent regulatory proteins and cytoskeletal fila-ments (Tasaka et al., 1986c, 1988). In 10^6 mast cells, there were 160 ng of calmodulin and a large portion of calmodulin is located in the cytosol fraction (Chakravarty and Nielsen, 1985). In vascular smooth muscle treated with Triton X-100, calmodulin inhibitors effectively suppressed Ca-dependent muscle contractions, but considerably higher concentrations of the compounds are required for such inhibition (Kreye et al., 1983). As shown in Fig. 29, W-7 inhibited approximately 50 % of histamine release at 10 μM, which is similar to reports of its 50 % inhibition of Ca^{2+}-dependent phosphodies-terase activity (Hidaka and Tanaka, 1983; Mio et al., 1991). These results suggest that calmodulin exists in permeabilized mast cells and plays some critical role in the Ca^{2+}-dependent process leading to histamine release. It has been suggested that Ca^{2+} also acts on the cytoskeletal systems, which may participate in the degranulation process (Tasaka et al., 1988). Mast cell granules are surrounded rather densely by actin filaments in a resting state, and are pushed or squeezed out of the cell membrane after appropriate stimulation (Tasaka et al., 1988). In accordance with this, cytochalasin D and colchicine effectively inhibited the histamine release induced by IP_3 or Ca^{2+}. These findings reinforce the hypothesis that the cytoskeletal system plays an important role in histamine release and degranulation.

Histamine release induced by IP_3 from permeabilized mast cells was inhibited dose-dependently by pretreatment with cAMP. This may indicate that cAMP inhibits IP_3-induced Ca^{2+} release from the ER (Ennis et al., 1980a; Yoshii et al., 1988). Furthermore, cAMP also inhibited Ca^{2+}-induced histamine release. The dose-response curves of cAMP in the histamine release inhibition induced by IP_3 and Ca^{2+} run parallel. This finding seems to indicate that cAMP effectively suppresses Ca^{2+} at a stage which might be close to the final step in histamine release. It has been reported that cAMP was not effective in altering histamine release from ATP^{4-}-permeabilized mast cells when it was added simultaneously with GppNHp, which is another GTP analogue and capable of inducing histamine release (Gomperts, 1983). However, in the present experiments permeabilized mast cells were pretreated with cAMP. During the incubation period, cAMP-dependent reactions may occur, inhibiting histamine release in the cytosol. In the skinned smooth muscle of rabbit mesenteric artery, Ca^{2+}-induced contraction was inhibited by cAMP and A-kinase pretreatment and Ca^{2+} uptake into the Ca^{2+} store was enhanced by these treatment (Itoh et al., 1985). In relation to this, it has been reported that the mitochondria isolated from dibutyryl-cAMP-treated PY815 cells take up $^{45}Ca^{2+}$ much faster than the mitochondria obtained from untreated cells. Also, $^{45}Ca^{2+}$-precharged mitochondria isolated from dibutyryl-cAMP-pretreated cells leaked $^{45}Ca^{2+}$ more slowly than those obtained from untreated cells (Ala'i and Ralph, 1986). In isolated smooth muscle preparations of the dog coronary artery, potassium contracture in Ca^{2+}-containing physiological salt solution was inhibited dose-dependently with procaine, which is known to suppress calcium-induced calcium release (Endo, 1977). However, the increase in Ca^{2+} uptake of the coronary artery was inhibited by procaine at the same concentration range. These findings suggest that the potassium contracture of the dog coronary artery is dependent on Ca^{2+}-induced Ca^{2+} release (Imai et al., 1984b). It is not known whether Ca^{2+}-induced Ca^{2+} release takes place in rat mast cells. If this is the case, the intracellular cAMP may effectively inhibit Ca^{2+} release from the Ca^{2+} store. Although it is not certain whether the mitochondrial uptake of Ca^{2+} is increased by cAMP, the effect of cAMP on calcium-induced calcium release seems plausible.

7. Role of ATP and activation of protein kinase C in Ca^{2+}-dependent histamine release from permeabilized rat mast cells

7.1. Introduction

It has been suggested that ATP is a prerequisite as an energy source of histamine release from mast cells (Johansen, 1980). However, no direct evidence to explain why ATP is necessary in histamine release has been obtained. It has been reported that in streptolysin-O-permeabilized rat mast cells, histamine release was dependent upon the presence of inosine 5'-triphosphate (ITP) in the medium (Howell and Gomperts, 1987), while ATP inhibited the reactions preceding the onset of exocytosis (Tatham and Gomperts, 1989). Koopmann and Jackson (1990) reported similar observations in digitonin-permeabilized rat mast cells. On the other hand, in catecholamine release from electropored chromaffin cells, the existence of ATP in the medium is indispensable (Baker and Knight, 1981), while in digitonin-permeabilized chromaffin cells cate-cholamine release partially depends on the presence of ATP (Wilson and Kirshner, 1983). Furthermore, an important role of protein kinase C in histamine release has been suggested on account of the synergic activation of 1-oleoyl-2-acetylglycerol (OAG) in A23187-induced histamine release (Katakami et al., 1984). The present study was performed to clarify the essential role of ATP and the activation of protein kinase C in histamine release using β-escin-permeabilized rat peritoneal mast cells (Izushi and Tasaka, 1991).

7.2. Content of ribonucleoside triphosphates in rat peritoneal mast cells

Although the ATP content of rat peritoneal mast cells has been reported (Johansen, 1980), in the case of other ribonucleoside triphosphates such as GTP, ITP, UTP and CTP, no such data are available. By means of an ion exchange column and HPLC system, the amounts of ribonucleoside triphosphates such as ATP, GTP, UTP, ITP and CTP were measured. ATP, GTP and UTP contents were $1,546 \pm 17$, 290 ± 4 and 172 ± 3 pmole/10^6 cells, respectively, while ITP and CTP were not detected. The ATP content measured by HPLC was similar to that reported by the luciferin-luciferase bioluminescence method (Howell and Gomperts, 1987). The GTP and UTP contents of mast cells were much lower than that of ATP.

7.3. Effects of nucleotides on histamine release from permeabilized mast cells

Rat peritoneal mast cells were permeabilized as described in the previous section (Izushi and Tasaka, 1991). After removal of ATP from KG buffer containing 640 nM of Ca^{2+}, no histamine release was elicited from permeabilized mast cells. On the other hand, when the buffer was devoid of Ca^{2+} in the presence of 3 mM ATP, histamine release was not provoked either (Fig. 32a). In this study, when mast cells were permeabilized, ATP in

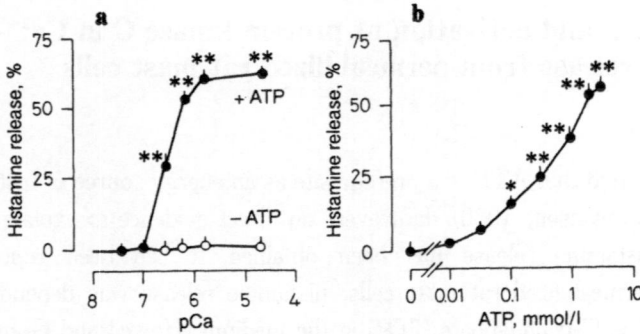

Fig. 32. ATP and Ca²⁺-dependent histamine release from β-escin-permeabilized rat mast cells. a. Influence of ATP on Ca²⁺-dependent histamine release from β-escin-permeabilized rat mast cells in the presence (●) or absence of ATP (○; 3mM). Ca²⁺ concentration (pCa) was changed from 7.5 to 4.5. b. ATP-dependent histamine release from β-escin-permeabilized mast cells in the presence of Ca²⁺. Permeabilized mast cells were incubated with 640 nM of Ca²⁺ in the presence of various concentrations of ATP for 5 min. Each value indicates the mean ± SEM (n = 4). * p < 0.05, ** p < 0.01; significantly different from the control. Reproduced from Izushi, K. and Tasaka, K.: Pharmacology, 42, 297-306 (1991), with permission.

Table 9. Effects of various nucleotides on Ca²⁺-dependent histamine release from β-escin-permeabilized mast cells

Nucleotide	Percent of control
ATP	100.0 ± 6.6
ADP	5.6 ± 1.3
AMP	4.3 ± 1.2
GTP	22.9 ± 3.7
UTP	21.6 ± 3.7
ITP	23.2 ± 3.9
CTP	22.5 ± 3.9
AMP-PNP	3.1 ± 0.8
β, γ-methylene ATP	4.6 ± 1.9

Rat mast cells were permeabilized by 7.5 μg/ml of β-escin in the absence of ATP. Subsequently, cells were exposed to Ca²⁺ (640 nM) and various nucleotides (3 mM) for 5 min. Data are expressed as a percentage of the histamine release induced by the concomitant existence of Ca²⁺ and ATP. Each value indicates the mean ± SEM (n = 4). Reproduced from Izushi, K. and Tasaka, K.: Pharmacology, 42, 297-306 (1991), with permission.

the cytoplasm was released within 2 min reaching a maximum of 50 %. Therefore, by replacing the medium repeatedly, almost none of the endogenous ATP remained. Actually, after the addition of 3 mM ATP to the KG buffer, Ca²⁺-induced histamine release took place with remarkable promptness. Histamine release was elicited concentration-dependently at concentrations ranging from 0.01 to 5 mM, reaching a maximum at 3 mM of ATP (Fig. 32a). However, when the other adenosine nucleotides, such as AMP and ADP, and non hydrolysable ATP analogues, such as AMP-PNP and β, γ-methylene ATP, were added in the place of ATP, no histamine release was elicited, though the Ca^{2+} concentration in the medium was fixed at the same level (Table 9). When some other nucleotides such as GTP, ITP, UTP and CTP were added to the medium instead of ATP, virtually no histamine release was brought about at the identical concentration of 3 mM.

7.4. Effects of Ca blockers and metabolic inhibitors on histamine release from permeabilized mast cells

As shown in Table 10, gallopamil, flunarizine and verapamil were not effective even at 10 μM to prevent Ca^{2+}/ATP-induced histamine release, while histamine release was slightly inhibited by 1 μM of Na_3VO_4. At higher concentrations (10 μM), Na_3VO_4 enhanced Ca^{2+}-independent histamine release (35.3 $\pm$ 4.6 %). Also, 1 nM of ouabain did not inhibit histamine release while N-ethylmaleimide inhibited Ca^{2+}/ATP-dependent histamine release concentration-dependently at concentrations higher than 10 μM. The inhibitory effect of N-ethylmaleimide seems to indicate that SH groups of functional protein(s) may exert some important role in histamine release.

7.5. Effects of protein kinase C activator on histamine release from permeabilized mast cells

When 12-O-tetradecanoylphorbol-13-acetate (TPA) was exposed to the intact mast cells

Table 10. Effects of some Ca blockers and metabolic inhibitors on Ca^{2+}- and ATP-dependent histamine release from β-escin-permeabilized mast cells

Drugs	Concentration (μM)	Percent of control
control		100.0 $\pm$ 5.1
gallopamil	10	98.4 $\pm$ 6.5
flunarizine	10	103.5 $\pm$ 4.4
verapamil	10	101.6 $\pm$ 6.2
ouabain	1	100.9 $\pm$ 5.5
Na_3VO_4	1	88.1 $\pm$ 6.8
	10	121.9 $\pm$ 4.9
N-ethylmaleimide	1	96.8 $\pm$ 4.5
	5	82.6 $\pm$ 4.3
	10	53.2 $\pm$ 5.1**
	50	31.3 $\pm$ 3.3**
	100	15.4 $\pm$ 4.6**

Permeabilized mast cells were stimulated by Ca^{2+} (640 nM) and ATP (3 mM) for 5 min. Data are expressed as a percentage of histamine released. Each value indicates the mean $\pm$ SEM (n = 4). ** $p < 0.01$: significantly different from control. Reproduced from Izushi, K. and Tasaka, K.: Pharmacology, 42, 297-306 (1991), with permission.

at 10 nM, histamine release took place very slowly and gradually as shown in Fig. 33. At the initial stage histamine release was almost negligible. Approximately 5 % of histamine release was brought about within 10 min. Subsequently, histamine release continued for a further 50 min, by which time about 50 % of the total histamine had been released. However, when 10 nM of TPA was added to KG buffer containing Ca^{2+} (0.1 μM) and ATP (3 mM), histamine release from permeabilized mast cells took place very

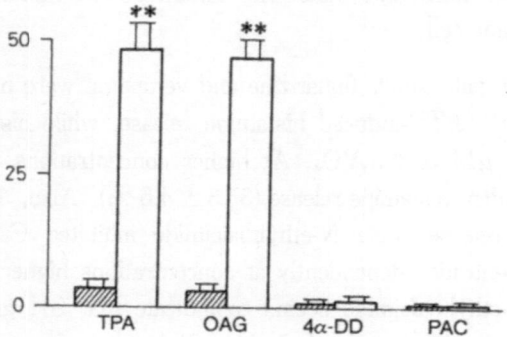

Fig. 33. Time course of histamine release from intact mast cells or β-escin-permeabilized mast cells elicited by TPA. Intact (●) or permeabilized (○) mast cells were stimulated by 10 nM of TPA, and histamine release into the medium was determined consequently. Permeabilized mast cells were incubated with 0.1 μM of Ca^{2+} and 3 mM ATP. Each value indicates the mean ± SEM (n = 4). * p < 0.05, ** p < 0.01: significantly different from the control. Reproduced from Izushi, K. and Tasaka, K.: Pharmacology, 42, 297-306 (1991), with permission.

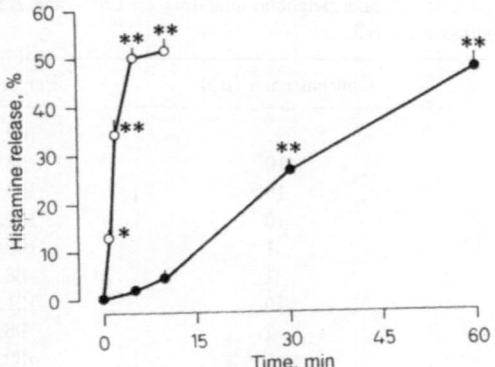

Fig. 34. Effects of phorbol esters and OAG on histamine release from β-escin-permeabilized mast cells. 4α-DD = 4α-phorbol 12, 13-didecanoate; PAC = 4β-phorbol 13-monoacetate. Hatched columns indicate the percentages of histamine release elicited from permeabilized mast cells stimulated with 10 nM phorbol esters or 30 μg/ml of OAG for 5 min in the absence of 0.1 μM Ca^{2+} and 3 mM ATP. Open columns indicate histamine release in the presence of Ca^{2+}. Each value indicates the mean ± SEM (n = 4). ** p < 0.01: significantly different from the control values. Reproduced from Izushi, K. and Tasaka, K.: Pharmacology, 42, 297-306 (1991), with permission.

promptly, reaching the maximum within 5 min (Fig. 33). However, when the permeabilized mast cells were suspended in KG buffer containing ATP (3 mM) and Ca^{2+} (0.1 μM) but not TPA, no histamine release was induced by ATP. This clearly indicates that TPA is effective to enhance the action of Ca^{2+}, since no histamine release was induced in the medium containing the corresponding concentrations of ATP and Ca^{2+} without TPA as shown in Fig. 32a. To examine whether or not histamine release stimulated by TPA is due to direct action on the mast cell granules, isolated membrane-intact granules were exposed to TPA in KG buffer for 10 min at 37°C. However, no histamine release was elicited at 10 nM of TPA. Furthermore, to make clear that TPA-induced histamine release from permeabilized mast cells is definitely related to the activation of protein kinase C, the effects of OAG and some other phorbol esters on histamine release from permeabilized cells were studied in the presence of Ca^{2+} (0.1 μM) and ATP (3 mM; Fig. 34). OAG (30 μg/ml), which is a synthetic diacylglyceride and capable of activating protein kinase C, also caused histamine release from permeabilized cells. By contrast, 4α-phorbol 12, 13-didecanoate (4α-DD) and 4β-phorbol 13-monoacetate (PAC), both phorbol esters having no property activating protein kinase C (Castagna et al., 1982), elicited no histamine release at the same concentration employed as TPA (10 nM). Actually, permeabilized mast cells retained 76.8 ± 5.6 % (n = 4) of the total protein kinase C activity.

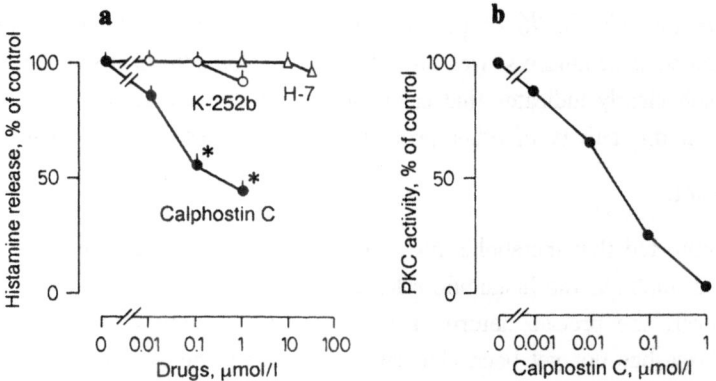

Fig. 35. Effects of protein kinase C inhibitors on histamine release and protein kinase C activity. a. Effects of protein kinase C inhibitors on histamine release from β-escin-permeabilized mast cells. Concentrations of protein kinase C inhibitors calphostin C (●), K-252b (○) and H-7 (△). Data are expressed as a percentage of histamine release elicited in the presence of Ca^{2+} and ATP. b. Effect of calphostin C on protein kinase C (PKC) activity. The source of crude enzyme was obtained from mast cell extract. Data are expressed as a percentage of the total protein kinase C activity. Each value indicates the mean $\pm$ SEM (n = 4). * p < 0.01: significantly different from the control. Reproduced from Izushi, K. and Tasaka, K.: Pharmacology, 42, 297-306 (1991), with permission.

7.6. Effects of some protein kinase C inhibitors on histamine release and protein kinase C activity of mast cells

As shown in Fig. 35a, protein kinase C inhibitors which interact with the catalytic domain (ATP-binding site) of protein kinase C, such as K-252b (Kase et al., 1987) and H-7 (Hidaka et al., 1984), have little effect on histamine release even at concentrations of 1 and 50 μM, respectively, whereas calphostin C, a new inhibitor of protein kinase C (Kobayashi et al., 1989), concentration-dependently inhibited Ca^{2+}/ATP-induced histamine release. More than 60 % of histamine release was inhibited by calphostin C at 1 μM.

In order to determine whether or not calphostin C actually inhibits protein kinase C activity in rat mast cells, the influence of the compound on Ca^{2+}- and phosphatidylserine/diglyceride-dependent protein kinase activity in the crude extract obtained from rat peritoneal mast cells was measured (Izushi and Tasaka, 1991). The enzyme activity measured at 37°C increased linearly with the incubation time (1-16 min) as well as with the protein content (0.5-5 μg of mast cell protein). Of the total protein kinase activities detected in rat mast cells, more than 90 % is dependent on Ca^{2+} and phospholipid/diglyceride, indicating that this specific type of protein kinase C has a crucial function in the mast cells. In this case, protein kinase C activity was defined as the difference between the activities measured with and without phospholipid/diglyceride combination in the presence of Ca^{2+}. As shown in Fig. 35b, calphostin C inhibited protein kinase C activity of rat mast cell extracts concentration-dependently. At 1 μM, calphostin C inhibited approximately 95 % of protein kinase C activity, while phosphatidylserine/diglyceride-independent kinase(s) was not affected by calphostin C at the same concentration. The result clearly indicates that calphostin C inhibits protein kinase C activity but does not inhibit the activity of other protein kinase(s) detected in mast cells.

7.7. Discussion

It has been reported that metabolic inhibitors decreased the intracellular ATP levels in mast cells and inhibited the histamine release from rat peritoneal mast cells (Johansen, 1980). However, the precise nature of the relationship between the ATP level and histamine release has not yet been elucidated. To clarify the role of ATP in histamine release, the study was carried out using permeabilized mast cells.

After β-escin treatment, intracellular ATP molecules were released into the medium. It has been reported that in digitonin-permeabilized chromaffin cells, Ca^{2+}-dependent catecholamine release seems to be partially dependent on exogenous ATP but in the remainder the release was carried out using endogenous ATP (Wilson and Kirshner, 1983). In a α-toxin-permeabilized rat pheochromocytoma cells (PC12 cells), the exocytotic release of dopamine was reported to be independent of exogenous ATP

(Ahnert-Hilger et al., 1985), while in digitonin-permeabilized PC12 cells (Peppers and Holz, 1986), catecholamine release was absolutely dependent on exogenous ATP. In streptolysin O-permeabilized mast cells, ITP gives rise to Ca^{2+}-induced histamine release, while ATP inhibited histamine release (Tatham and Gomperts, 1989; Koopmann and Jackson, 1990). However, in the present study, ITP was not detected in rat peritoneal mast cells. In accordance with this, it was also reported that ITP was not detected in L1210 murine leukemia cells (Pogolotti and Santi, 1982). As shown in Fig. 32b, at concentrations ranging from 0.05 to 5 mM, ATP elicited Ca^{2+}-dependent histamine release. When ATP concentration was fixed at 3 mM, the maximum histamine release was induced by Ca^{2+} at 1 μM. This Ca^{2+} concentration is almost the same as that determined at the peak of histamine release in RBL-2H3 cells (Beaven et al., 1984). To confirm whether or not ATP hydrolysis is a prerequisite in histamine release, the effects of nonhydrolyzable ATP analogues such as AMP-PNP and β, γ-methylene ATP were tested. However, none of them was effective (Table 9). These results indicate that in Ca^{2+}-dependent histamine release the hydrolysis of ATP is inevitable. As previously shown, cAMP inhibited IP_3-induced Ca^{2+} release from the intracellular Ca store (Yoshii et al., 1988), and it was also effective in inhibiting histamine release from β-escin-permeabilized mast cells (Izushi and Tasaka, 1989). These results seem to indicate that hydrolysis of ATP taking place in association with histamine release may not be related to the changes in intracellular cAMP levels.

It is known that ATP is hydrolyzed by various types of ATPase. It has been reported that vanadate inhibited dog kidney Na-K-ATPase (Ki = 40 nM) (Cantley et al., 1977) and rabbit skeletal muscle Ca-Mg-ATPase (Ki = 5 μM) (Wang et al., 1979). Na-K-ATPase is not involved in Ca^{2+}/ATP-induced histamine release, since the histamine release was not affected by treatment with 1 mM of ouabain or 1 μM of Na_3 VO_4. Ca-Mg-ATPase seems to play no crucial role in Ca^{2+}/ATP-dependent histamine release. N-Ethylmaleimide inhibited histamine release from permeabilized mast cells concentration-dependently (Table 10), and this may suggest that some proteins having the SH group may be important in the process leading to histamine release.

It became apparent that protein kinase C seems to be the intracellular target of TPA. Extracellular TPA caused a very slow histamine release from intact mast cells (Fig. 33). In accordance with this, Cantwell and Foreman (1987) reported that histamine release from intact mast cells induced by TPA was very slow, achieving a maximum release at 90 min. However, in permeabilized mast cells, TPA induced a very prompt histamine release even in the presence of 0.1 μM of Ca^{2+}. As shown in Fig. 32a, no histamine release took place in the medium containing ATP (3 mM) and Ca^{2+} (0.1 μM) but not TPA. This finding clearly indicates that TPA is effective to promote the action of Ca^{2+} and the site of action of TPA is located in the intracellular organelle; the site of action of TPA is definitely related to protein kinase C, since OAG is also effective, but phorbols which are not capable of activating protein kinase C (4α-DD and PAC) were ineffective in inhibiting the histamine release. Knight and Scrutton (1984) reported that TPA caused ^{14}C-serotonin release in the presence of Ca^{2+} in electroporred human platelets and that TPA increased the secretory response to Ca^{2+} by shifting the concentration-response curve towards the left. In digitonin-permeabilized chromaffin cells, Ca^{2+}-induced catecholamine release was enhanced by pretreatment with TPA before cell permeabilization (Pocotte et al., 1985). However, in the present experiment, a much more potent histamine release took place after cell permeabilization.

In the present experiment, K-252b and H-7 had little effect in the presence of 3 mM of ATP. Since H-7 and K-252b act competitively with ATP on the catalytic domain of protein kinase C in the presence of excessive amounts of ATP (3 mM), these inhibitions might not be effective. Calphostin C inhibits protein kinase C activity by interacting with the regulatory domain of the enzyme (Kobayashi et al., 1989), but it does not interact directly with Ca^{2+}, phospholipids or ATP-binding sites. Calphostin C is approximately 1,000 times more potent in inhibiting the enzyme activity of protein kinase C than any other protein kinases (Kobayashi et al., 1989). In this study, phospholipid/diglyceride-independent protein kinase activity was not inhibited by the same concentration of calphostin C. Histamine release from permeabilized mast cells was inhibited at 1 μM of calphostin C; 95 % of protein kinase C activity was suppressed at the same concentration. Synergic actions of the IP_3-induced Ca^{2+}-mobilization system and diglyceride-protein kinase C system in rat peritoneal mast cells have been reported (Katakami et al., 1984). However, our results suggest that protein kinase C activated by Ca^{2+} plays some crucial role in histamine release, and the increment in intracellular Ca^{2+} concentrations offers a synergistic effect on protein kinase C activation and other Ca^{2+}-dependent reactions. TPA, compound 48/80 and concanavalin A are supposed to induce the translocation of protein kinase C from the cytosol to the membrane, and subsequent phosphorylation of the membrane protein occurs in the cell membrane (Nago et al., 1987). Identification of phosphorylated proteins induced by protein kinase C is described in the next chapter.

8. Ca^{2+}-induced translocation of protein kinase C during Ca^{2+}-dependent histamine release from β-escin-permeabilized mast cells

8.1. Introduction

Recently, it has been found that the translocation of protein kinase C (PKC) from the cytosol to the plasma membrane takes place as a result of proper stimulation in a variety of cells (Nishizuka, 1984). However, the causal relation between cell activation and PKC translocation is not yet clearly understood. In rát basophilic leukemia cells, the translocation of PKC occurs in the early stage after stimulation of IgE receptors, but it may not be requisite for the process leading to serotonin release (White and Metzger, 1988). Recently, it was found that the translocation of PKC in human neutrophils was augmented in proportion to an increase in cytosolic Ca^{2+} concentrations (O'Flaherty et al., 1990). However, in the methods employed for measuring the translocation of PKC in these reports, the PKC extraction from the cell fractions was performed in a medium containing a high concentration of EGTA in the absence of Ca^{2+}. Under such conditions, EGTA inevitably chelates Ca^{2+}, and this detaches PKC from the membrane fraction (Kikkawa et al., 1982); consequently, even when the PKC activity was increased in the membrane fraction in response to the stimulation, PKC may return to the cytosol fraction during the procedures extracting PKC. In this study, we employed β-escin-treated permeabilized mast cells in which the extent of histamine release could be adjusted by altering Ca^{2+} concentrations in the medium. Moreover, when the PKC activity in the membrane fraction was measured, the extraction of PKC could be performed in the presence of various Ca^{2+} concentrations. The presence of Ca^{2+} is essential to prevent the detachment of PKC from the membrane fraction and to evaluate the role of Ca^{2+} in inducing the PKC activity. This allows the accurate assessment of PKC translocation in histamine release.

8.2. PKC in the permeabilized mast cells

As previously shown, when mast cells were permeabilized with 7.5 μg/ml of β-escin, cytosolic ATP was readily released from the cells, while only 15 % of the total LDH was released (Izushi and Tasaka, 1989). Similar results were repeatedly obtained and this seems to indicate that even after permeabilization, only slight damage was induced in the cells. In accord with this, only 20 % of the total PKC activity was released in the medium during 10 min of permeabilization. This may further indicate that the cells were only slightly damaged and that PKC was present mainly in the cytosolic fraction.

8.3. Translocation of PKC in relation to histamine release

To elucidate the role of Ca^{2+} in the translocation of PKC, the determination of the enzyme activity was performed dividing the whole process of Ca^{2+}-induced stimulation into the incubation stage and the sonication stage. Permeabilized cells were incubated at 37 °C in Mg^{2+}/ATP-free potassium glutamate (KG) buffer (in mM; potassium glutamate 137, EDTA 5, β-escin 7.5 μg/ml, PIPES 10; pH 6.8) in the presence and in the absence of Ca^{2+} (incubation period); subsequently, the incubation medium was renewed, and the cells were sonicated at 4 °C in the same medium both in the presence and absence of Ca^{2+} (sonication stage). When the changes in PKC activity in relation to Ca^{2+} concentration were determined in these two stages (A and B in Table 11), in which the Ca^{2+} concentration of KG buffer was increased from 0 (A) to 1.4 μM (B) both in the incubation and sonication stages, the PKC activity in the membrane fraction increased more than 10 times (from 5 to 54 % of the total activity) with an accompanying decrease in the enzyme activity of the cytosol fraction. When Ca^{2+} was added to the KG buffer only in the incubation period (C in Table 11), Ca^{2+}-dependent translocation of PKC was not induced. Also, when permeabilized cells were sonicated in the Ca^{2+}-containing KG-buffer after incubation in the absence of Ca^{2+} (D in Table 11), no increase in PKC activity was found in the membrane fraction, indicating that no translocation took place. Thus, it becomes evident that when the translocation of PKC is studied, all procedures

Table 11. Influence of Ca^{2+} on PKC activity in the membrane fraction of permeabilized mast cells.

Stage	Ca^{2+} concentration		PKC activity (%)	
	incubation	sonication	soluble	membrane
A	0	0	94.5 ± 1.5	5.5 ± 1.2
B	1.4	1.4	$45.8 \pm 2.1^{**}$	$54.2 \pm 2.1^{**}$
C	1.4	0	91.4 ± 1.2	8.6 ± 1.2
D	0	1.4	93.2 ± 1.4	6.8 ± 1.4

Data are expressed as percentage of the total PKC activity. Each value is the mean $\pm$ SEM (n = 5). ** p < 0.01: significantly different from control (A). Reproduced from Izushi,. and Tasaka, K.: Pharmacology, 44, 61-70 (1992), with permission.

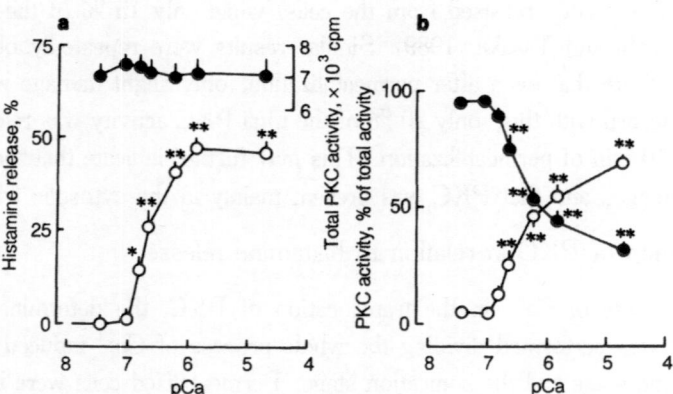

Fig. 36. Effects of Ca^{2+} on histamine release and translocation of PKC. a. Histamine release (○) and the total PKC activity (●) in β-escin-permeabilized mast cells induced by Ca^{2+}. Permeabilized mast cells were incubated in KG buffer containing various concentrations of Ca^{2+} for 5 min at 37°C. b. Translocation of PKC activity from soluble fraction to membrane fraction induced by Ca^{2+}. ● = PKC activity in the soluble fraction. ○ = PKC activity in the membrane fraction. Permeabilized mast cells were incubated for 5 min in Mg^{2+}/ATP-free KG buffer containing various concentrations of Ca^{2+}. The PKC activity was expressed as percentage of the total PKC activity. Values indicate means $\pm$ SEM (n = 4). * p < 0.05, ** p < 0.01: significantly different from control. Reproduced from Izushi, K. and Tasaka, K.: Pharmacology, 44, 61-70 (1992), with permission.

should be carried out in the presence of Ca^{2+} (B in Table 11).

When permeabilized mast cells were exposed to Ca^{2+}, histamine release was elicited dose-dependently at concentrations ranging from 10^{-7} to 10^{-6} M, reaching the maximum at 10^{-6} M of Ca^{2+} (Fig. 36a). However, the total PKC activity in permeabilized cells did

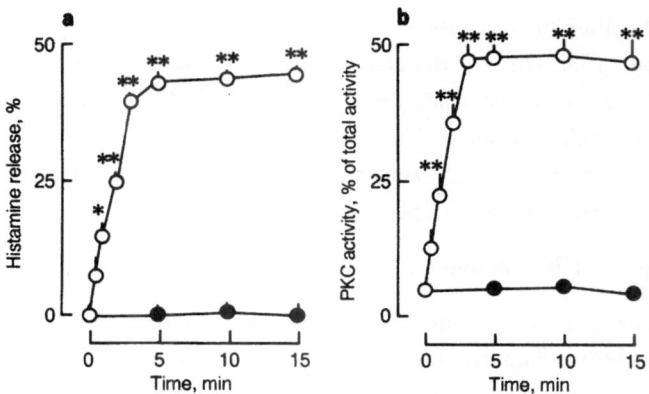

Fig. 37. Time course of histamine release and translocation of PKC. a. Time course of histamine release from permeabilized mast cells induced by 0.6 μM of Ca^{2+}. Permeabilized mast cells were incubated in the presence ($\bigcirc$) and in the absence ($\bullet$) of 0.6 μM Ca^{2+} at 37°C. b. Time course of translocation of PKC to the membrane fraction. Permeabilized mast cells were incubated with Mg^{2+}/ATP-free KG buffer in the presence ($\bigcirc$) and in the absence ($\bullet$) of 0.6 μM Ca^{2+}. After sonication and subsequent centrifugation, PKC activity in soluble and membrane fractions was measured. Data are expressed as percentage of the total activity. Each value indicates the mean $\pm$ SEM (n = 4). * p < 0.05, ** p < 0.01: significantly different from control. Reproduced from Izushi, K. and Tasaka, K.: Pharmacology, **44**, 61-70 (1992), with permission.

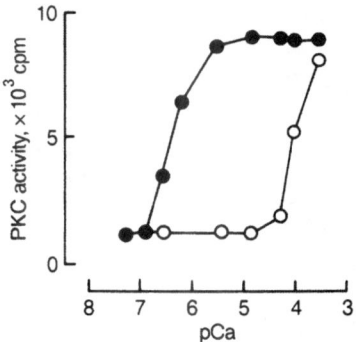

Fig. 38. Effect of Ca^{2+} on PKC activity in mast cell extract in the presence and in the absence of diolein. PKC activity was measured in the presence ($\bullet$) and in the absence ($\bigcirc$) of diolein at various Ca^{2+} concentrations with a fixed concentration of phosphatidylserine. Each value indicates the mean $\pm$ SEM (n = 4). Reproduced from Izushi, K. and Tasaka, K.: Pharmacology, **44**, 61-70 (1992), with permission.

not actually change in a variety of pCa (Fig. 36a). In the resting state, 95 % of PKC activity of permeabilized cells was detected in the soluble fraction which contained Ca^{2+} below 0.1 μM (Izushi and Tasaka, 1991). As Ca^{2+} concentration in the medium increased from 10^{-7} to 10^{-6} M, translocation of PKC to the membrane fraction was elicited with an accompanying decrease in PKC activity of the soluble fraction (Fig. 36b). At 1 μM of Ca^{2+} (pCa = 6), more than 50 % of PKC was translocated to the membrane fraction. As shown in Fig. 37a, Ca^{2+}-dependent histamine release from permeabilized cells was initiated within 30 sec of the addition of Ca^{2+} (0.6 μM) and achieved nearly the maximum at 3 min (Fig. 37a). In the absence of Ca^{2+}, no such histamine release was elicited (Fig. 37a). On the other hand, the translocation of PKC started within 30 sec of the addition of Ca^{2+} (0.6 μM) and achieved the maximum within 3 min (Fig. 37b). It is evident that the time course of Ca^{2+}-induced translocation of PKC was almost identical to that of Ca^{2+}-induced histamine release.

8.4. Ca^{2+}-dependent PKC activation in mast cell crude extract

PKC activity of the mast cell extract increased in accordance with an increase in Ca^{2+} concentrations in the medium as shown in Fig. 38. In the absence of diolein, a high concentration of Ca^{2+} (pCa = 4) was necessary for the activation of PKC. However, when 2 μg/ml of diolein and 20 μg/ml of phosphatidylserine were added to the medium, the PKC activity of the extracts increased in accordance with increases in Ca^{2+} concentra-

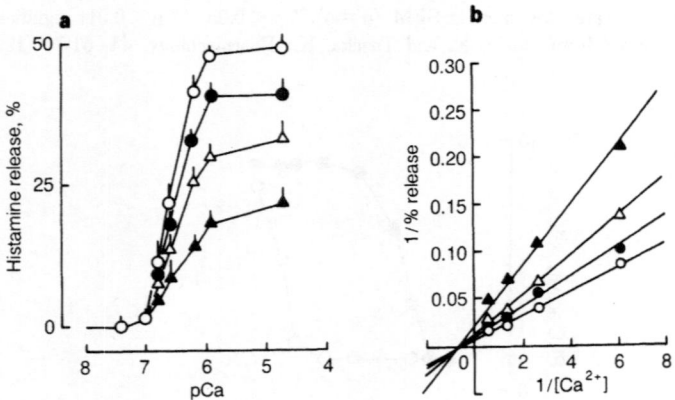

Fig. 39. Effect of calphostin C on Ca^{2+}-dependent histamine release. a. Inhibitory effect of calphostin C on Ca^{2+}-dependent histamine release from permeabilized mast cells. b. Dixon plot: permeabilized mast cells were incubated for 5 min at 37 °C in the presence of various concentrations of Ca^{2+} and calphostin C. ○ = control; ● = 0.01 μM calphostin C; △ = 0.1 μM calphostin C; ▲ = 1 μM calphostin C. Each value indicates the mean ± SEM (n = 4). Reproduced from Izushi, K. and Tasaka, K.: Pharmacology, 44, 61-70 (1992), with permission.

tions in the assay medium, and the enzyme activity reached a plateau at approximately 1 μM of Ca^{2+} (Fig. 38). The calculated Km value of PKC in the crude extract for Ca^{2+} was about 0.33 μM.

8.5. Effect of calphostin C on histamine release, translocation of PKC and the binding to phospholipid vesicles

Calphostin C, a specific PKC inhibitor, dose-dependently inhibited Ca^{2+}-induced histamine release from permeabilized cells (Fig. 39a). With 1 μM of calphostin C, approximately 60 % of Ca^{2+}-dependent histamine release was inhibited. As shown in the Dixon plot analysis (Fig. 39b), no competitive inhibition was observed between the effects of calphostin C and Ca^{2+} concentrations. Furthermore, to investigate whether or not calphostin C interferes with the translocation of PKC, permeabilized cells were incubated in the presence of Ca^{2+} and calphostin C for 5 min. Thereafter, the medium was replaced by one devoid of calphostin C, and the cells were destroyed to extract PKC. As shown in Fig. 40, when PKC activity in the membrane fraction was measured in the presence of calphostin C, PKC activity was slightly lower than that determined in the control cells. This may indicate that the Ca^{2+}-dependent translocation of PKC was not inhibited by calphostin C at 1 μM. Since calphostin C inhibits PKC activity by competing for the binding site of DG in the cell membrane (Kobayashi et al., 1989), it seems possible that calphostin C itself does not interfere with the translocation of PKC. These findings indicates that the site of action of calphostin C may be located in the cell membrane and that PKC translocation takes place in association with the stimulation of rat mast cells.

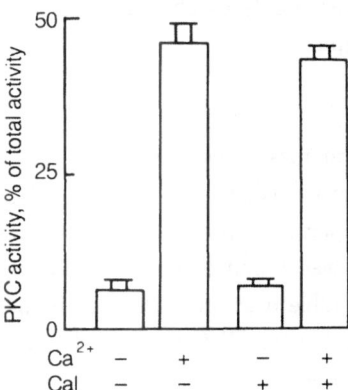

Fig. 40. Effect of calphostin C on translocation of PKC to the membrane in permeabilized mast cells. Permeabilized cells were incubated in various combinations of calphostin C (Cal; 1 μM) and Ca^{2+} (0.6 μM) in Mg^{2+}/ATP-free KG buffer. After sonication and centrifugation, PKC activity in the membrane fractions was measured (Izushi and Tasaka, 1992). Data are expressed as percentages of the total PKC activity. Each value is the mean ± SEM (n = 4). Reproduced from Izushi, K. and Tasaka, K.: Pharmacology, 44, 61-70 (1992), with permission.

152

Table 12. Binding of the crude PKC in the mast cell extracts to phospholipid vesicles

Ca²⁺ (μM)	Control (cpm)	+ Phospholipid vesicle (cpm)
0	6,762 ± 248	6,882 ± 242
0.16	6,968 ± 350	5,564 ± 290*
0.32	7,178 ± 199	3,625 ± 171**
0.64	6,961 ± 371	424 ± 88**
1.4	6,992 ± 291	200 ± 43**

Mast cell extract was incubated with and without (control) phospholipid vesicles (200 μg/ml phosphatidylserine) for 10 min at 37°C in the presence and in the absence of Ca^{2+}. After incubation, the reaction mixture was centrifuged. The PKC activity in the supernatant was measured (Izushi and Tasaka, 1992). Each value indicates the mean ± SEM (n = 4). * p < 0.05, ** p < 0.01: significantly different from control. Reproduced from Izushi, K. and Tasaka, K.: Pharmacology, 44, 61-70 (1992), with permission.

However, it is not certain whether PKC actually binds to phospholipids in the cell membrane after translocation. To make this point clear the binding of PKC to phospholipid vesicles was studied.

The crude extract was incubated with phospholipid vesicles, prepared with phosphatidylserine alone, in the presence and in the absence (control) of Ca^{2+} at 37°C for 10 min. After incubation, the reaction mixture was centrifuged at 100,000 × g for 10 min at 4°C to precipitate the vesicles (Izushi and Tasaka, 1992). When the PKC activity in the supernatant was measured, it was less than 10 % of the control as the Ca^{2+} concentration was increased beyond 0.64 μM (Table 12). This result clearly indicates that PKC in the crude extract binds to phospholipid vesicles in the presence of Ca^{2+} depending on its concentration.

8.6. Discussion

It was reported that when rat mast cells were challenged with antigen, the total PKC activity increased significantly (White et al., 1985; Kurosawa and Parker, 1986). However, in the present experiment the total PKC activity was not altered by Ca^{2+} stimulation. By contrast, it was found that Ca^{2+}-induced stimulation caused the translocation of PKC from the soluble fraction to the membrane fraction in a Ca^{2+}-dependent manner. The amount of translocated PKC runs parallel to that of histamine release from permeabilized mast cells. Moreover, when PKC was incubated with phospholipid vesicles in the presence of Ca^{2+}, the binding of PKC to phospholipid vesicles markedly increased (Table 12). The reason why PKC binds to the membrane fraction in the presence of Ca^{2+} is assumed to be as follows: when PKC interacts with Ca^{2+}, the hydrophobic domains of PKC may be exposed, and consequently, a hydrophobic interaction may occur between PKC and phospholipids, especially with phosphatidylserine, in the cell membrane as shown between calmodulin and lipids (Tanaka and Hidaka, 1980).

In connection with this, the 'E-F hand' domain, which can be found in calcium-binding proteins such as calmodulin, was also found in the V_3 region of PKCα and PKCβ (Parker et al., 1986; Coussens et al., 1986).

It is known that DG, another activator of PKC, is formed from cell membrane phospholipids after physiological stimulations. In the resting state of rat peritoneal mast cells, the content of DG is about 100 pmol/10^6 cells (Kennerly, 1987). Based on this finding, it was calculated that the intracellular concentration of DG is about 12.5 μg/ml given that the total cell volume of 10^6 cells is about 4 μl and that the molecular weight of DG is 500. In the presence of diolein (2 μg/ml), PKC activity in the crude extract increased in accordance with increases in Ca^{2+} concentrations, and the Km value for Ca^{2+} was 0.33 μM as indicated in Fig. 38. Kurosawa and Parker (1986) reported that the apparent Km value of PKC purified from rat mast cells was 10^{-3} M. However, this value is very much larger than that determined in PKC obtained from rat brain (Kosaka et al., 1988) and that measured in the present experiment. The Km value determined in the present experiment corresponds to the Ca^{2+} concentration which can induce the half-maximum effect of histamine release from permeabilized mast cells. Calphostin C, a specific PKC inhibitor which is more potent than other PKC inhibitors, inhibits PKC activity by competing for the binding site in the regulatory domain of PKC with DG (Kobayashi et al., 1989). As shown in Fig. 39b, the inhibitory mechanism of calphostin C is not competitive for Ca^{2+}, and the translocation of PKC is not significantly inhibited by calphostin C even at a concentration of 1 μM (Fig. 40). However, calphostin C dose-dependently inhibited Ca^{2+}-induced histamine release from permeabilized mast cells (Fig. 39a). It was assumed that calphostin C probably inhibits the interaction of membrane-bound PKC with DG, and this might be the reason for the histamine release inhibition.

In connection with this, it is known that when PKC is activated directly or indirectly, the phosphorylation of phosphatidylinositol and its conversion to phosphatidylinositol monophosphate are stimulated. Thus, a potential increase in the availability of the substrate for phospholipase C specific for phosphatidylinositol-4, 5-bisphosphate occurs (Cockcroft et al., 1985). In the stimulated mast cells, phosphatidylinositol-4, 5-bisphosphate may degraded to inositol 1, 4, 5-trisphosphate (IP_3) and DG in the presence of phospholipase C (Nakamura and Ui, 1985). IP_3 stimulates Ca^{2+} release from the intracellular Ca^{2+} store (endoplasmic reticulum), and this remarkably triggers histamine release (Yoshii et al,. 1988). Thus, it can be concluded that translocation of PKC takes place in the early stage of histamine release, and this triggers the subsequent process(es) in histamine release.

9. Phosphorylation of smg p21B in rat peritoneal mast cells in association with histamine release inhibition by dibutyryl-cAMP

9.1. Introduction

It is known that protein phosphorylation plays some critical roles in the regulation of various cellular responses and functions. So far, many protein kinases have been identified, and protein kinase A and C in particular have been shown to be involved in a wide variety of cell functions (Nishizuka, 1986; Edelman et al., 1987). We first demonstrated the existence of vimentin, an intermediate filament protein, in rat peritoneal mast cells, and this protein was phosphorylated after stimulation with some histamine releasers. Further, we found that this phosphorylation proceeded in the presence of protein kinase C (Izushi et al., 1992a). By contrast, the substrate protein(s) of protein kinase A and its physiological functions in rat mast cells remains to be identified. In platelets, prostacyclin has been shown to increase cAMP level through stimulation of adenylate cyclase, which is effective in leading the inhibition of platelet functions (Watson et al., 1984). Moreover, the activation of protein kinase A elicited by increased intracellular cAMP increases the phosphorylation of several proteins having molecular weights of 240, 50, 24 and 22 kDa (Haslam et al., 1978; Kawata et al., 1989).

In rat peritoneal mast cells, it is known that an increase in cAMP level is effective in inhibiting histamine release (Alm and Bloom, 1982; Tasaka et al., 1986a). One possible mechanism of cAMP in inhibitory reaction is assumed to be prevention of Ca^{2+} release from intracellular Ca store (Yoshii et al., 1988). Although the inhibitory effect of cAMP on histamine release from rat mast cells is exerted in association with the activation of protein kinase A, the exact mechanism of the interaction between cAMP and protein kinase A is not known. To clarify this point, the investigation was carried out (Izushi et al., 1992b).

9.2. Histamine release inhibition by cAMP and protein kinase A

Rat peritoneal mast cells were incubated with dibutyryl cAMP (Bt_2cAMP) and then stimulated by compound 48/80. In control cells, compound 48/80 (0.5 μg/ml) released about 50 % of the total histamine, and the amount of IP_3 formed within 5 sec increased to approximately 9 times that of the resting cells. As shown in Fig. 41A, Bt_2cAMP induced a dose-dependent inhibition of histamine release from rat peritoneal mast cells. At a concentration of 1 mM, Bt_2cAMP inhibited histamine release by approximately 70 %. In order to examine whether or not this inhibitory effect of cAMP is exerted in association with the activation of protein kinase A, the effect of H-8, a protein kinase A inhibitor (Hidaka et al., 1984), was investigated. When various concentrations of H-8 were added with 1 mM of Bt_2cAMP during the preincubation period, Bt_2cAMP-induced histamine release inhibition was dose-dependently restored by H-8, and the maximum

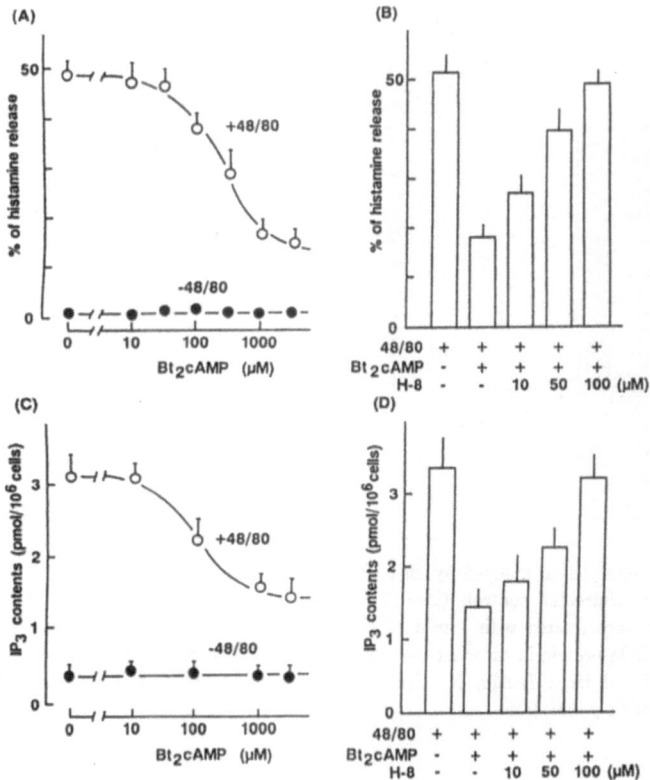

Fig. 41. Effects of Bt$_2$cAMP and H-8, a protein kinase A inhibitor, on histamine release and IP$_3$ formation in rat peritoneal mast cells. Histamine release was elicited by compound 48/80 (0.5 μg/ml) for 10 min. (A) The inhibitory effect of Bt$_2$cAMP on histamine release from rat mast cells. (B) The restoring effect of H-8 on histamine release from rat mast cells. (C) The inhibitory effect of Bt$_2$cAMP on IP$_3$ formation in rat mast cells. (D) The restoring effect of H-8 on Bt$_2$cAMP-induced inhibition of IP$_3$ formation. The results indicate the mean ± SEM (n = 4). Reproduced from Izushi, K., Shirasaka, T., Chokki, M. and Tasaka, K.: FEBS Lett., 314, 241-245 (1992), with permission.

effect of H-8 was reached at 100 μM (Fig. 41B). When mast cells were exposed to compound 48/80 for 5 sec, the intracellular content of IP$_3$ increased to approximately 9 times that of the control. The amount of IP$_3$ formed by compound 48/80 was also inhibited by pretreatment with Bt$_2$cAMP at the same concentration range (Fig. 41C). Moreover, Bt$_2$cAMP-induced inhibition of IP$_3$ formation was dose-dependently restored by the presence of H-8 as in the case of histamine release (Fig. 41D).

These results seem to suggest that the inhibitory effect of cAMP in mast cells might be related to both the amount of activated protein kinase A and the amount of target protein phosphorylated by protein kinase A.

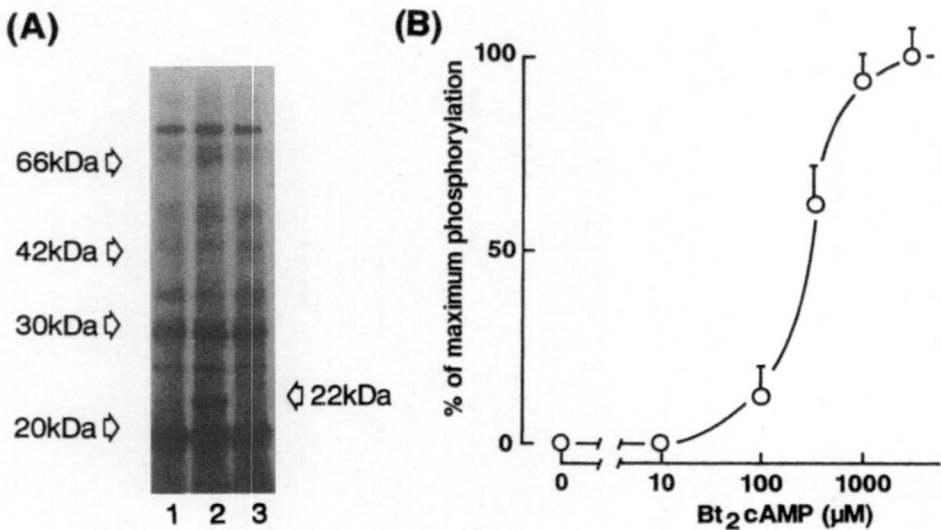

Fig. 42. Protein phosphorylation induced by Bt$_2$cAMP in rat mast cells. (A) Protein phosphorylation in rat mast cells. (Lane 1) Untreated control; (lane 2) rat mast cells were treated with 1 mM of Bt$_2$cAMP; (lane 3) rat mast cells were treated with 1 mM of Bt$_2$cAMP plus 100 μM of H-8. (B) Dose-dependent phosphorylation of 22 kDa protein in rat mast cells induced by Bt$_2$cAMP. The results indicate the mean $\pm$ SEM (n = 4). Reproduced from Izushi, K., Shirasaka, T., Chokki, M. and Tasaka, K.: FEBS Lett., 314, 241-245 (1992), with permission.

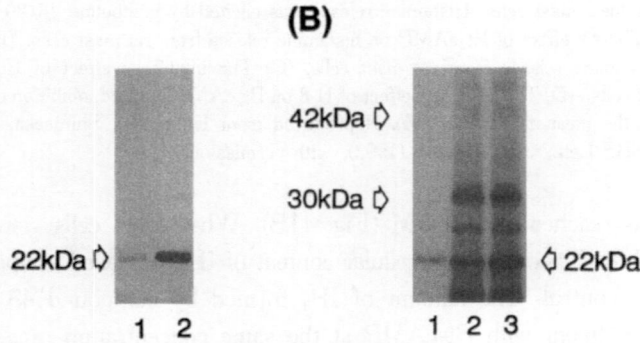

Fig. 43. Immunoblot analysis of smg p21B and Bt$_2$cAMP-induced protein phosphorylation of rat mast cells. (A) Lane 1, rat mast cell proteins; lane 2, purified human platelet smg p21B. (B) Lane 1, phosphorylated smg p21B by protein kinase A; lane 2, phosphorylated proteins in rat mast cells elicited by 1 mM of Bt$_2$cAMP; lane 3, phosphorylated proteins in rat mast cells plus phosphorylated smg p21B. Reproduced from Izushi, K., Shirasaka, T., Chokki, M. and Tasaka, K.: FEBS Lett., 314, 241-245 (1992), with permission.

9.3. Phosphorylation of smg p21B in rat mast cells

Protein phosphorylation in rat peritoneal mast cells exposed to Bt_2cAMP was investigated. As shown in Fig. 42A, when rat mast cells were incubated with 1 mM of Bt_2cAMP, several proteins were phosphorylated. In particular, 22 kDa protein was markedly phosphorylated, as shown in Fig. 42A (lane 2). However, phosphorylation of 22 kDa protein was clearly inhibited in the presence of 100 μM of H-8 (Fig. 42A, lane 3). Bt_2-cAMP-induced phosphorylation of 22 kDa protein increased in a dose-dependent manner (Fig. 42B) and reached a maximum at 1 mM of Bt_2cAMP. The concentration of Bt_2-cAMP necessary for protein phosphorylation was the same as the concentration of Bt_2-cAMP required in inhibiting histamine release and IP_3 formation, as shown in Fig. 41.

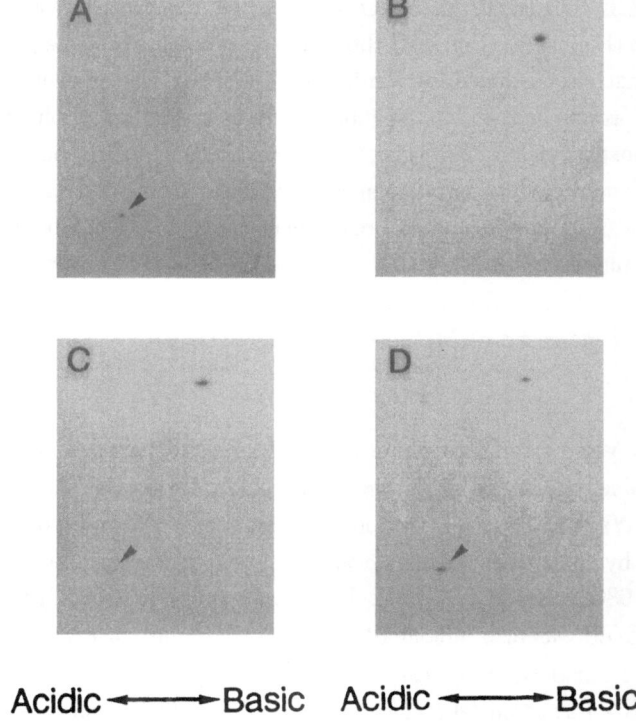

Acidic ◄────► Basic   Acidic ◄────► Basic

Fig. 44. Two-dimensional PAGE analysis of protein phosphorylation by Bt_2cAMP in rat mast cells. (A) smg p21B phosphorylated by protein kinase A. (B) Protein phosphorylation in rat mast cells in the absence of Bt_2cAMP. (C) Protein phosphorylation in rat mast cells induced by 1 mM of Bt_2cAMP. (D) Phosphorylated protein in rat mast cells elicited by Bt_2cAMP plus phosphorylated smg p21B by protein kinase A. Arrows indicate phosphorylated smg p21B (A), the 22 kDa phosphorylated protein in rat mast cells (C) and the mixture (D: A plus C). Reproduced from Izushi, K., Shirasaka, T., Chokki, M. and Tasaka, K.: FEBS Lett., **314**, 241-245 (1992), with permission.

The same molecular weight (22 kDa) of phosphorylated protein has been reported in platelet (Haslam et al., 1978; Kawata et al., 1989) and this was identified as smg p21B, a ras-related small molecular weight G protein (Kawata et al., 1989). But, it is not known whether or not smg p21B exists in rat peritoneal mast cells. As shown in Fig. 43A, smg p21B was detected as a single protein band (lane 1) when rat mast cell proteins were tested in the immunoblot analysis using anti-smg p21B antiserum. This protein band had the same molecular weight as that of purified human platelet smg p21B (Fig. 43A, lane 2). The 22 kDa protein phosphorylated in response to Bt_2cAMP in rat mast cells had the same molecular weight as the smg p21B phosphorylated by protein kinase A in cell-free system (Fig. 43B, lane 1 and 2). When these two were mixed and subjected to SDS-PAGE, two proteins comigrated to the same position (Fig. 43B, lane 3).

To confirm whether or not these two proteins are identical, the 22 kDa phosphorylated protein in rat mast cells and phosphorylated smg p21B were subjected to two-dimensional PAGE (O'Farrell, 1975). One radioactive spot was observed as phosphorylated smg p21B (Fig. 44A). Fig. 44C shows the autoradiogram of the protein phosphorylation in rat mast cells induced by 1 mM of Bt_2cAMP. The phosphorylated 22 kDa protein appeared as one spot and its position was the same as that of phosphorylated smg p21B. When phosphorylated 22 kDa protein in mast cells and phosphorylated smg p21B were mixed and subjected to two-dimensional PAGE, these two proteins comigrated together showing exactly the same molecular weight and the same isoelectric point (Fig. 44D). These results clearly indicate that smg p21B is located in rat peritoneal mast cells, and that 22 kDa protein, which is phosphorylated by protein kinase A in rat mast cells in the presence of increased cAMP level, is 'smg p21B'.

9.4. Discussion

It is known that when rat mast cells are treated so as to increase intracellular cAMP contents, histamine release is significantly inhibited (Tasaka et al., 1986a; Alm and Bloom, 1982). We also reported that intracellular Ca^{2+} concentrations and histamine release induced by histamine releasers were inhibited by pretreatment with Bt_2cAMP (Tasaka et al., 1986a; Mio et al., 1991). Moreover, an increase in IP_3 formation elicited by compound 48/80 was also inhibited by pretreatment with Bt_2cAMP (Fig. 41C). As shown in Fig. 41B and D, it is clear that the inhibitory effects of cAMP on histamine release and on IP_3 formation take place in association with protein kinase A activation, since H-8 was effective in reversing these inhibitory reactions. So far, in rat peritoneal mast cells the direct substrate protein(s) for protein kinase A remain to be identified. We found that 22 kDa protein was markedly phosphorylated by treatment with Bt_2cAMP (Fig. 42A and B). However, this phosphorylated protein was not detected in either compound 48/80- or substance P-stimulated rat peritoneal mast cells (Izushi et al., 1992a). In human platelets, four proteins of 240, 50, 24 and 22 kDa have been shown to be phosphorylated in response to PGE$_1$, a cAMP-elevating agent, and 22 kDa phosphorylated protein was estimated as smg p21B (Kawata et al., 1989). As shown in one- and two-dimensional PAGE (Fig. 43B and 44), 22 kDa protein phosphorylated in response to Bt$_2$cAMP in rat peritoneal mast cells comigrates with purified smg p21B; this seems to indicate that these two proteins have the same molecular weights and isoelectric points. Moreover, smg p21B exists in rat peritoneal mast cells as indicated by immunoblot analysis using anti-smg p21B antiserum (Fig. 43A). Protein kinase A phosphorylates smg p21B mainly at Ser^{179}, which is located between the polybasic region and the geranyl-geranylated cystein residue in the C-terminal region (Hata et al., 1991). The C-terminal region is essential to interact with the stimulatory GDP/GTP exchange protein named smg GDS (Yamamoto et al., 1990). When smg p21B was phosphorylated by protein kinase A, it markedly increased the action of smg GDS and initiated the conversion from GDP-bound inactive form to the GTP-bound active form (Hata et al., 1991). It is known that smg p21B has the same putative effector domain and consensus C-terminal sequences as the ras gene products (Ohmori et al., 1989; Yamamoto et al., 1990). Therefore, it is possible to assume that smg p21B exerts actions similar to, or antagonistic to, those of the ras gene products (Kawata et al., 1989; Ohmori et al., 1988).

Wakelam et al. (1986) showed that normal p21^{N-ras} might be involved in the regulation of receptor-linked phospholipase C activation without affecting the increase of receptor numbers. Lapetina and Reep (1987) also reported that ras-like 29 kDa G protein might regulate phospholipase C activity in human platelets. Moreover, the activation of PIP$_2$-specific phospholipase C was prompted by the addition of small molecular weight G protein (Baldassare et al., 1988; Wang et al., 1989). It has been reported that IP$_3$ formation in human platelets was inhibited by pretreatment with prostacyclin, a potent adenylate cyclase activator (Watson et al., 1984). IP$_3$ formation, induced by compound 48/80 in rat mast cells, was also inhibited by the pretreatment with Bt$_2$cAMP (Fig. 41C), and this inhibitory action was restored by H-8 in a dose-dependent fashion (Fig. 41D). IP$_3$ promotes Ca^{2+} release from intracellular Ca^{2+} store and released Ca^{2+} is essential for triggering histamine release from rat peritoneal mast cells (Yoshii et al., 1988; Izushi and Tasaka, 1991; Mio et al., 1991). The inhibitory effects (to histamine release and IP$_3$ formation) of Bt$_2$cAMP and 22 kDa protein phosphorylation were induced by at the same concentration range of Bt$_2$cAMP. It is possible that smg p21B is phosphorylated by protein kinase A in response to increased intracellular cAMP levels, and that at least some parts of the inhibitory actions of protein kinase A system in rat mast cells are exerted by smg p21B. Although it is not known whether phospholipase C in mast cells is also regulated by small molecular weight G protein, it can be assumed that phosphorylated smg p21B is useful to convert GDP-bound inactive form to GTP-bound active form, and consequently, it may inhibit phospholipase C activation by the antagonistic action to phospholipase C-activated G protein.

In this experiment, we first demonstrated that smg p21B exists in rat peritoneal mast cells and that it is phosphorylated by activation of protein kinase A. Further, it was assumed that smg p21B may play some important role in inhibiting histamine release from rat peritoneal mast cells.

References

Abdel-Latif A.A. (1986) Calcium-mobilizing receptors, polyphosphoinositides, and the generation of second messengers. *Pharmacol. Rev.*, 38: 227-272

Ahnert-Hilger G., Bhakdi S. and Gratzl (1985) Minimal requirements for exocytosis: a study using PC12 cells permeabilized with staphylococcal α-toxin. *J. Biol. Chem.*, 260: 12730-12734

Akagi M., Mio M., Tasaka K. and Kiniwa S. (1983) Mechanism of histamine release inhibition induced by azelastine. *Pharmacometrics*, 26: 191-198

Alm P.E. and Bloom G.D. (1982) Cyclic nucleotide involvement in histamine release from mast cells. A reevaluation. *Life Sci.*, 30: 213-218

Amende L.M. and Donlon M.A. (1985) Isolation of cellular membranes from rat mast cells. *Biochim. Biophys. Acta*, 812: 713-720

Ala'i R. and Ralph R.K. (1986) Cyclic AMP and Ca^{2+} uptake by mastocytoma mitochondria. *Cell Calcium*, 7: 13-27

Baker P.F. and Knight D.E. (1981) Calcium control of exocytosis and endocytosis in bovine adrenal medullary cells. *Phil. Trans. R. Soc. Lond. B. Biol. Sci.*, 296: 83-103

Baldassare J.J., Knipp M.A., Henderson P.A. and Fisher G.J. (1988) GTPγS-stimulated hydrolysis of phosphatidylinositol-4, 5-bisphosphate by soluble phospholipase C from human platelets requires soluble GTP-binding protein. *Biochem. Biophys. Res. Commun.*, 154: 351-357

Barrett K.E. and Pearce F.L. (1983) A comparison of histamine secretion from isolated peritoneal mast cells of the mouse and rat. *Int. Arch. Allergy Appl. Immunol.*, 72: 234-238

Beaven M.A., Roger J., Moore J.P. Hesketh T.R., Smith G.A. and Metcalfe J.C. (1984) The mechanism of the calcium signal and correlation with histamine release in 2H3 cells. *J. Biol. Chem.*, 259: 7129-7136

Bennet J.P., Cockcroft S. and Gomperts B.D. (1981) Rat mast cells permeabilized with ATP secrete histamine in response to calcium ions buffered in the micromolar range. *J. Physiol.*, 317: 335-345

Berridge M.J. (1984) Inositol trisphosphate and diacylglycerol as second messengers. *Biochem. J.*, 220: 345-360

Berridge M.J. and Irvine R.F. (1984) Inositol trisphosphate, a novel second messenger in cellular signal transduction. *Nature*, 312: 315-321

Borgers M., Thone F.J.M., Xhonneux B.J.M. and de Clerck F.F.P. (1983) Localization of calcium in red blood cells. *J. Histochem. Cytochem.*, 31: 1109-1116

Brass L.F. and Joseph S.K. (1985) A role for inositol triphosphate in intracellular Ca^{2+} mobilization and granule secretion in platelets. *J. Biol. Chem.*, 260: 15172-15179

Burgess G.M., McKinney J.S., Fabiato A., Leslie B.A. and Putney J.W. (1983) Calcium pools in saponin-permeabilized guinea pig hepatocytes. *J. Biol. Chem.*, 258: 15336-15345

Burgoyne R.D. (1987) Control of exocytosis. *Nature*, 328: 112-113

Cantley L.C., Josephson L., Warner R., Yanagisawa M., Lechene C. and Guidotti G. (1977) Vanadate is a potent (Na, K)-ATPase inhibitor found in ATP derived from muscle. *J. Biol. Chem.*, 252: 7421-7423

Cantwell M.E. and Foreman J.C. (1987) Phorbol esters induced a slow, non-cytotoxic release of histamine from rat peritoneal mast cells. *Agents Actions*, 20: 165-168

Castagna M., Takai Y., Kaibuchi K., Sano K., Kikkawa U. and Nishizuka Y. (1982) Direct activation of calcium-activated, phospholipid-dependent protein kinase by tumor-promoting phorbol esters. *J. Biol. Chem.*, 257: 7847-7851

Chakravarty N. and Nielsen E.M. (1985) Calmodulin in mast cells and its role in histamine release. *Agents Actions*, 16: 122-125

Chiou C.Y. and Malagodi M.H. (1975) Studies on the mechanism of action of a new Ca^{2+} antagonist, 8-(N, N-diethylamino)-octyl 3, 4, 5-trimethoxybenzoate hydrochloride in smooth and skeletal muscles. *Br. J. Pharmacol.*, 53: 279-285

Church M.K. and Gradidge C.F. (1980) Oxatomide: inhibition and stimulation of histamine release from human lung and leukocytes *in vitro*. *Agents Actions*, 10: 4-7

de Clark F., van Reempts J. and Borgers M. (1981) Comparative effects of oxatomide on the release of histamine from rat peritoneal mast cells. *Agents Actions*, 11: 184-192

Cockcroft S., Barrowman M.M. and Gomperts B.D. (1985) Breakdown and synthesis of polyphosphoinositides in fMet-Leu-Phe stimulated neutrophils. *FEBS Lett.*, 181: 259-263

Coussens L., Parker P.J., Rhee L., Yang-Feng T.L., Chen E., Waterfield M.D., Francke U. and Ullrich A. (1986) Multiple, distinct forms of bovine and human protein kinase C suggest diversity in cellular signaling pathways. *Science*, 233: 859-866

Curtis B.M. and Catterall W.A. (1985) Phosphorylation of the calcium antagonist receptor of the voltage-sensitive calcium channel by cAMP-dependent protein kinase. *Proc. Natl. Acad. Sci. USA*, 82: 2528-2532

de Duve C., Pressman B.C., Gianetto R., Wattiaux R. and Appelmans F. (1955) Tissue fractionation studies. 6. Intracellular distribution patterns of enzymes in rat-liver tissue. *Biochem. J.*, 60: 604-617

Edelman A.M., Blumenthal D.K. and Krebs E.G. (1987) Protein serine-threonine kinase. *Ann. Rev. Biochem.*, 56: 567-614

Endo M. (1977) Calcium release from the sarcoplasmic reticulum. *Physiol. Rev.*, 57: 71-108

Ennis M., Atkinson G and Pearce F.L. (1980a) Inhibition of histamine release induced by compound 48/80 and peptide 401 in the presence and absence of calcium. Implication for the mode of action of antiallergic compounds. *Agents Actions*, 10: 222-228

Ennis M., Truneth A., White J.R. and Pearce F.L. (1980b) Calcium pools involved in histamine release from rat mast cells. *Int. Arch. Arch. Allergy Appl. Immunol.*, 62: 467-471

Fewtrell C.M.S., Foreman J.C., Jordan C.C., Oehme P., Renner H. and Stewart J.M. (1982) The effects of substance P on histamine release and 5-hydroxytryptamine release in the rat. *J. Physiol.*, 330: 393-411

Fox P.C., Basciano L.K. and Siraganian R.P. (1982) Mouse mast cell activation and desensitization for immune aggregate-induced histamine release. *J. Immunol.*, 129: 314-319

Garteiz D.A., Hook R.H., Walker B.J. and Okerholm R.A. (1982) Pharmacokinetics and biotransformation studies of terfenadine in man. *Arzneim. -Forsch.*, 32: 1185-1190

Gennis R.B. (1989) Biomembranes. In Springer Advanced Texts in Chemistry. New York: Springer-Verlag

Gomperts B.D. (1983) Involvement of guanine nucleotide-binding protein in the gating of Ca^{2+} by receptors. *Nature*, 306: 64-66

Grosman N. (1986) Effects of TMB-8 on histamine release from isolated rat mast cells. *Int. Arch. Allergy Appl. Immunol.*, 79: 253-258

Haslam R.J., Davidson M.M.L., Davies T., Lynham J.A. and McClenagham M.D. (1978) Regulation of blood platelet function by cyclic nucleotides. *Adv. Cyclic Nucleotide Res.*, 9: 533-552

Hata Y., Kaibuchi K., Kawamura S., Hiroyoshi M., Shirataki H. and Takai Y. (1991) Enhancement of the actions of smg p21 GDP/GTP exchange protein by protein kinase A-catalyzed phosphorylation of smg p21. *J. Biol. Chem.*, 266: 6571-6577

Hayashi H., Ichikawa A., Saito T. and Tomita K. (1976) Inhibitory role of cyclic adenosine 3: 5-monophosphate in histamine release from rat peritoneal mast cells *in vitro. Biochem. Pharmacol.*, 25: 1907-1913

Hidaka H. and Tanaka T. (1983) Naphthalenesulfonamides as calmodulin antagonists. *Methods Enzymol.*, 102: 185-194

Hidaka H., Inagaki M., Kawamoto S. and Sasaki Y. (1984) Isoquinoline-sulfonamide, novel and potent inhibitors of cyclic nucleotide dependent protein kinase and protein kinase C. *Biochemistry*, 23: 5036-5041

Hirata M., Suematsu E., Hashimoto T., Hamachi T. and Koga T. (1984) Release of Ca^{2+} from a non-mitochondrial store site in peritoneal macrophages treated with saponin by inositol 1, 4, 5-trisphosphate. *Biochem. J.*, 223: 229-236

Hirata M., Kukita M., Sasaguri T., Suematsu E., Hashimoto T. and Koga T. (1985) Increase in Ca^{2+} permeability of intracellular Ca^{2+} store membrane of saponin-treated guinea pig peritoneal macrophage by inositol 1, 4, 5-trisphosphate. *J. Biochem.*, 97: 1575-1582

Howell T.W. and Gomperts B.D. (1987) Rat mat cells permeabilized with streptolysin O secrete histamine in response to Ca^{2+} at concentrations buffered in the micromolar range. *Biochim. Biophys. Acta*, 927: 177-183

Imai A., Ishizuka Y., Nakashima S. and Nozawa Y. (1984a) Differential activation of membrane phospholipid turnover by compound 48/80 and A23187 in rat mast cells. *Arch. Biochem. Biophys.*, 232: 259-268

Imai S., Nakazawa H., Imai H. and Nabata H. (1984b) Effects of procaine on the isolated dog coronary artery. *Arch. Int. Pharmacodyn.*, 271: 98-105

Itoh T., Kanmura Y., Kuriyama H. and Sasaguri T. (1985) Nitroglycerine- and isoprenaline-induced vasodilatation: assessment from actions of cyclic nucleotides. *Br. J. Pharmacol.*, 84: 393-406

Izushi K. and Tasaka K. (1989) Histamine release from β-escin-permeabilized rat peritoneal mast cells and its inhibition by intracellular Ca^{2+} blockers, calmodulin inhibitors and cAMP. *Immunopharmacology*, 18: 177-186

Izushi K. and Tasaka K. (1991) Essential role of ATP and possibility of activation of protein kinase C in Ca^{2+}-dependent histamine release from permeabilized rat peritoneal mast cells. *Pharmacology*, 42: 297-308

Izushi K. and Tasaka K. (1992) Ca²⁺-induced translocation of protein kinase C during Ca²⁺-dependent histamine release from beta-escin-permeabilized rat mast cells. *Pharmacology*, 44: 61-70

Izushi K., Fujiwara Y. and Tasaka K. (1992a) Identification of vimentin in rat peritoneal mast cells and its phosphorylation in association with histamine release. *Immunopharmacology*, 23: 153-161

Izushi K., Shirasaka T., Chokki M. and Tasaka K. (1992) Phosphorylation of smg p21B in rat peritoneal mast cells in association with histamine release by dibutyryl-cAMP. *FEBS Lett.*, 314: 241-245

Johansen T. (1980) Adenosine triphosphate level during anaphylactic histamine release in rat mast cells in vitro. Effects of glycolytic and respiratory inhibitors. *Eur. J. Pharmacol.*, 58: 107-115

Kase H., Iwahasi K., Nakanishi S., Matsuda Y., Yamada K., Takahashi M., Murakata C., Sato A. and Kaneko M. (1987) K-252 compounds, novel and potent inhibitors of protein kinase C and cyclic nucleotide-dependent protein kinases. *Biochem. Biophys. Res. Commun.*, 142: 436-440

Katakami Y., Kaibuchi K., Sawamura M., Takai Y. and Nishizuka Y. (1984) Synergic action of protein kinase C and calcium for histamine release from rat peritoneal mast cells. *Biochem. Biophys. Res. Commun.*, 12: 573-578

Kawata M., Kikuchi A., Hoshijima M., Yamamoto K., Hashimoto E., Yamamura H and Takai Y. (1989) Phosphorylation of smg p21, a ras p21-like GTP-binding protein, by cyclic AMP-dependent protein kinase in a cell-free system and in response to prostaglandin E₁ in intact human platelets. *J. Biol. Chem.*, 264: 15688-15695

Kennerly D.A. (1987) Diacylglycerol metabolism in mast cells. *J. Biol. Chem.*, 262: 16305-16313

Kennerly D.A., Sullivan T.J. and Parker C.W. (1979) Activation of phospholipid metabolism during mediator release from stimulated rat mast cells. *J. Immunol.*, 122: 152-159

Kikkawa U., Takai Y., Minakuchi R., Inohara S. and Nishizuka Y. (1982) Calcium-activated, phospholipid-dependent protein kinase from rat brain. *J. Biol. Chem.*, 257: 13341-13348

Knight D.E. and Scrutton M.C. (1984) Cyclic nucleotides control a system which regulates Ca²⁺ sensitivity of platelet secretion. *Nature*, 309: 66-68

Kobayashi E., Nakano H., Morimoto M., Tamaoki T. (1989) Calphostin C (UCN-1028C), a novel microbial compound, is a highly potent and specific inhibitor of protein kinase C. *Biochem. Biophys. Res. Commun.*, 159: 548-553

Koopmann W.R. and Jackson R.C. (1990) Calcium- and guanine-nucleotide-dependent exocytosis in permeabilized rat mast cells. *Biochem. J.*, 265: 363-373

Kosaka Y., Ogita K., Ase K., Nomura H., Kikkawa U. and Nishizuka Y. (1988) The heterogeneity of protein kinase C in various rat tissues. *Biochem. Biophys. Res. Commun.*, 15: 973-981

Kreye V.A.W., Ruegg J.C. and Hofmann F. (1983) Effects of calcium-antagonist and calmodulin antagonist drugs on calmodulin dependent contractions of chemically skinned vascular smooth muscle from rabbit renal arteries. *Naunyn-Schmied. Archs. Pharmacol.*, 323: 85-89

Kurosawa M. and Parker C.W. (1986) Characterization of calcium-activated, phospholipid-dependent protein kinase from rat serosal mast cells and RBL-1 cells. *Cell. Immunol.*, 103:

381-393

Lapetina E.G. and Reep B.R. (1987) Specific binding of (α-^{32}P)GTP to cytosolic and membrane-bound proteins of human platelets correlates with the activation of phospholipase C. *Proc. Natl. Acad. Sci. USA*, 84: 2261-2265

Lindau M. and NüBe O. (1987) Pertussis toxin does not effect the time course of exocytosis in mast cells stimulated by intracellular application of GTP-γ-S. *FEBS Lett.*, 222: 317-321

Martonosi A.N. (1984) Mechanisms of Ca^{2+} release from sarcoplasmic reticulum of skeletal muscle. *Physiol. Rev.*, 64: 1240-1320

Mio M., Izushi K. and Tasaka K. (1991) Substance P-induced histamine release from rat peritoneal mast cells and its inhibition by antiallergic agents and calmodulin inhibitors. *Immunopharmacology*, 22: 59-66

Mitchell R.H. (1975) Inositol phospholipids and cell surface receptor function. *Biochim. Biophys. Acta*, 415: 81-147

Nago S., Nagata K., Kohmura Y., Ishizuka T. and Nozawa Y. (1987) Redistribution of phospholipid/Ca^{2+}-dependent protein kinase in mast cells activated by various agonists. *Biochem. Biophys. Res. Commun.*, 142: 645-653

Nakamura T. and Ui M. (1985) Simultaneous inhibitions of inositol phospholipid breakdown, arachidonic acid release, and histamine secretion in mast cells by islet-activating protein, pertussis toxin. *J. Biol. Chem.*, 260: 3584-3593

Nishizuka Y. (1984) The role of protein kinase C in cell surface signal transduction and tumour promotion. *Nature*, 308: 693-698

Nishizuka Y. (1986) Studies and perspectives of protein kinase C. *Science*, 233: 305-312

O'Farrell P.H. (1975) High resolution two dimensional electrophoresis of proteins. *J. Biol. Chem.*, 250: 4007-4021

O'Flaherty J.T., Jacobson D.P., Redman J.F. and Rossi A.G. (1990) Translocation of protein kinase C in human polymorphonuclear neutrophils. *J. Biol. Chem.*, 265: 9146-9152

Ohmori K., Ishii H., Takei Y., Shuto K. and Nakamizo N. (1982a) Pharmacological studies of oxatomide. 3. Effect on experimental asthma and Schultz-Dale response in rats and guinea pigs. *Folia Pharmacol. Jpn.*, 80: 481-493

Ohmori K., Ishii H., Takei Y., Shuto K. and Nakamizo N. (1982b) Pharmacological studies of oxatomide. 4. Effect on the histamine release from rat isolated peritoneal mast cells (PEC) and lung slices. *Folia Pharmacol. Jpn.*, 80: 441-449

Ohmori T., Kikuchi A., Yamamoto K., Kawata M., Kondo J. and Takai Y. (1988) Identification of a platelet Mr 22,000 GTP-binding protein as the novel smg-21 gene product having the same putative effector domain as the ras gene products. *Biochem. Biophys. Res. Commun.*, 157: 670-676

Ohmori T., Kikuchi A., Yamamoto K., Kim S. and Takai Y. (1989) Small molecular weight GTP-binding proteins in human platelet membranes. Purification and characterization of a novel GTP-binding protein with a molecular weight of 22,000. *J. Biol. Chem.*, 264: 1877-1881

Ohsako S. and Deguchi T. (1983) Phosphatidic acid mimics the muscarinic action of acetylcholine in cultured bovine chromaffin cells. *FEBS Lett.*, 152: 62-66

Pang D.C. and Sperelakis N. (1983) Nifedipine, diltiazem, bepridil and verapamil uptakes into

165

cardiac and smooth muscles. *Eur. J. Pharmacol.*, 87: 199-207

Parker P.J., Coussens L., Totty N., Rhee L., Young S., Chen E., Stabel S., Waterfield M. D. and Ullrich A. (1986) The complete primary structure of protein kinase C - The major phorbol ester receptor. *Science*, 233: 853-859

Patel K.R. (1981) The effect of calcium antagonist, nifedipine in exercise-induced asthma. *Clin Allergy*, 11: 429-432

Peppers S.C. and Holz R.W. (1986) Catecholamine secretion from digitonin-treated PC12 cells: effects of Ca^{2+}, ATP and protein kinase C activators. *J. Biol. Chem.*, 261: 14665-14669

Pershadsingh H.A. and McDonald J.M. (1980) A high affinity calcium-stimulated magnesium-dependent adenosine triphosphatase in rat adipocyte plasma membrane. *J. Biol. Chem.*, 255: 4087-4093

Pocotte S.L., Frye R.A., Senter R.A., TerBush D.R., Less S.A. and Holz R.W. (1985) Effects of phorbol esters on catecholamine secretion and protein phosphorylation in adrenal medullary cell culture. *Proc. Natl. Acad. Sci. USA*, 82: 930-934

Pogolotti A.L. and Santi D.V. (1982) High-pressure liquid chromatography-ultraviolet analysis of intracellular nucleotides. *Anal. Biochem.*, 126: 335-345

Pointer R.H., Butcher F.R. and Fain J.N. (1976) Studies on the role of cyclic guanosine 3, 5-monophosphate and extracellular Ca^{2+} in the regulation of glycogenesis in rat liver cells. *J. Biol. Chem.*, 251: 2987-2992

Prentki M., Wollheim C.B. and Lew P.D. (1984) Ca^{2+} homeostasis in permeabilized human neutrophils. *J. Biol. Chem.*, 259: 13777-13782

Ritchie D.M., Sierchio J.N., Bishop C.M., Hedli C.C., Levinson S.L. and Capetola R.J. (1984) Evaluation of calcium entry blockers in several models of immediate hypersensitivity. *J. Pharmacol. Exp. Ther.*, 229: 690-695

Spat A., Fabiat A. and Rubin R.P. (1986) Binding of inositol trisphosphate by a liver microsomal fraction. *Biochem. J.*, 233: 929-932

Spearman T.N. and Butcher F.R. (1983) The effect of calmodulin antagonists on amylase release from the rat parotid gland in vitro. *Pflugers Arch.*, 397: 220-224

Sullivan T.J., Parker K.L., Eisen S.A. and Parker C.W. (1975) Modulation of cyclic AMP in purified rat mast cells. II. Studies on the relationship between intracellular cyclic AMP concentrations and histamine release. *J. Immunol.*, 114: 1480-1485

Tanaka T. and Hidaka H. (1980) Hydrophobic regions function in calmodulin-enzyme(s) interactions. *J. Biol. Chem.*, 255: 11078-11080

Tasaka K. (1986) Anti-allergic drugs. *Drugs Today*, 22: 101-133

Tasaka K., Mio M. and Okamoto M. (1986a) Intracellular calcium release induced by histamine releasers and its inhibition by some antiallergic drugs. *Ann. Allergy*, 56: 464-469

Tasaka K., Mio M. and Okamoto M. (1986b) Changes in intracellular Ca^{2+} distribution of rat peritoneal mast cells before and after histamine release. *Agents Actions*, 18: 61-64

Tasaka K., Akagi M. and Miyoshi K. (1986c) Distribution of actin filaments in rat mast cells and its role in histamine release. *Agents Actions*, 18: 49-52

Tasaka K., Akagi M., Mio M., Miyoshi K. and Nakaya N. (1987a) Inhibitory effect of oxatomide on intracellular Ca mobilization, Ca uptake and histamine release, using rat peritoneal mast cells. *Int. Arch. Allergy Appl. Immunol.*, 83: 348-353

Tasaka K., Mio M. and Okamoto M. (1987b) The role of intracellular Ca^{2+} in the degranulation of skinned mast cells. *Agents Actions*, 20: 157-160

Tasaka K., Akagi M., Miyoshi K. and Mio M. (1988) Role of microfilaments in the exocytosis of rat peritoneal mast cells. *Int. Arch. Allergy Appl. Immunol.*, 87: 213-221

Tasaka K. and Mio M. (1989) Microfilament-associated degranulation of sensitized guinea-pig lung mast cells. *Agents Actions*, 27: 79-82

Tasaka K., Sugimoto Y. and Mio M. (1990) Sequential analysis of histamine release and intracellular Ca^{2+} release from murine mast cells. *Int. Arch. Allergy Appl. Immunol.*, 91: 211-213

Tasaka K., Mio M., Fujisawa K. and Aoki I. (1991) Role of microtubules on Ca^{2+} release from the endoplasmic reticulum and associated histamine release from rat peritoneal mast cells. *Biochem. Pharmacol.*, 14: 1031-1037

Tatham P.E.R. and Gomperts B.D. (1989) ATP inhibits onset of exocytosis in permeabilized mast cells. *Biosci. Rep.*, 9: 99-109

Tsein R.Y., Pozzan T. and Rink T.J. (1982) Calcium homeostasis in intact lymphocytes: cytoplasmic free calcium monitored with a new, intracellularly trapped fluorescence indicator. *J. Cell Biol.*, 94: 325-334

Wakelam M.J.O., Davies S.A., Houslay M.D., McKay I., Marshall C.J. and Hall A. (1986) Normal p21(N-ras) couples bombesin and other growth factor receptors to inositol phosphate production. *Nature*, 322: 173-176

Wang P., Nishihata J., Takabori E., Yamamoto K., Toyoshima S. and Osawa T. (1989) Purification and partial amino acid sequences of a phospholipase C-associated GTP-binding protein from calf thymocytes. *J. Biochem.*, 105: 461-466

Wang T., Tsei L.I., Solaro J., Frassi de Gende A.O. and Schwartz A. (1979) Effects of potassium on vanadate inhibition of sarcoplasmic reticulum Ca^{2+}-ATPase from dog cardiac and rabbit skeletal muscle. *Biochem. Biophys. Res. Commun.*, 91: 356-361

Watson S.P., McConell R.T. and Lapetina E.G. (1984) The rapid formation of inositol phosphates in human platelets by thrombin is inhibited by prostacyclin. *J. Biol. Chem.*, 259: 13199-13203

Weiss G.B. (1974) Cellular pharmacology of lanthanum. *Ann. Rev. Pharmacol.*, 14: 343-354

White J.R., Ishizaka T., Ishizaka K. and Sha'afi R.I. (1984) Direct demonstration of increased intracellular concentration of free calcium as measured by quin-2 in stimulated rat peritoneal mast cell. *Proc. Natl. Acad. Sci. USA*, 81: 3978-3982

White J.R., Pluznik D.H., Ishizaka K. and Ishizaka T. (1985) Antigen-induced increase in protein kinase C activity in plasma membrane of mast cells. *Proc. Natl. Acad. Sci. USA*, 82: 8193-8197

White K.N. and Metzger H. (1988) Translocation of protein kinase C in rat basophilic leukemic cells induced by phorbol ester or by aggregation of IgE receptors. *J. Immunol.*, 141: 942-947

Wick S.M. and Hepler P.K. (1982) Selective localization of intracellular Ca^{2+} with potassium antimonate. *J. Histochem. Cytochem.*, 30: 1190-1204

Wilson S.P. and Kirshner N. (1983) Calcium-evoked secretion from digitonin-permeabilized adrenal medullary chromaffin cells. *J. Biol. Chem.*, 258: 4994-5000

Wolf B.A., Florholmen J., Colca J.R. and McDaniel M.L. (1987) GTP mobilization of Ca^{2+} from the endoplasmic reticulum of islets. *Biochem. J.*, 242: 137-141

Yamamoto T., Kaibuchi K., Mizuno T., Hiroyoshi M., Shirataki H. and Takai Y. (1990) Purification and characterization from bovine brain cytosol of proteins that regulate the GDP/GTP exchange reaction of smg p21s, ras p21-like GTP-binding proteins. *J. Biol. Chem.*, 265: 16626-16634

Yoshii N., Mio M. and Tasaka K. (1988) Ca uptake and Ca releasing properties of the endoplasmic reticulum in rat peritoneal mast cells. *Immunopharmacology*, 16: 107-113

Yoshii N., Mio M., Akagi M. and Tasaka K. (1991) Role of endoplasmic reticulum, an intracellular Ca^{2+} store, in histamine release from rat peritoneal mast cell. *Immunopharmacology*, 21: 13-22

Chapter 5

Role of cytoskeleton in the histamine release from mast cells

1. Role of microfilaments in the exocytosis of rat peritoneal mast cells

1.1. Introduction

It has been supposed by several investigators that cytoskeleton is involved in the exocytosis of rat mast cells, though conclusive evidence has not been submitted (Gillespie et al., 1968; Rölich, 1975). Recently, three-dimensional images of a subplasmalemmal network and granules surrounded by a small-meshed network in the same cell have been shown by Nielsen and Jahn (1984). In addition, the existence of actin filaments in the mast cells was reported by Tasaka et al. (1986a). In resting cells, actin immunofluorescence appeared as a net-like formation surrounding each granule. After stimulation with secretagogues, the distribution of the actin filaments became very irregular and disordered. In immunoelectron microscopy, patches of anti-actin immunogold particles were observed in the perigranular, nucleus and cell membranes. Curiously enough, a dense distribution of immunogold particles was often observed in the microvilli and on the surface of extruded granules (Tasaka et al., 1986a). This finding indicates that the actin located on the cell surface probably participates in exocytosis. In order to study more precisely the events leading to degranulation, this study was performed.

1.2. Fluorescence microscopy of mast cells

1.2.1. Mast cell labeling with tetramethylrhodamine-conjugated actin (TMR-actin)

When normal rat mast cells were incubated with tetramethylrhodamine-conjugated actin (TMR-actin) $(0.5 \mu g/ml)$, no fluorescence was observed. However, when mast cells were treated with compound 48/80 $(0.2 \mu g/ml)$ and the resulting degranulation were elicited, remarkable TMR-actin fluorescence was observed on the extruded granules and on the cell surface (Fig. 1). The fluorescence intensity was roughly proportional to the degree of degranulation. These findings indicate that F-actin or its fragments appear on the extruded granules and cell surface as the binding sites of G-actin in association with the process leading to degranulation. After stimulation with compound 48/80 $(0.2 \mu g/ml)$ at $25°C$ for 60 sec, the histamine released in the medium was $37.8 \pm 1.4\%$ (n = 5).

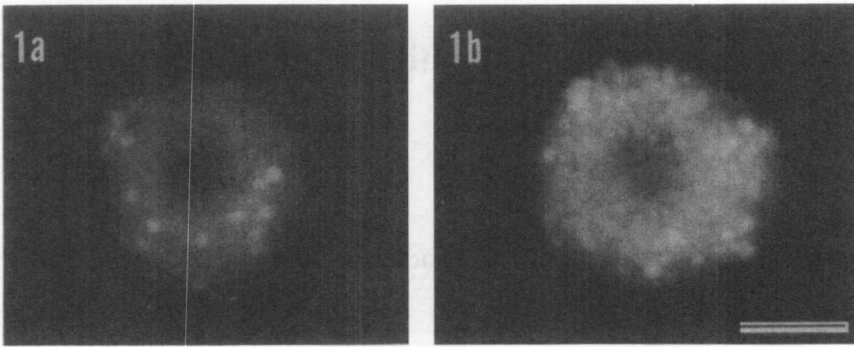

Fig. 1. Distribution of TMR-actin on rat peritoneal mast cells. Mast cells were exposed to 0.2 μg/ml of compound 48/80 for 30 sec at 25 °C. Intense fluorescence is seen on the extruded granules and on the cell surface in a and b. Bar = 10 μm. Reproduced from Tasaka, K., Akagi, M., Miyoshi, K. and Mio, M.: Int. Archs. Allergy Appl. Immunol., 87, 213-221 (1988), with permission.

1.2.2. Mast cell labeling with rhodamine-phalloidin

Perfused with β-escin (7.5 μg/ml) dissolved in a cytosol-like solution (in mM; KCl 110, NaCl 10, MgCl$_2$ 4, ATP 3, EGTA 0.1, Tris-HCl 5; pH 6.8), mast cells were permeabilized and subsequently rhodamine-phalloidin (1 μg/ml) was incorporated into the cytoplasm through the micropores. Fig. 2a shows the fluorescence image detected on the cell surface. The serpentine ridges appeared to anastomose all over the cell surface. SEM observations of microfolds on the rat mast cell surface with similar appearance have been reported (Kessler and Kuhn, 1975; Burwen and Satir, 1977). When the cell was perfused with a solution containing 0.1 mM of free Ca^{2+} for 15 sec (Fig. 2b), a network formation stained with rhodamine-phalloidin appeared just beneath of the cell surface. The meshes seemed to be composed of actin filaments and were similar to those of the subplasmalemmal network seen by SEM (Nielsen and Jahn, 1984), but somewhat rougher. In some cases, circular fluorescence structures were observed on the cell surface and were considered to be the pore through which the granules are extruded (Fig. 2c). Perfusion with Ca^{2+} (0.1 mM) for 15 sec, induced the release of 22.5 ± 1.3% of total histamine (n = 5).

1.3. SEM appearance of polymerized actin on the mast cell surface

Fig. 3a shows an SEM image of the compound 48/80-stimulated mast cell surface. Many extruded granules were observed on the cell surface, but only few filaments were seen. However, with the addition of G-actin (3 μg/ml) to the cells suspension prior to that of compound 48/80, numerous filaments which formed a complicated dense network on the cell surface became apparent (Fig. 3b). SEM revealed that both the Y-shaped

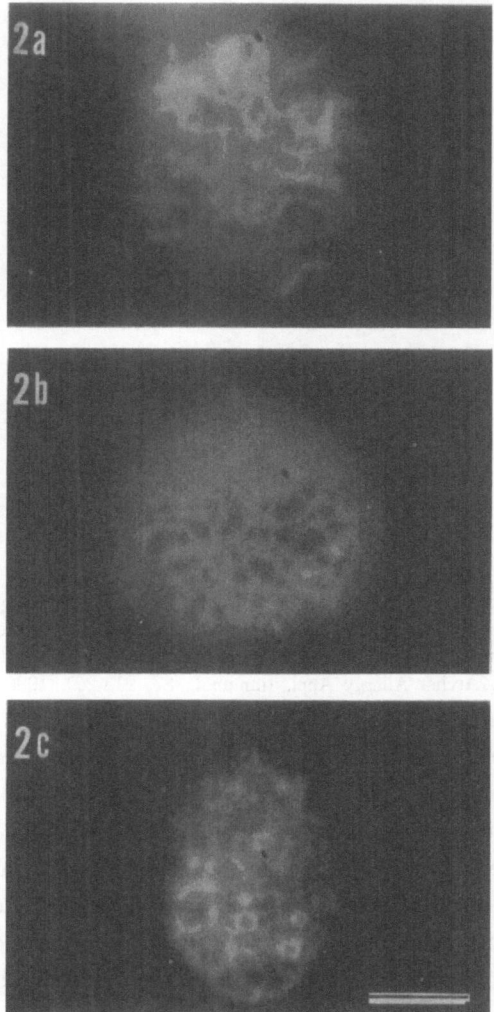

Fig. 2. Fluorescence images of the cell surfaces of permeabilized mast cells stained with rhodamine-phalloidin. The video micrographs of fluorescent images of mast cells were taken from a video cassette tape recorder (Tasaka et al., 1988). a. Mast cell before Ca^{2+} perfusion. b. Mast cell perfusion with 0.1 mM of free Ca^{2+} for 15 sec. c. Mast cell perfused with 0.1 mM of free Ca^{2+} for 30 sec. Bar $= 5 \mu$m. Reproduced from Tasaka, K., Akagi, M., Miyoshi, K. and Mio, M.: Int. Archs. Allergy Appl. Immunol., 87, 213-221 (1988), with permission.

'branches' of actin filaments and the connecting filaments run between the extruded granules and the cell surface. These findings indicate that actin filaments are exposed on the cell surface according to the progress of exocytotic process.

A granule surrounded by dense filaments protruding from the cell surface can be

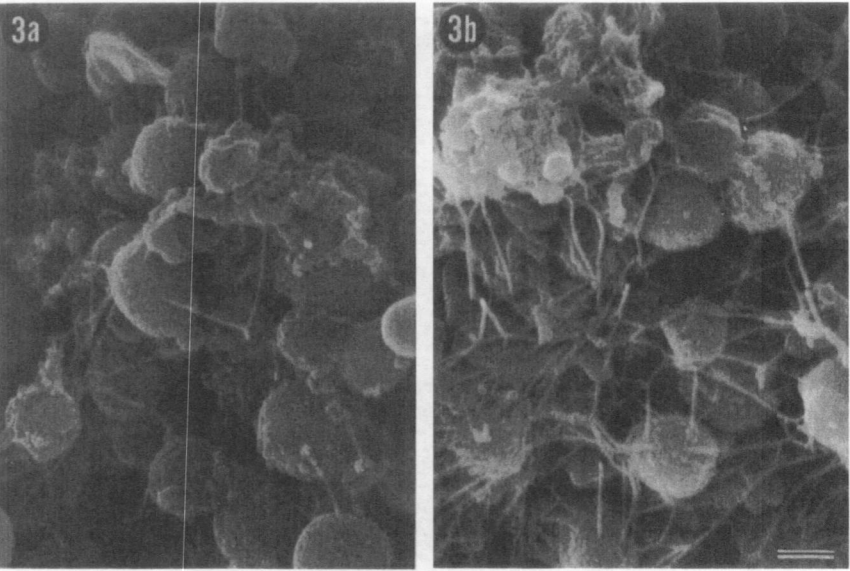

Fig. 3. Appearance of actin filaments in mast cell stimulated with 0.2 μg/ml of compound 48/80 in the absence (a) or presence (b) of exogenous G-actin (3 μg/ml). The comparison of the two figures clearly shows an increase in actin filaments. x 55,250. Bar = 200 nm. Reproduced from Tasaka, K., Akagi, M., Miyoshi, K. and Mio, M.: Int. Archs. Allergy Appl. Immunol., 87, 213-221 (1988), with permission.

observed in Fig. 4. As shown in Fig. 4a, nearly half of the granule which has passed through the cell membrane is exposed to the medium. However, the rest is restrained by many filaments. In Fig. 4b, the granule is definitely out of the cell but the lower part facing the cell surface is connected to many filaments. On the surface of the extruded granule, small particles can be seen. These particles are thought to be the fragments of disrupted microfilaments. Moreover, in Fig. 5., circular filaments can be observed in the periphery of the pore through which the granule seems to have been extruded. These observations may support the assumption that the filaments play a role in inducing degranulation by squeezing the granules out of the cell.

1.4. Whole-mount preparation

Fig. 6a and b show a TEM image of the dense intracellular networks surrounding the granules in normal mast cells. The filaments, having various widths, seem to consist of closely interwoven microfilaments, microtubules and intermediate filaments. Some microfilaments were connected side to side with microtubules or intermediate filaments; others formed end-to-side connections with different types of filaments. Y-shaped branches were also observed. It seems likely that these filaments are related not only to maintaining the cell shape as a whole but also to holding the secretory granules in position.

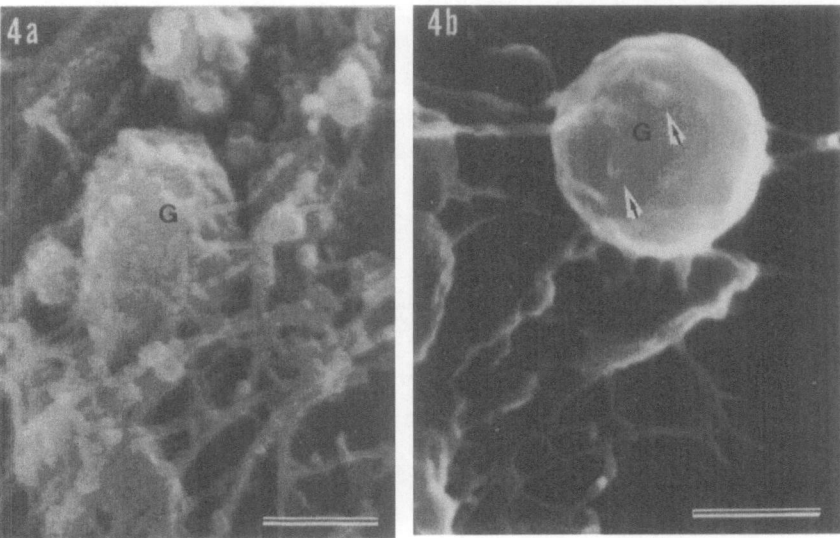

Fig. 4. SEM views of whole-mount preparations of mast cells stimulated with 0.2 μg/ml of compound 48/
80. The protruding granules (G) are surrounded by masses of microfilaments. a. On the protruding surface,
no connection with filaments is seen. b. Small particles (arrows) are seen on the extruded granule.
x 114,700. Bars = 200 nm. Reproduced from Tasaka, K., Akagi, M., Miyoshi, K. and Mio, M.: Int.
Archs. Allergy Appl. Immunol., 87, 213-221 (1988), with permission.

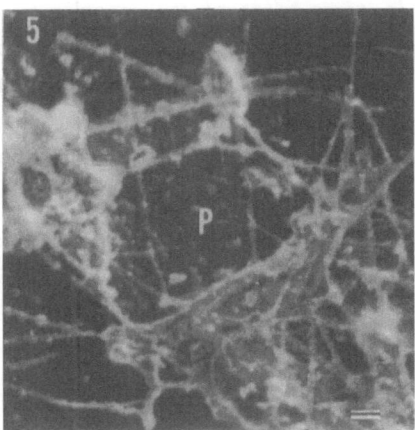

Fig. 5. SEM view of whole-mount preparations of mast cells stimulated with 0.2 μg/ml of compound 48/
80. A circular microfilament structure through which the granule has been discharged is seen. P = pore.
x 22,500. Bar = 200 nm. Reproduced from Tasaka, K., Akagi, M., Miyoshi, K. and Mio, M.: Int.
Archs. Allergy Appl. Immunol., 87, 213-221 (1988), with permission.

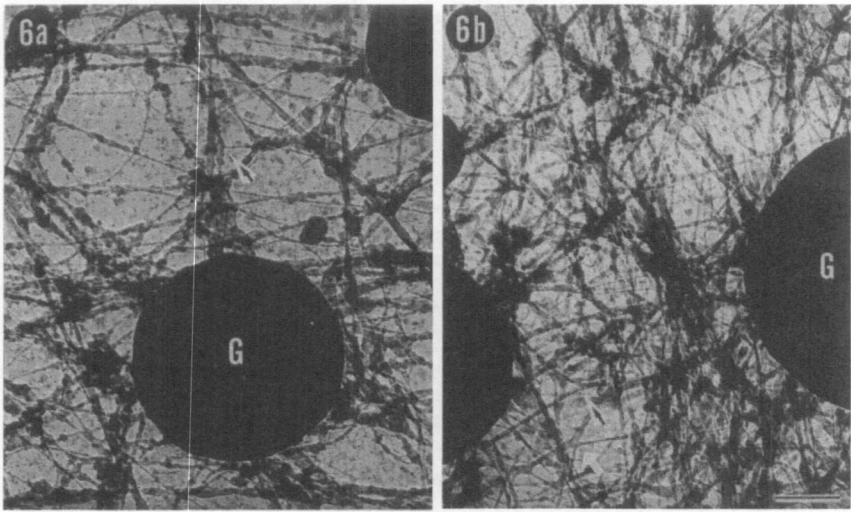

Fig. 6. Whole-mount preparations of normal, unstimulated mast cells viewed with a TEM. Many types of intertwined filaments form the complicated network. G = Secretory granule. a. End-to-side contact of microfilament to microtubule (arrow). x 57,000. b. Lateral contact (white arrow) and Y-shaped branch (black arrow). x 61,200. Bar = 200 nm. Reproduced from Tasaka, K., Akagi, M., Miyoshi, K. and Mio, M.: Int. Archs. Allergy Appl. Immunol., 87, 213-221 (1988), with permission.

Fig. 7a shows whole-mount preparations of 48/80-stimulated mast cells (0.2μg/ml). The extruded granules observed in the periphery of the cell were connected to the cell surface by thin filaments. In some areas of the cytoplasm, the distribution of thin filaments was less dense, indicating that disruption of microfilaments had taken place. Under high magnification (Fig. 7b), several filaments were seen to connect the extruded granules to the cell surface, and a few of them were disrupted in the middle. These findings seem to be related to those of the SEM preparations and may indicate that the connecting filaments are necessary for the extrusion of granules (Fig. 4b). Fig. 8 shows a granule which is about to detach from the cell surface. Many filaments were assembled very densely in the cytoplasmic area around the bottom of the protruding granule, but were not connected with the outer surface of the extruded granule. Such a dense distribution of microfilaments at the bottom of the granule may be useful to push the granule out of the cell.

1.5. Changes in actin contents before and after treatment with compound 48/80

As shown in Fig. 9, SDS-PAGE analysis revealed that a 43 kDa protein, which was supposed to be actin, significantly increased in the plasma membrane fraction within 1 min after exposure to compound 48/80 (0.2μg/ml). Immunoblotting provided confirmation that the 43

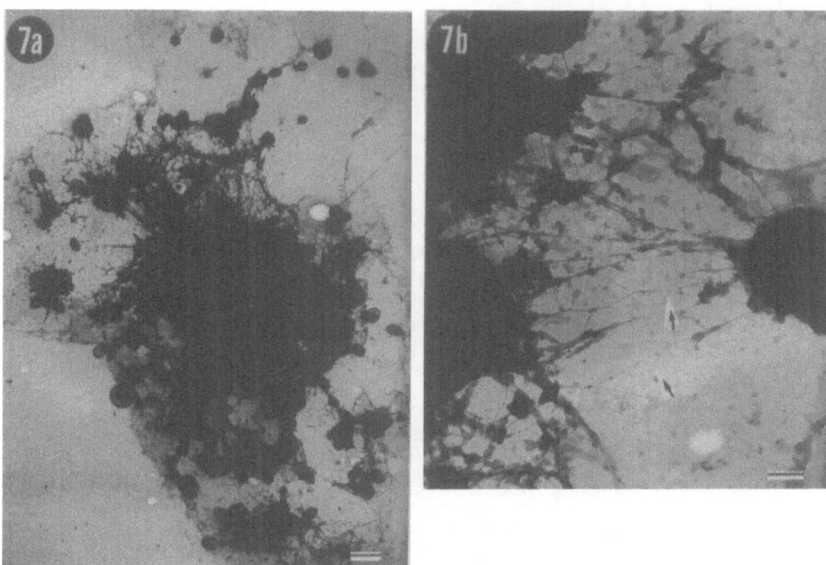

Fig. 7. Whole-mount preparations of mast cells stimulated with 0.2 μg/ml of compound 48/80, viewed in TEM. a. Whole cell image. Extruded granules in contact with filaments are seen at the periphery of the cell. b. The extruded granule is still connected to the cell surface by microfilaments. A few filaments are disrupted in the middle (arrows). x 26,860. Bar = 200 nm. Reproduced from Tasaka, K., Akagi, M., Miyoshi, K. and Mio, M.: Int. Archs. Allergy Appl. Immunol., **87**, 213-221 (1988), with permission.

kDa protein is actually actin. A slight increase in the actin protein of the granule fraction was observed and the amount of actin decreased in the cytosol fraction. Since G-actin probably binds firmly to the fragmented F-actin appearing on the cell membranes, this finding seems to indicate that the actin content increased in the plasma membrane as a consequence of the outward mobilization of actin filaments associated with degranulation.

Treatment with compound 48/80 caused a remarkable decrease in the amount of Triton X-100 insoluble (Triton-insoluble) fractions and a correspondent increase in the Triton-soluble fractions (Fig. 10). Since the F-actin in the cytoskeleton is thought to be insoluble to Triton X-100, these results suggest that treatment with compound 48/80 decreased the F-actin content in the cytoskeleton of mast cells. Concomitantly, the Triton-soluble fractions considered to be G-actin increased. It seems that with degranulation in rat mast cells there is a simultaneous disruption of F-actin in the cytoplasm (Fig. 7a).

1.6. Discussion

It was confirmed that the cytoskeleton of the rat mast cell participates in the process leading to exocytosis. The existence of actin filaments in unstimulated mast cells has been reported previously (Tasaka et al., 1986a). Since phalloidin cannot pass through the cell

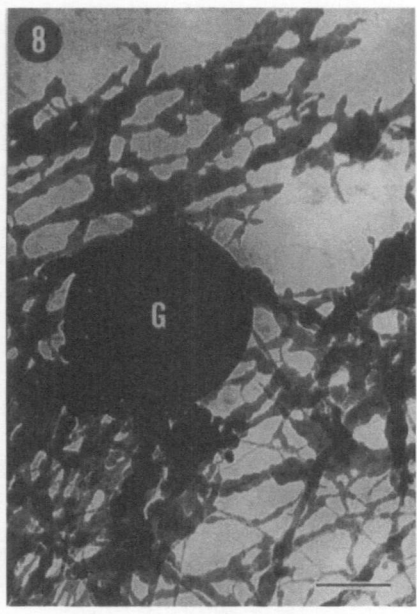

Fig. 8. Whole-mount preparation of mast cells stimulated with $0.2\,\mu g/ml$ of compound 48/80. On the cytoplasmic side (at the bottom of the granule, G), the filaments are gathered densely as if they are pushing the granule out. x 59,400. Bar = 200 nm. Reproduced from Tasaka, K., Akagi, M., Miyoshi, K. and Mio, M.: Int. Archs. Allergy Appl. Immunol., 87, 213-221 (1988), with permission.

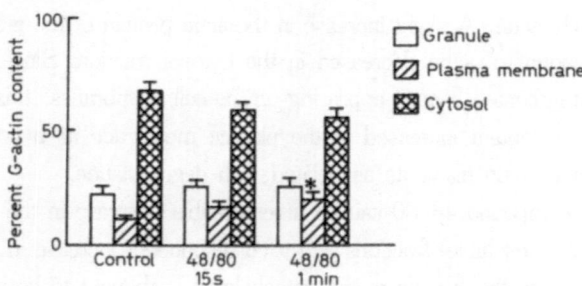

Fig. 9. Percent changes in 43 kDa protein (G-actin) contents in three fractions obtained from rat peritoneal mast cells before and after stimulation with compound 48/80 $(0.2\,\mu g/ml)$. n = 6, * p < 0.05. Reproduced from Tasaka, K., Akagi, M., Miyoshi, K. and Mio, M.: Int. Archs. Allergy Appl. Immunol., 87, 213-221 (1988), with permission.

membrane, permeabilization is prerequisite to incorporate phalloidin into the mast cell without fixation (Barak et al., 1981). It has been reported that Ca^{2+} perfusion into permeabilized mast cells is an efficient way to induce degranulation (Tasaka et al., 1987). Because phalloidin binds very selectively to F-actin, its fluorescence image, showing

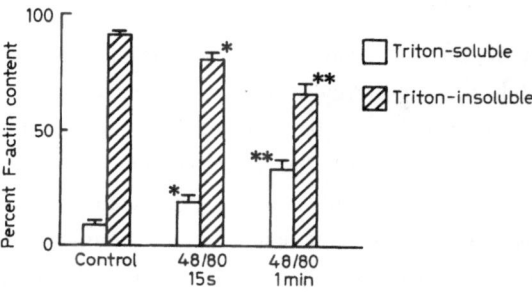

Fig. 10. Percent changes in 43 kDa protein (F-actin) contents extracted from rat peritoneal mast cells before and after stimulation with compound 48/80 (0.2 μg/ml). n = 6. * p < 0.05; ** p < 0.01. Reproduced from Tasaka, K., Akagi, M., Miyoshi, K. and Mio, M.: Int. Archs. Allergy Appl. Immunol., 87, 213-221 (1988), with permission.

serpentine ridges on the cell surface, indicates that actin filaments exist in the cell membrane of just beneath it. Using SEM, folds in the plasma membrane of rat mast cell, which are similar in appearance to the ridges, have been observed and reported repeatedly (Kessler and Kuhn, 1975; Burwen and Satir, 1977). Also, it has been shown very clearly that dense patches of anti-actin immunogold particles exist in the microvilli of rat peritoneal mast cells exposed to compound 48/80 (Tasaka et al., 1986a). Since the localization of the particles coincides with that of actin and the folds of the cell surface in the SEM image correspond to the microvilli in that of TEM, it seems reasonable to assume that these two structures are identical. Moreover, rhodamine-phalloidin fluorescence observed on the surface of stimulated mast cells appears to surround the degranulation pores or cover the surface of extruded granules (Fig. 1a, 2c) and the extruded granules, stained with TMR-actin, emitted strong fluorescence. Since G-actin has a high affinity for binding actin fragments, these observations indicate that actin filaments (or their fragments) become exposed on the cell surface and on the extruded granules. The evidence yielded by fluorescence microscopy seems to coincide with that of electron microscopy (Fig. 3). In association with this, it has been reported that chromaffin granule membranes isolated from bovine adrenal medulla contain a polypeptide with the same electrophoretic mobility as that observed in the actin on SDS-PAGE (Burridge and Phillips, 1975). The presence of actin in the latter preparation has been confirmed by the antibody precipitation technique (Meyer and Burger, 1979).

Electron microscopy showed that the extruded granules located near the cell surface are connected to it by many filaments (Fig. 4, 7, 8) and some of these filaments may be actin filaments as shown in Fig. 3. Similar observations were made with whole-mount preparations (Fig. 7, 8). Clearly, these results indicate that immediately after degranulation the granules are still connected to the cell surface. Thereafter, the filaments are disrupted (Fig. 7b) and, consequently, the granules will be dissociated from the cell surface. In accordance with these morphological changes, when rat mast cells were exposed to compound 48/80, the actin content significantly increased in the plasma membrane fraction, while that in the cytosol fraction decreased. It appears that after stimulation with compound 48/80, F-actin filaments decompose and move toward the plasma membrane. As a consequence, the actin content in the ˙plasma membrane increases (Fig. 9) while the F-actin content in the cytoplasm decreases.

In the whole-mount preparations, filaments formed complicated networks. These networks may be necessary, on the one hand, to maintain the cytoplasmic consistency and, on the other hand, to facilitate some cellular activities such as locomotion and ruffling. Granules were interconnected by these filaments and trapped in the networks. It has been reported that calcium-triggered actomyocin interactions may be useful in pushing secretory granules toward the plasma membrane (Durham, 1974). Chromaffin granule membrane-actin interactions in bovine adrenal glands are inhibited by increasing free calcium ion concentrations up to those corresponding to the activation of exocytosis, and the association of actin filaments with membrane vesicles may be significant in the maintenance of cytoplasmic structure and consistency. The Ca^{2+}-dependent dissociation of the chromaffin granules from a network of cytoskeletal actin filaments may be necessary for their movement toward the plasma membrane during exocytosis (Fowler and Pollard, 1982). Although it can be assumed that a similar mechanism regulates both the association and dissociation of the mast cell granule membrane and the surrounding actin filaments, it is necessary to elucidate the crucial event inducing granule dissociation prior to histamine release. The dense wrapping of filaments which covered the granules in the resting state (Fig. 6) was disconnected during the final stage of degranulation on the cell surface (Fig. 7b). Moreover, a concentrated mass of filaments was seen at the degranulation site (Fig. 4, 8) and, prior to degranulation, the filaments located close to the cell surface were in a circular configuration, surrounding the periphery of the granules. This structure may be effective in pushing or squeezing the granules out of the cell. Such an organization of filaments does not simply depend on the Ca^{2+} concentration in the cytoplasm. It is assumed that factors other than the presence of Ca^{2+} participate in the unidirectional movement of granules toward the cell membrane and the circular configuration of (actin) filaments of the cell surface.

2. Microfilament-associated degranulation

2.1. Introduction

It has been shown in electron microscopic observation that complicated filamentous networks exist in the mast cells connecting individual granules, and connecting the granules to the cell membrane (Nielsen and Jahn, 1984; Tasaka et al., 1988). In immunoelectron microscopy, the complex distribution of actin filaments in the mast cells has also been shown (Tasaka et al., 1986a). These observations suggest that cytos-keletons play some roles in the histamine release (Nielsen and Jahn, 1984; Tasaka et al., 1988; Orr et al., 1972; Gillespie et al., 1968). Actually, it has been shown that pretreatment with colchicine or cytochalasins inhibits histamine release from mast cells (Orr et al., 1972; Gillespie et al., 1968). Previously, we reported that F-actin content in the membrane fractions of the mast cell increased in association with histamine release, suggesting that a polymerization of G actin may occur during the degranulation process (Tasaka et al., 1988). Although, there has been some evidence that cytoskeletal elements may participate in expelling the granules out of the cell (Tasaka et al., 1988), the whole process of the degranulation was not followed in detail. It was the purpose of this study to investigate the movement of actin filaments in association with degranulation.

2.2. Microfilament-associated local degranulation of rat peritoneal mast cells

When compound 48/80 was applied locally to the mast cell surface via a microelectrode filled with $100\,\mu\text{g/ml}$ of compound 48/80, localized extrusion of granules and the filaments were seen close to the site of stimulation within a few minutes (Fig. 11). The

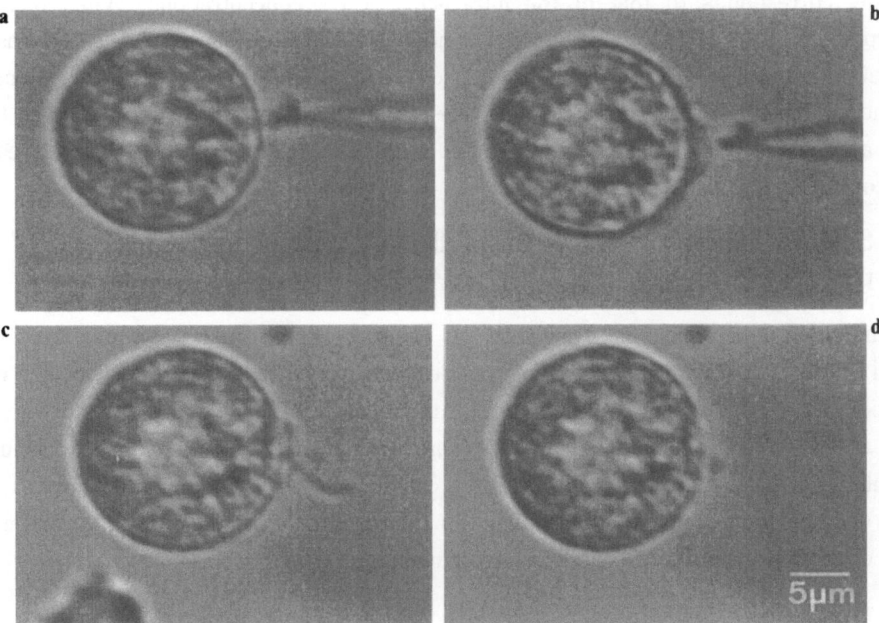

Fig. 11. Local degranulation and its repair. Extrusion of the microfilaments from a rat peritoneal mast cell was induced simultaneously with localized protrusion of the granules after local application of compound 48/80. a. Before application. b. Initiation of microfilament extrusion. c. Elongated filaments are observed. d. Extruded filaments are completely reincorporated. Reproduced from Mio, M. and Tasaka, K.: Int. Archs. Allergy Appl. Immunol., 88, 369-371 (1989), with permission.

extrusion of the granule was intimately related to the elongation of the filaments. The extruded granules were connected to the cell surface via thin filaments, and the filaments were elongated rather radially. In some instances, the length of the filaments were as long as $5\,\mu$m. When the stimulation ceased, the filament became gradually shorter, and finally the extruded filaments were reincorporated into the cell with attached granules, indicating that a repair process occurs after degranulation (a post-degranulation repair process).

2.3. Microfilament-associated degranulation of sensitized guinea-pig lung mast cells

When sensitized guinea-pig lung mast cells were stimulated with antigen ($10\,\mu$g/ml), degranulation was observed within 2-3 min. The extruded granules were connected to the cell surface by thin filaments, as shown in Fig. 12a. In some cases, the filaments projected to a length of $15\,\mu$m. Some of the extruded granules were much larger than those seen in cytoplasm, indicating that they became swollen in the medium. Thereafter, the elongated filaments became progressively shorter and, within 7-8 min, the extruded granule were reincorporated into the cytoplasm. This phenomenon may represent a post-degranulation repair process. The time course of such morphological changes roughly corresponds to that of the intracellular Ca^{2+} concentrations. When quin 2-incorporated mast cells were exposed to antigen, the fluorescence intensity increased, indicating an increase in Ca^{2+} concentration (Fig. 12b). The fluorescence intensity reached a maximum approximately 2 min after addition of antigen, having increased to a level approximately 10 times higher than that of the control, and gradually diminished to the control level 7-8 min later.

2.4. Selective binding of rhodamine-phalloidin to the extruded filaments of rat peritoneal mast cells

When $3\,\mu$g/ml of rhodamine-phalloidin, an F-actin specific fluorescent dye, was perfused in the microchamber containing rat peritoneal mast cells stimulated locally with compound 48/80, the site corresponding to the local degranulation was stained with the dye, as shown in Fig. 13. Intense fluorescence was observed on the extruded granules, the filaments connecting the granules to the cell surface, and the limited area where the local degranulation took place. However, no fluorescence was observed in normal mast cells exposed to the fluorescent dye for 15 min at 37°C.

2.5. Discussion

From these experiments, it became evident that the filaments connecting the extruded granules to the cell surface consist of actin filaments. This seems to indicate that degranulation takes place in association with the outward elongation of actin filaments. Microfilaments probably reinforce to expel the granules out of the cell (Tasaka et al.,

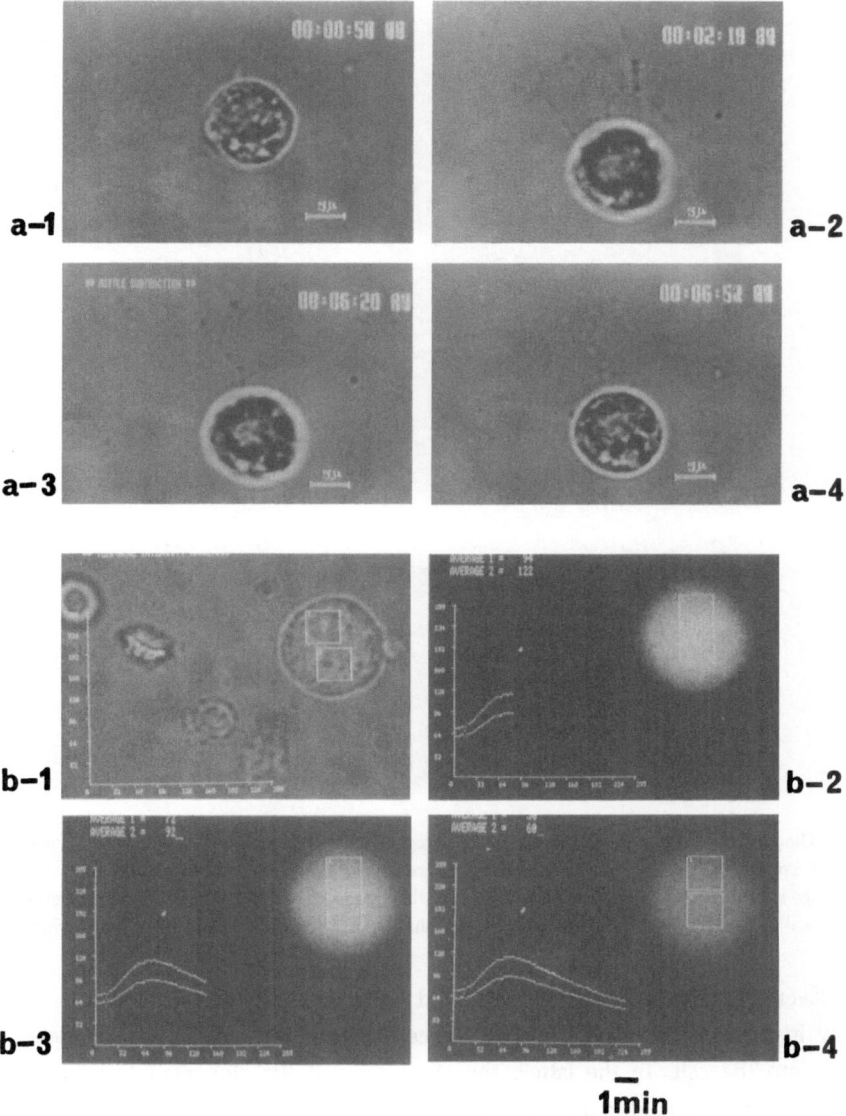

Fig. 12. Morphological changes in sensitized guinea pig lung mast cells and sequential changes in intracellular Ca^{2+} concentrations after exposure to antigen. (a) Changes in the length of extruded microfilaments to granules (1) 50 sec, (2) 2 min 18 sec, (3) 6 min 20 sec, and (4) 6 min 52 sec, after addition of antigen to the medium. (b) Changes in fluorescence intensity derived from the quin 2-Ca complex. The target areas for fluorescence measurement are indicated as two squares on the cell surface. The fluorescence intensity in each area is integrated and expressed as the Y-axis. The X-axis represents the number of video-frames. The scale is in minutes. (1) Before, (2) 2 min 18 sec, (3) 4 min 20 sec, and (4) 7 min after antigen exposure. Reproduced from Tasaka, K. and Mio, M.: Agents Actions, 27, 79-82 (1989), with permission.

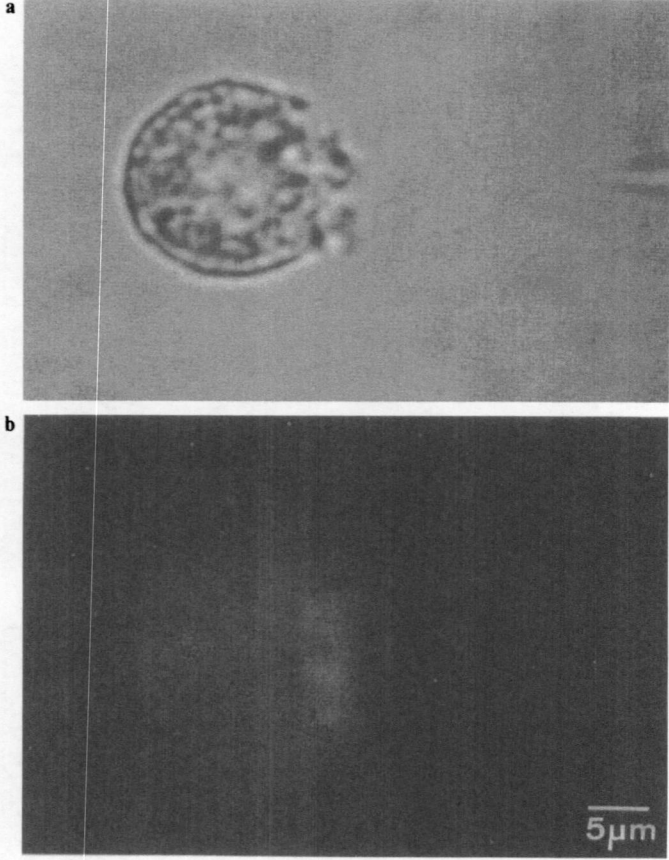

Fig. 13. Bindings of rhodamine-phalloidin at the sites of local degranulation of rat peritoneal mast cells induced by compound 48/80. a. Phase-contrast microscopy: extruded granules are observed locally. b. Fluorescence microscopy: fluorescence is observed in the area of local degranulation. Reproduced from Mio, M. and Tasaka, K.: Int. Archs. Allergy Appl. Immunol., 88, 369-371 (1989), with permission.

1988). Even after the granules are dislocated from the cell, the actin filaments are active not only in keeping the granules near the cell surface but also in the reuptake of the granules into the cell. In the latter, the shortening of the elongated filaments may be caused by either the contraction or depolymerization of actin filaments inside the cell.

As indicated in the degranulation of sensitized guinea pig lung mast cells induced by antigen exposure, the time courses of these morphological changes and of the intracellular Ca^{2+} concentrations corresponded. It appeared that the projection created by the outward movement of microfilaments may be related to an increased intracellular Ca^{2+} concentration. Although it has been strongly proposed that an increase of the intracellular concentration of Ca^{2+} is an essential trigger for the histamine release (Tasaka et al.,

1986b), neither the site of action of Ca^{2+}, nor how Ca^{2+} activates the mast cell so as to induce histamine release and degranulation has been reported. It is known that actomyosin requires Ca^{2+} for the contraction (Korn, 1978) and that the length of the actin filaments is regulated by actin-regulating proteins such as calmodulin and cytocalbins (Yazawa et al., 1987) in the presence of Ca^{2+}. Moreover, it is also known that Ca^{2+} plays an important role in regulating the functions of various cytoskeletal proteins such as actin-binding and microtubule-associated proteins. In the cases of macrophages and platelets, it has been shown that gelsolin is a Ca^{2+}-sensitive regulatory factor of conformational changes in microfilaments, especially as it decreases actin filament lengths (Hartwig and Stossel, 1982; Tellam and Friden, 1982). Its shortening effect on the actin filament is very rapid and reversible. By contrast, platelets develop numerous thin projections during the formation of blood clots and these projections contain a larger number of actin filaments that quickly form from a pool of unpolymerized actin. Although it is not clear at the present time how Ca^{2+} activates the actin filaments in the mast cells, the actin filaments might be a most pertinent factor in inducing degranulation in the presence of proper Ca^{2+} concentration.

Fig. 14 shows a putative scheme for microfilament-associated degranulation of mast cell (Tasaka et al., 1989). The granules were surrounded by very densely with microfilaments and microtubules of various widths. When intracellular Ca^{2+} concentration was raised with the proper stimuli, microfilaments located near the periphery may pull the granules close to the plasma membrane. When the granule attached to the cell membrane,

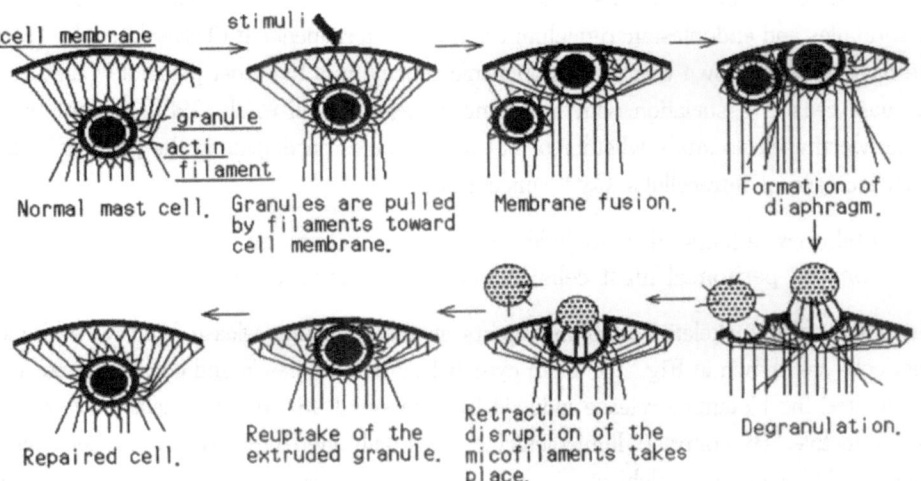

Fig. 14. Putative scheme for the microfilament-associated degranulation. Reproduced from Tasaka K., Mio M. and Akagi M.: In 'Bioinformatics', (eds.) O. Hatase and J. H. Wang, Elsevier, Amsterdam, 1989, 195-203, with permission.

fusion between the cellular and granular membranes may take place. Subsequently, a "diaphragm" was formed at the fusion site. When the diaphragm was disrupted, the granule was expelled out of the cell (exocytosis) in association with concomitant movement of actin filaments. Elongation of actin filaments may be induced by polymerization of G-actin in accordance with the degranulation process. On some occasion, disruption of the distended filaments occurs and a resulting degranulation takes place. After degranulation, shortening of the actin filaments takes place from time to time and finally, the extruded granules are reincorporated into the cell. In the cytoplasm, not only the retraction but also depolymerization of F-actin may occur, resulting in the shortening of filaments.

3. Role of microtubules on Ca^{2+} release from the endoplasmic reticulum and associated histamine release from rat peritoneal mast cells

3.1. Introduction

So far, it has been repeatedly mentioned that microfilaments in the mast cell play a leading role in the process of exocytosis (Tasaka et al., 1988; Tasaka and Mio, 1989; Mio and Tasaka, 1989). By contrast, the function of microtubules in histamine release is not recognized as clearly as in the case of microfilaments. In addition, although it is known that an increase in intracellular Ca^{2+} concentration is prerequisite for the histamine release from mast cells, the role of cytoskeleton in association with the increase in intracellular Ca^{2+} concentrations is not known. It has been shown that the structures of microtubules and endoplasmic reticulum are highly interdependent (Terasaki et al., 1986). We have already shown that the endoplasmic reticulum is the most plausible Ca store in the mast cells in association with histamine release (Yoshii et al., 1988). Therefore, it seems worthwhile to study whether or not microtubules participate in the process leading to an increase of intracellular Ca^{2+} concentration.

3.2. Inhibitory effects of cytoskeleton-inhibiting agents on the histamine release from rat peritoneal mast cells induced by compound 48/80

The effects of cytoskeleton-inhibiting agents on the histamine release from rat peritoneal mast cells are shown in Fig. 15. Both cytochalasin D, vinblastin and colchicine effectively inhibited the histamine release induced by compound 48/80; cytochalasin D was the most effective. By contrast, lumicolchicine was totally ineffective even at higher concentrations. Although lumicolchicine is an analog of colchicine, it is effective in neither interfering with the polymerization of the microtubules nor in affecting the function of the microtubules. Consequently, it was assumed that the inhibitory effect of colchicine may be exerted by inhibiting the polymerization of microtubules.

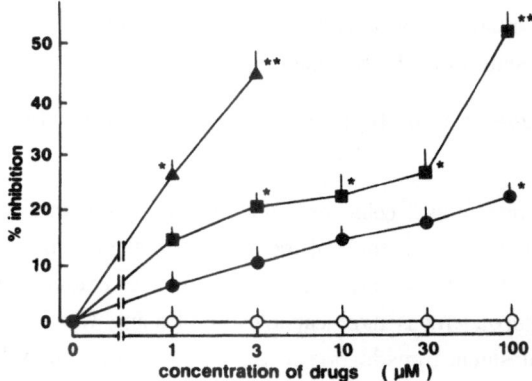

Fig. 15. Effects of cytoskeleton-inhibiting agents on the histamine release from rat peritoneal mast cells induced by compound 48/80 (0.5 μg/ml). (●) colchicine, (■) vinblastin, (○) lumicolchicine, (▲) cytochalasin D. * and ** represent $p < 0.05$ and $p < 0.01$, respectively. Each point represents the mean ± SEM (n = 5). Histamine release in the control experiments was 71.4 ± 1.9% (n = 9). Reproduced from Tasaka, K., Mio, M., Fujisawa, K. and Aoki, I.: Biochem. Pharmacol., 41, 1031-1037 (1991a), with permission.

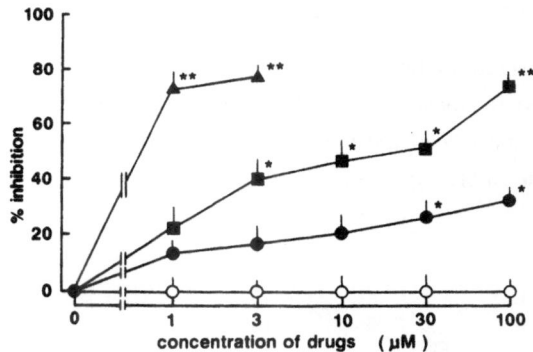

Fig. 16. Effects of cytoskeleton-inhibiting agents on the ^{45}Ca uptake of rat peritoneal mast cells induced by compound 48/80 (0.5 μg/ml). (●) colchicine, (■) vinblastin, (○) lumicolchicine, (▲) cytochalasin D. * and ** represent $p < 0.05$ and $p < 0.01$, respectively. Each point represents the mean ± SEM (n = 5). Reproduced from Tasaka, K., Mio, M., Fujisawa, K. and Aoki, I.: Biochem. Pharmacol., 41, 1031-1037 (1991a), with permission.

3.3. Effects of cytoskeleton-inhibiting agents on the ^{45}Ca uptake of rat peritoneal mast cells induced by compound 48/80

In order to study the inhibitory mechanism of cytoskeleton-inhibiting agents on histamine release from mast cells, the effects of these compounds on ^{45}Ca uptake were investigated.

As shown in Fig. 16, the ^{45}Ca uptake into rat mast cells elicited by compound 48/80 was dose-dependently inhibited by separate pretreatments of cytochalasin D, vinblastin and colchicine, at concentrations higher than $1\,\mu$M.

3.4. Effect of cytoskeleton-inhibiting agents on the IP$_3$ formation of rat peritoneal mast cells

It is known that when mast cells are stimulated by histamine releasers, phosphatidylinositol (PI) breakdown in the cell membrane is elicited so as to liberate inositol 1, 4, 5-trisphosphate (IP$_3$) at the early stage of the cell activation. Since IP$_3$ acts as a phospholipid-derived second messenger capable of releasing Ca^{2+} from intracellular Ca store, the activation of signal transduction in the cell membrane may be reflected by the IP$_3$ formation. In order to study the influence of cytoskeletons on the IP$_3$ formation, rat mast cells were stimulated with compound 48/80. In the resting stage, the IP$_3$ content of mast cells was 0.12 ± 0.05 pmol/10^6 cells. However, when the cells were stimulated with compound 48/80 for 5 sec, the IP$_3$ content markedly increased to 3.78 ± 0.07 pmol/10^6 cells, which is more than a 30-fold increase compared to the control level. Neither of the cytoskeleton-inhibiting agents tested was effective in inhibiting the IP$_3$ formation (Fig. 17).

3.5. Changes in intracellular Ca^{2+} concentration of mast cells

The effects of cytoskeleton-inhibiting agents on the Ca^{2+} release from the intracellular Ca store were determined by means of a video-intensified microscopy system. When the quin 2 ($5\,\mu$M)-loaded mast cells were stimulated with compound 48/80 ($0.35\,\mu$g/ml) in a Ca-free medium, the fluorescence intensity of the cell increased promptly and reached

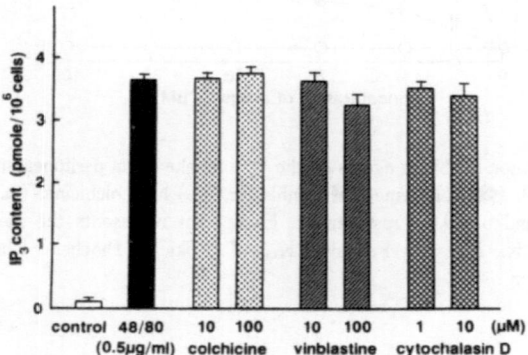

Fig. 17. Effects of cytoskeleton-inhibiting agents on the IP$_3$ formation of rat peritoneal mast cells induced by compound 48/80 ($0.5\,\mu$g/ml). Each column represents the mean $\pm$ SEM (n $=$ 5). Reproduced from Tasaka, K., Mio, M., Fujisawa, K. and Aoki, I.: Biochem. Pharmacol., 41, 1031-1037 (1991a), with permission.

Table 1. Inhibitory effects of cytoskeleton-inhibiting agents on the Ca^{2+} release from intracellular Ca store induced by compound 48/80 (0.35 μg/ml)

Compounds (μM)	Fluorescence intensity (arbitrary units)	% inhibition
control	41.2 ± 2.2	—
Cytochalasin D		
1	40.3 ± 1.3	2.18
10	39.6 ± 1.7	3.88
Colchicine		
1	$20.3 \pm 1.8^*$	50.73
10	$14.7 \pm 1.4^*$	64.32
100	$7.6 \pm 1.5^*$	81.55
Vinblastin		
1	$25.7 \pm 1.6^*$	37.62
10	$13.9 \pm 2.9^*$	66.26
100	$6.4 \pm 0.8^*$	84.47

The fluorescence intensity of quin 2-loaded mast cells stimulated with compound 48/80 were measured at 7 sec after stimulation. Each value represents the mean $\pm$ SEM (n = 50). * indicates statistical significance in $p < 0.01$. Reproduced from Tasaka, K., Mio, M., Fujisawa, K. and Aoki, I.: Biochem. Pharmacol., 41, 1031-1037 (1991a), with permission.

maximum level within a few seconds, indicating that Ca^{2+} was released from the intracellular Ca store. Thereafter, the fluorescence intensity gradually decreased to the control level within 30 sec after stimulation. The inhibitory effects of cytoskeleton-inhibiting agents on the Ca^{2+} release from the intracellular Ca store are summarized in Table 1. Although cytochalasin D did not alter the Ca^{2+} release from the intracellular Ca store, both vinblastin and colchicine effectively inhibited it at concentrations higher than 1 μM.

3.6. Effects of colchicine and compound 48/80 on the fluorescence intensity of tubulin-stained mast cell

The fluorescence image of rat mast cells stained with an anti-tubulin antibody is shown in Fig. 18. Intense fluorescence was observed at the cell periphery and filamentous structures were seen inside the cell. When the mast cells were treated with compound 48/80, the microtubules seemed to be disrupted in various places and a cluster of microtubule fragments were observed (Fig. 18b).

Fig. 19 shows the changes in the amount of microtubules in mast cells exposed to colchicine or compound 48/80. In this experiment, after permeabilization with 0.1% of Triton X-100, the cells were loaded with an anti-tubulin antibody. The process of permeabilization leads to a drastic loss of tubulin monomers, so that the fluorescence intensity of the cell stained with anti-tubulin antibody corresponds to the amount of

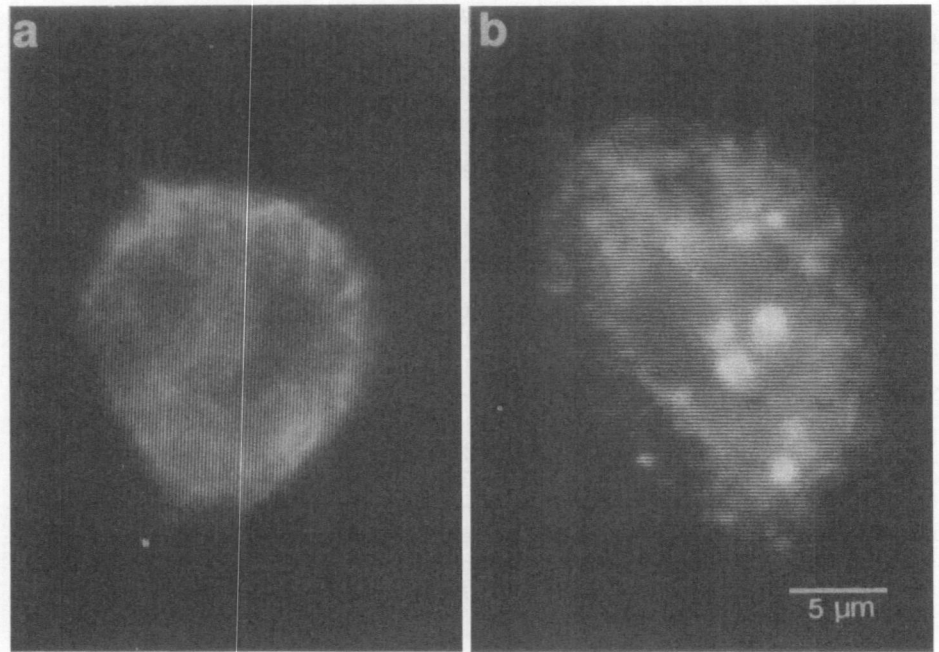

Fig. 18. Immunofluorescence microscopic image of rat peritoneal mast cells stained with anti-tubulin antibody. (a) control, (b) stimulated with compound 48/80 (0.5 μg/ml). Reproduced from Tasaka, K., Mio, M., Fujisawa, K. and Aoki, I.: Biochem. Pharmacol., 41, 1031-1037 (1991a), with permission.

polymerized form of tubulin. The fluorescence intensity of mast cells treated with colchicine decreased in a dose-dependent fashion, indicating that colchicine acts to inhibit the polymerization of the microtubules. By contrast, in the mast cells treated with compound 48/80, the fluorescence intensity increased significantly and dose-dependently, indicating that an increase in polymerization was elicited. When the mast cells treated with 10 μM of colchicine were stimulated with 0.5 μg/ml of compound 48/80, the fluorescence intensity of the tubulin-stained mast cell was slightly lower than that of the control level, but higher than that of the cells treated with colchicine alone.

3.7. TEM observation of the localization of intracellular Ca and microtubules

When permeabilized mast cells were exposed to potassium antimonate solution, potassium antimonate entered into the cytoplasm and interacted with Ca^{2+} to precipitate it as Ca antimonate. After this treatment, the particles of Ca antimonate appeared as black dots on the endoplasmic reticulum (Fig. 20).

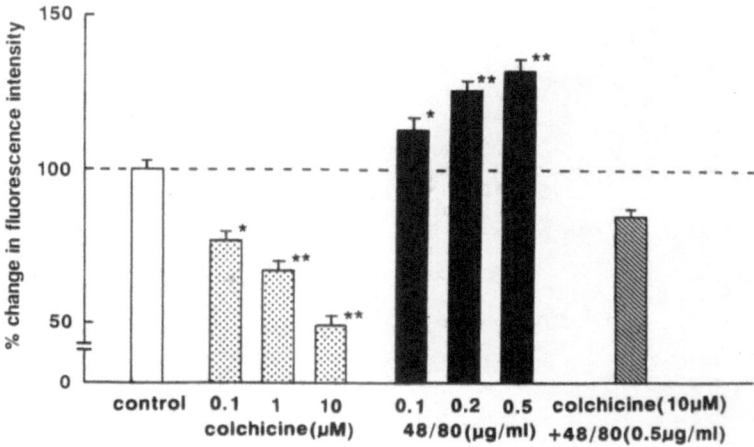

Fig. 19. Effects of compound 48/80 on the fluorescence intensity of tubulin-stained rat peritoneal mast cells. In each column, over 500 cells were measured. * and ** represent $p < 0.05$ and $p < 0.01$, respectively. Reproduced from Tasaka, K., Mio, M., Fujisawa, K. and Aoki, I.: Biochem. Pharmacol., 41, 1031-1037 (1991a), with permission.

3.8. Discussion

In this experiment, it was shown that cytochalasin D, colchicine and vinblastine effectively inhibited the histamine release from mast cells due to compound 48/80. The results are in agreement with those reported previously (Orr et al., 1972; Gillespie et al., 1968). Since lumicolchicine did not affect the histamine release, it was assumed that the inhibitory effect of colchicine was exerted by its specific action on microtubules. When the concentration of vinblastin was increased from 30 to 100μM, the effect of vinblastine in inhibiting histamine release increased remarkably as shown in Fig. 15. Since it is known that vinblastin promotes formation of tubulin paracrystals very efficiently at 50 μM, the potent histamine release inhibition elicited by vinblastine at 100μM may be caused by the paracrystalline formation of cytoplasmic tubulin (Weber and Osborn, 1981).

At least two pathways are proposed to increase intracellular Ca^{2+} concentrations: an increase in Ca^{2+} transport from an extracellular medium, or an increment of Ca^{2+} release from the intracellular Ca store (Ennis et al., 1980). In the former, it is suggested that a receptor-operated Ca channel may be of importance (Corcia et al., 1988), while in the latter, a rapid formation of IP_3 and the resulting Ca^{2+} release from the endoplasmic reticulum may lead to histamine release (Yoshii et al., 1988; Tasaka et al., 1990). Since none of the cytoskeleton-inhibiting agents used in the present study affected the IP_3 production of mast cells induced by compound 48/80, it became clear that both

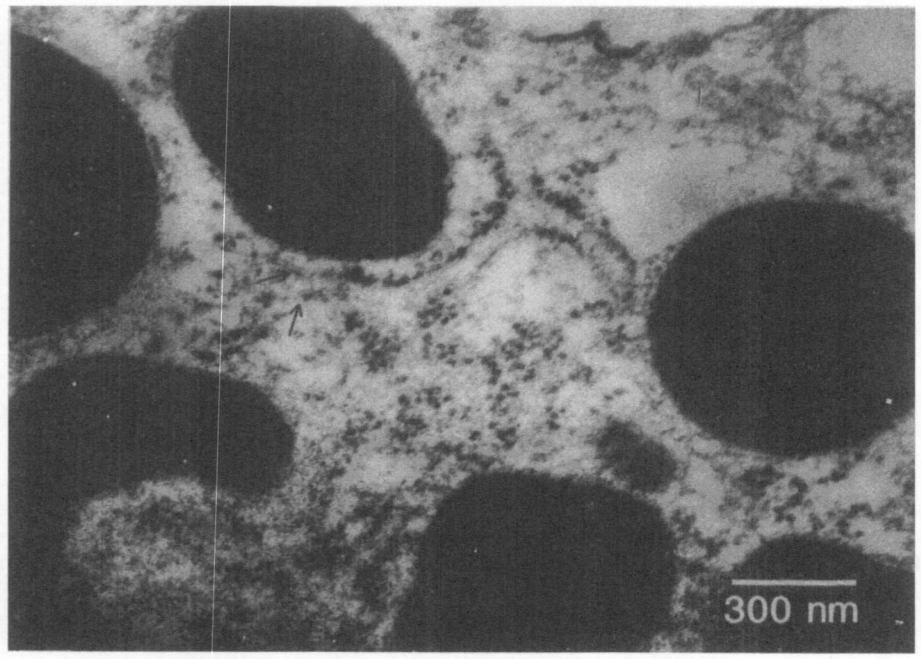

Fig. 20. TEM appearance of the inside structures and Ca stores of the rat peritoneal mast cells. The precipitates of Ca antimonate were observed as small black dots on the surface of endoplasmic reticulum, where microtubules are attached (arrows). Reproduced from Tasaka, K., Mio, M., Fujisawa, K. and Aoki, I.: Biochem. Pharmacol., 41, 1031-1037 (1991a), with permission.

microfilaments and microtubules may not participate in the process(es) of IP_3 production in the early stage of the cell activation.

For the measurement of intracellular Ca^{2+} concentrations, Fura 2/AM is widely used in many types of cells. However, it has been reported that ester-loaded Fura 2 was accumulated significantly in the secretory granules and that the fluorescence was lost during exocytosis (Almers and Neher, 1985). Furthermore, Highsmith et al. (1986) reported that Fura 2/AM may be converted to Fura 2 via the intermediate form called Fura 2'. Fura 2' is a compound which is lipophilic and fluorescent even in the absence of an interaction with Ca^{2+}. Since these peculiar fluorescences unrelated to intracellular Ca^{2+} were noticed, Fura 2/AM was not employed in the present experiment. In connection with this, when ^{45}Ca uptake into mast cells under stimulation with compound 48/80 was tested preliminarily according to the silicon oil method (Foreman et al., 1973), it was noticed that mast cells spun through silicon oil were often damaged and the values were scattered diversely.

In the present experiment, mast cells were stimulated with compound 48/80.

However, it is known that in IgE-dependent histamine release the increase in intracellular Ca^{2+} concentration is exclusively dependent on the Ca uptake from extracellular medium. As shown in Fig. 16, anti-microtubule agents inhibit the ^{45}Ca influx significantly and dose-dependently in the histamine release induced by compound 48/80. The identical mechanism may be exerted more effectively in inhibiting the histamine release from sensitized mast cells.

On the other hand, Ca^{2+} release from the intracellular Ca store elicited by exposure to compound 48/80 was effectively inhibited by both colchicine and vinblastine, but not by cytochalasin D. Since these two compounds were not effective in inhibiting IP_3 production, it was assumed that the endoplasmic reticulum of mast cells treated with any of the anti-microtubule agents became irresponsible to IP_3, and that microtubules may be important in regulating the function of the intracellular Ca store. From morphological observation it has been suggested that the structures of the endoplasmic reticulum and the microtubules are highly interdependent (Terasaki et al., 1986). Also, it was demonstrated that in electron microscopy, the microtubules are linked to the endoplasmic reticulum, where many black dots of Ca antimonate were observed as indicated in Fig. 20. From these findings, it was proposed that microtubules play a critical role in histamine release by releasing Ca^{2+} into the cytosol from the intracellular Ca store: the endoplasmic reticulum.

Since anti-microtubule agent (colchicine) is not easily incorporated into the cells, the incubation should be extended up to 2 hr. Thus, in the study of the histamine release from mast cells, the experiment must be carried out in the presence of extracellular Ca^{2+} (RPMI-1640 medium). Therefore, colchicine definitely inhibited intracellular Ca^{2+} release, certain amounts of Ca were incorporated from the extracellular medium. Ca uptake and histamine release were inhibited similarly by colchicine, as shown in Figs. 15 and 16.

As shown in Fig. 19, the fluorescence intensity of the microtubules in the mast cells treated with colchicine decreased in a dose-dependent fashion, clearly indicating that colchicine acts to depolymerize the microtubules. By contrast, the fluorescence intensity of microtubules in the mast cells treated with compound 48/80 increased dose-dependently. However, as shown in Fig. 18, disruption of the microtubules was elicited by stimulation with compound 48/80. These observations seem to suggest that after exposure to compound 48/80, a transient fragmentation of the microtubules takes place, while tubulin polymerization occurs sequentially. In accordance with this, the colchicine pretreatment prevented an increase in the fluorescence intensity of mast cells induced by compound 48/80, but the fluorescence intensity was higher than that of the cells treated with colchicine alone. This may indicate that depolymerization may be reversed as a consequence of an excessive incorporation of Ca^{2+} into mast cells in association with the histamine release induced by compound 48/80. It was reported that monomeric tubulin became polymeric in the presence of millimolar order of ATP and Ca^{2+} at the micromolar range (Nishida et al., 1979); such changes in ATP and Ca^{2+} concentrations are frequently encountered in the cytosol of the activated mast cells. (Tasaka et al., 1986b).

4. Identification of vimentin in rat peritoneal mast cells and its phosphorylation in association with histamine release

4.1. Introduction

In many cells, protein phosphorylation is one of the important pathways of signal transduction, eliciting changes in enzyme activity, cellular responses and functions (Nishizuka, 1984; Edelman et al., 1987). It has been reported that when rat peritoneal mast cells are stimulated with histamine releasers, some proteins in the mast cells are rapidly phosphorylated (Sieghart et al., 1978; Wells and Mann, 1983). However, none of the phosphorylated proteins have been identified, and the relation of these proteins to histamine release from rat mast cells has not been elucidated. In platelets, phosphorylation of myosin light chains has been implicated with the initiation of the cell shape-change response (Daniel et al., 1984). Although the general process of histamine release from mast cells has been well documented, the molecular basis for histamine release is not yet understood.

It has been shown that microfilaments and microtubules play some important roles in the process of histamine release from rat peritoneal mast cells (Tasaka et al., 1988, 1989, 1991a, b, c). However, the existence of an intermediate filament, which is known as the third type of cytoskeletal element in many cells, has not yet been verified in rat peritoneal

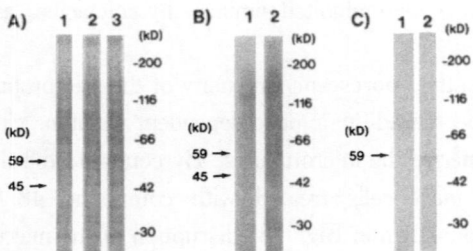

Fig. 21. Protein phosphorylation in rat mast cells and Triton insoluble fraction of mast cells. (A) ^{32}P-labeled rat mast cells were stimulated with $0.5 \mu g/ml$ compound 48/80 or $20 \mu M$ substance P for 10 sec. Phosphorylated proteins were subjected to SDS-PAGE and autoradiography. 1: control; 2: compound 48/80; 3: substance P. (B) β-Escin-permeabilized mast cells were exposed to $0.6 \mu M$ Ca^{2+} in the presence of $[\gamma$-^{32}P$]$ ATP for 2 min. 1: control; 2: $0.6 \mu M$ Ca^{2+}. (C) Mast cells were stimulated with $0.5 \mu g/ml$ compound 48/80 for 10 sec and extracted with 1% Triton X-100. Phosphorylated proteins in Triton insoluble fraction were analyzed by SDS-PAGE and autoradiography. 1: control; 2: compound 48/80. Reproduced from Izushi, K., Fujiwara, Y. and Tasaka, K.: Immunopharmacology, 23, 153-161 (1992), with permission.

mast cells.

4.2. Protein phosphorylation of non-permeabilized and permeabilized rat mast cells

As shown in Fig. 21A, when rat mast cells were stimulated for 10 sec with either 0.5 μg/ml compound 48/80 or 20 μM substance P, some proteins were rapidly phosphorylated. In particular, 45 and 59 kDa proteins were markedly phosphorylated in both cases. When permeabilized rat mast cells were stimulated by 0.6 μM Ca^{2+} simultaneously with 1 mM [γ-^{32}P] ATP, several proteins including 45 kDa and 59 kDa proteins, were phosphorylated as shown in Fig. 21B. However, in the absence of Ca^{2+}, no phosphorylation was apparent in permeabilized mast cells. Among the phosphorylated proteins, the most remarkable phosphorylation was seen in the 59 kDa band. This result seems to indicate that phosphorylation in both 45 kDa and 59 kDa proteins takes place in association with an increased intracellular Ca^{2+} concentration and simultaneous activation of Ca^{2+}-dependent protein kinase during the process of histamine release. Of these two proteins, only 59 kDa protein was detected in Triton X-100 insoluble fraction when the cells were stimulated with compound 48/80. By contrast, the 45 kDa protein was not detected in Triton X-100 insoluble fraction (Fig. 21C).

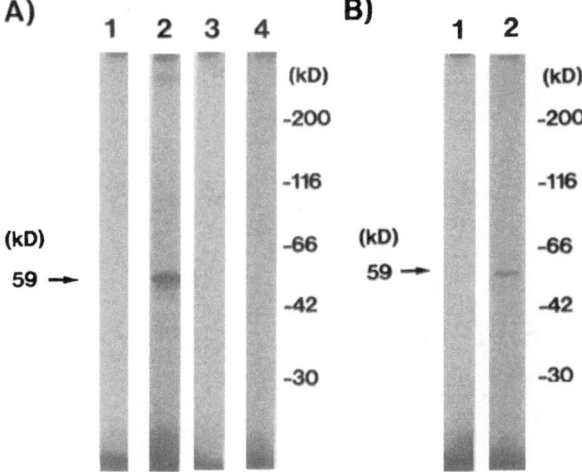

Fig. 22. Radioimmunoprecipitation of phosphorylated protein in non-permeabilized and permeabilized mast cells. (A) Rat mast cells were stimulated for 10 sec with 0.5 μg/ml compound 48/80 and extracted proteins were incubated with various antibodies. The immunoprecipitated protein was analyzed by SDS-PAGE and autoradiography. 1: normal mouse serum; 2: anti-vimentin antibody; 3: anti-desmin antibody; 4: anti-tubulin antibody. (B) Permeabilized mast cells were stimulated for 2 min with 0.6 μM Ca^{2+}. After that, immunoprecipitation was carried out by means of anti-vimentin antibody. 1: control; 2: in the presence of 0.1 μM Ca^{2+}. Reproduced from Izushi, K., Fujiwara, Y. and Tasaka, K.: Immunopharmacology, 23, 153-161 (1992), with permission.

4.3. Identification of 59 kDa phosphorylated protein in mast cells

After ^{32}P-incorporated cells were incubated with compound 48/80 (0.5 μg/ml) for 10 sec, cellular proteins were extracted with radioimmunoprecipitation assay buffer (containing NaCl, NaF, sodium pyrophosphate, sodium orthovanadate, EDTA, phenylmethylsl-fonylfluoride, Triton X-100, SDS, sodium deoxycholate, Tris-H_3PO_4; pH 7.5: RIPA buffer) and subsequent centrifugation was carried out. The supernatant was incubated with one of several antibodies, and the antigen-antibody complex was precipitated with protein A beads. Thereafter, the precipitated specimens were washed with RIPA buffer and the pellets were boiled in SDS-sample buffer. Subsequently, the sample were analyzed by SDS-PAGE and autoradiography. Neither anti-desmin, anti-tubulin anti-body nor normal mouse serum were precipitated with phosphorylated protein. Only anti-vimentin mouse monoclonal antibody bonded to phosphorylated 59 kDa protein to precipitate it (Fig. 22A). In the case when rat mast cells were stimulated with substance P (20 μM), the 59 kDa phosphorylated protein was precipitated by anti-vimentin anti-body. Moreover, when permeabilized mast cells were stimulated with 0.6 μM Ca^{2+}, the 59 kDa phosphorylated protein interacted with anti-vimentin antibody as seen in the case of non-permeabilized mast cells (Fig. 22B). This results clearly indicate that the 59 kDa phosphorylated protein in mast cells corresponds to vimentin, and that vimentin was phosphorylated by Ca^{2+}-dependent protein kinase.

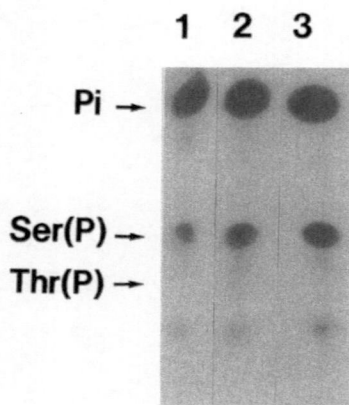

Fig. 23. Phosphoamino acid analysis of phosphorylated vimentin. Phosphorylated vimentin was obtained from rat peritoneal mast cells stimulated with 0.5 μg/ml compound 48/80 for 0 (lane 1), 5 (lane 2) and 10 sec (lane 3). Phosphoamino acid analysis was performed with cellulose TLC plate. The positions of phosphoserine (Ser(P)), phosphothreonine (Thr(P)), inorganic phosphate (Pi) are indicated. Reproduced from Izushi, K., Fujiwara, Y. and Tasaka, K.: Immunopharmacology, 23, 153-161 (1992), with permission.

Furthermore, as shown in Fig. 23, it became clear that serine residue is, indeed, the only phosphoamino acid in phosphorylated vimentin. When the ratio of Triton X-100 soluble/insoluble vimentin was determined, it became clear that almost 100% of vimentin detected in the resting rat mast cells was Triton X-100 insoluble. However, after the mast cell stimulation with compound 48/80, Triton insoluble vimentin decreased to $49.6 \pm 5.4\%$ (n = 3) of the total vimentin within 10 sec.

The time courses of vimentin phosphorylation and histamine release from rat mast cells stimulated with either compound 48/80 or substance P are shown in Fig. 24. Histamine release from rat peritoneal mast cells induced by compound 48/80 or substance P takes place very rapidly, reaching a maximum within 10 sec (Fig. 24). Similarly, the vimentin phosphorylation induced by either compound 48/80 or substance P was clearly observed within 5 sec and reached a maximum within 10 sec (Fig. 24A (inset) and Fig. 24B (inset)). This result clearly indicates that the time course of vimentin phosphorylation is almost the same as that of histamine release.

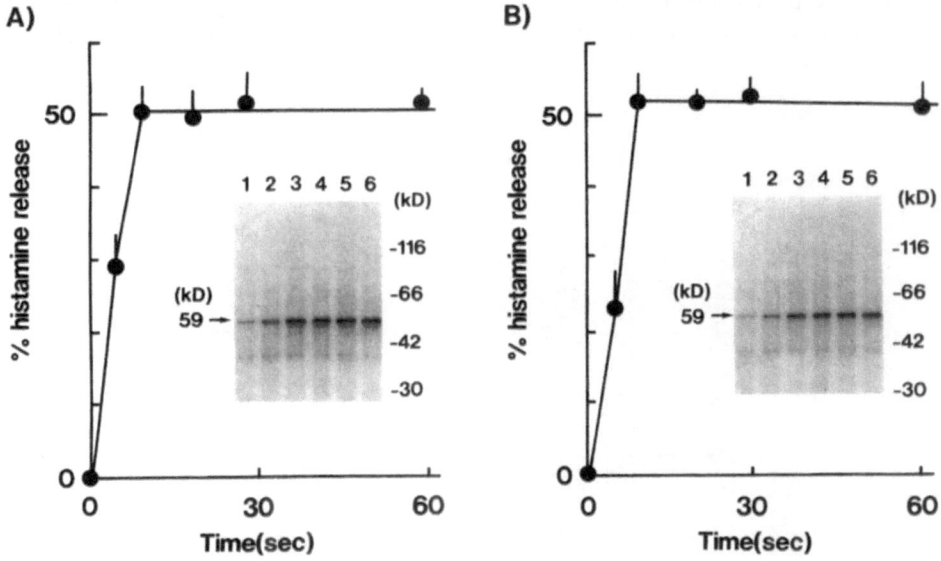

Fig. 24. Time courses of histamine release and vimentin phosphorylation of rat mast cells induced by compound 48/80 ($0.5 \mu g/ml$) or substance P ($20 \mu M$). Each value of histamine release indicates the mean $\pm$ SEM (n = 4). Inset: Phosphorylation of vimentin induced by compound 48/80 (A) and substance P (B). 1: 0 sec, 2: 5 sec, 3: 10 sec, 4: 20 sec, 5: 30 sec, 6: 60 sec. Reproduced from Izushi, K., Fujiwara, Y. and Tasaka, K.: Immunopharmacology, 23, 153-161 (1992), with permission.

4.4. Effects of calphostin C and cAMP on the phosphorylation of vimentin in permeabilized mast cells

Fig. 25 shows a representative autoradiogram of phosphorylated vimentin detected in β-escin permeabilized rat peritoneal mast cells. In permeabilized mast cells, Ca^{2+}-induced phosphorylation of vimentin was dose-dependently inhibited by calphostin C, a specific protein kinase C inhibitor (lanes 3, 4). On the other hand, trifluoperazine and W-7, which are calmodulin inhibitors, also inhibited vimentin phosphorylation in permeabilized mast cells (lanes 5, 6). When permeabilized mast cells were incubated with $10\,\mu M$ cAMP in the presence of $[\gamma\text{-}^{32}P]$ ATP, no phosphorylation of vimentin was detected (lane 7).

4.5. Immunofluorescent microscopic observation of mast cells

The immunofluorescence imaging of mast cells stained with an anti-vimentin antibody is shown in Fig. 26. In control cells, the filamentous structures were seen inside the cells, and each granule seemed to be surrounded by vimentin filaments (Fig. 26A). After stimulation with compound 48/80 for 20 sec, the filamentous structures promptly disappeared from the periphery of the cells, and the fluorescence intensity decreased (Fig.

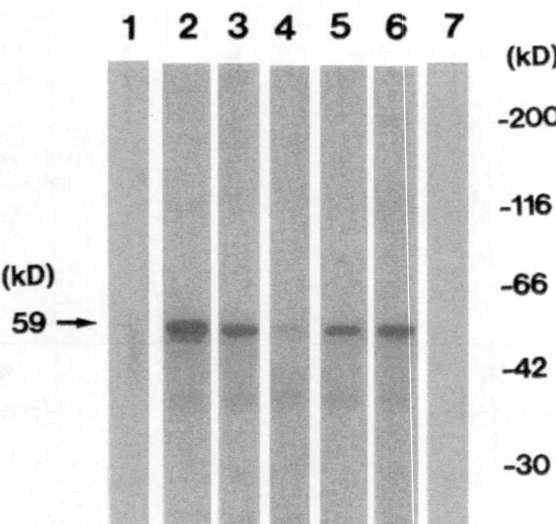

Fig. 25. Effects of calphostin C, calmodulin inhibitors and cAMP on vimentin phosphorylation in permeabilized mast cells. 1: resting cells; 2: Ca^{2+} $0.6\,\mu M$; 3: Ca^{2+} + calphostin C $0.5\,\mu M$; 4: Ca^{2+} + calphostin C $1\,\mu M$; 5: Ca^{2+} + trifluoperazine $10\,\mu M$; 6: Ca^{2+} + W-7 $25\,\mu M$; 7: cAMP $10\,\mu M$. Reproduced from Izushi, K., Fujiwara, Y. and Tasaka, K.: Immunopharmacology, 23, 153-161 (1992), with permission.

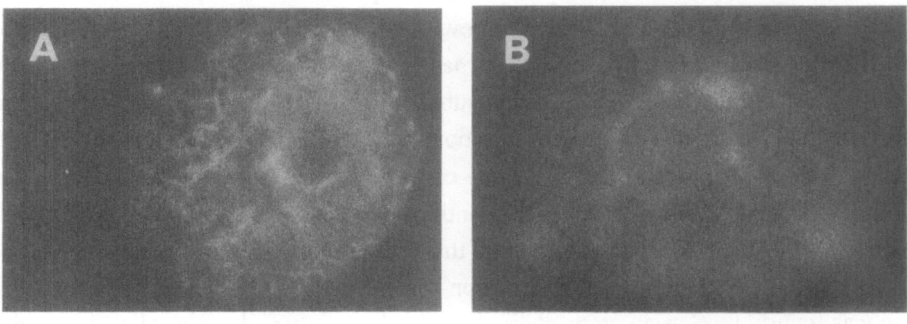

Fig. 26. Immunofluorescent microscopic images of rat mast cells stained by anti-vimentin antibody. (A) Control cell. (B) Stimulated with compound 48/80 (0.5 μg/ml) for 20 sec. Bar = 10 μm. Reproduced from Izushi, K., Fujiwara, Y. and Tasaka, K.: Immunopharmacology, 23, 153-161 (1992), with permission.

Table 2. Changes in fluorescence intensity of vimentin filaments in rat peritoneal mast cells stimulated with compound 48/80 (0.5 μg/ml).

Time (sec)	Fluorescence intensity (% of control)
0	100.0
5	78.8 ± 7.1
10	52.3 ± 5.8
30	54.9 ± 6.7
60	51.2 ± 5.1

In one experiment, the fluorescence intensities of 100 cells were measured (n = 4). Each value indicates the mean ± SEM. Reproduced from Izushi, K., Fujiwara, Y. and Tasaka, K.: Immunopharmacology, 23, 153-161 (1992), with permission.

26B). A dim and homogeneous fluorescence covered each cell. Table 2 indicates the changes in the amount of vimentin filaments in mast cells stimulated with compound 48/80 (0.5 μg/ml). Vimentin filaments are stable in the presence of Triton X-100 (Bormann et al., 1986). Therefore, permeabilization with Triton X-100 leads to almost complete leakage of non-filamentous (monomer form) vimentin from the cytoplasm, so that the fluorescence intensity of stained mast cells reflects the amount of vimentin filament in the mast cells. After the cell stimulation, phosphorylation of vimentin filaments in the mast cells and vimentin depolymerization take place very rapidly matching each other during the initial stage of histamine release.

4.6. Discussion

In this study, it was revealed that when rat mast cells were stimulated with compound

48/80 or with substance P, phosphorylation of proteins takes place very rapidly, especially in 45 and 59 kDa proteins as shown in Figs. 21 and 24. Therefore, it was assumed that such a rapid phosphorylation may be related in some way to the initiation process of histamine release. Concerning substance P, it is totally unknown, so far, whether or not such a protein phosphorylation occurs in relation to histamine release. However, it became apparent that this is the case: substance P induces phosphorylation of 45 and 59 kDa proteins which are identical to those found after treatment with compound 48/80. Several other proteins than 45 and 59 kDa proteins were also phosphorylated. However, no direct relation between these proteins and histamine release was found. In β-escin permeabilized mast cells, protein phosphorylation was also induced in the presence of $0.6 \mu M$ Ca^{2+}. Neither histamine release (Izushi and Tasaka, 1991) nor protein phosphorylation was provoked without Ca^{2+} as shown in Fig. 21B. The protein phosphorylation found in rat peritoneal mast cells stimulated with histamine releasers may be intimately related to an increase of intracellular Ca^{2+} concentration which occur after the stimulations.

Phosphorylated 59 kDa protein was also detected in Triton X-100 insoluble fraction as shown in Fig. 21C. Since the cytoskeletal proteins are insoluble in Triton X-100 (Jennings et al., 1981; Bormann et al., 1986; Tasaka et al., 1991a), the 59 kDa protein seem to be one of the cytoskeletal elements. In this study, we found that the 59 kDa phosphorylated protein is, indeed, vimentin as shown in Fig. 22. So far, vimentin is known as one of the proteins of the intermediate filament. As shown in Fig. 24, vimentin phosphorylation took place very rapidly in accordance with the time course of histamine release. Furthermore, phosphoamino acid analysis revealed that only serine residue of vimentin was phosphorylated.

The cytoskeletal proteins are likely to be the potential targets for protein kinases; microfilaments and microtubules are substrate of protein kinases *in vivo* and *in vitro* (Adelstein and Conti, 1975; Daniel et al., 1984; Kawamoto and Hidaka 1984; Tsuyama et al., 1986). In addition to these cytoskeletal elements, cells contain intermediate filaments, another class of cytoplasmic filaments. Intermediate filaments are major components of the cytoskeleton of the cells and appear to play a significant role maintaining organization of the cytoplasmic space (Lazarides, 1980). In rabbit ciliary processes or human umbilical vein endothelial cells, vimentin has been identified as phosphorylated protein (Bormann et al., 1986; Yoshimura et al., 1989). Furthermore, it has been reported that phosphorylation of vimentin may be induced by protein kinase C, protein kinase A, as well as Ca^{2+}-calmodulin dependent protein kinase II *in vitro* (Inagaki et al., 1987, 1988; Tokui et al., 1990). As shown in Fig. 25, Ca^{2+}-induced vimentin phosphorylation in permeabilized mast cells was suppressed by calphostin C and calmodulin inhibitors, while cAMP was not effective at all in eliciting phosphorylation. These results seem to suggest that vimentin is phosphorylated by activated protein kinase

C or Ca²⁺-calmodulin dependent protein kinase, but not by protein kinase A. Most calmodulin inhibitors are known to inhibit not only Ca²⁺-calmodulin dependent protein kinase but also protein kinase C (Schatzman et al., 1981). However, it is known that calphostin C causes very selective inhibition on protein kinase C without affecting calmodulin or other protein kinases (Kobayashi et al,. 1989). Therefore, vimentin in rat mast cells might be phosphorylated exclusively by protein kinase C rather than Ca²⁺-calmodulin dependent protein kinase.

As shown in Fig. 26, network formation of vimentin filaments was seen inside cells in the resting state. In contrast, in stimulated mast cells, vimentin filaments became quite disordered and a dim FITC-fluorescence was observed homogeneously on the whole cell area. This probably indicates that a marked depolymerization of vimentin filaments takes place. At the present time, the role of intermediate filaments in histamine release is not known. Vimentin filaments probably hold the secretory granules tightly in the resting state. After exposure to compound 48/80, depolymerization of vimentin filaments occurs in stimulated cells, and consequently, (1) granules may become easily movable and (2) in the loosened condition, a series of reactions leading to histamine release may be activated promptly inside the granules. From this study, it became clear that vimentin in mast cells may be phosphorylated by protein kinase C, and subsequent depolymerization of vimentin filaments may be the reason for degranulation and histamine release from rat peritoneal mast cells.

References

Adelstein R.S. and Conti M.A. (1975) Phosphorylation of platelet myosin increases actin-activated myosin ATPase activity. *Nature*, 256: 597-598

Almers W. and Neher E. (1985) The Ca signal from fura-2 loaded mast cells depends strongly on the method of dye-loading. *FEBS Lett.*, 192: 13-18

Barak L.S., Nothnagel E.A., DeMarco E.F. and Webb W.W. (1981) Differential staining of actin in metaphase spindles with 7-nitrobenz-2-oxa-1, 3-diazole-phallacidin and fluorescent DNase: Is actin involved in chromosomal movement? *Proc. Natl. Acad. Sci. USA*, 78: 3034-3038

Bormann B.-J., Huang C.-K., Lam G.F. and Jaffe E.A. (1986) Thrombin-induced vimentin phosphorylation in cultured human umbilical vein endothelial cells. *J. Biol. Chem.*, 261: 10471-10474

Burridge K. and Phillips J.H. (1975) Association of actin and myosin with secretory granule membranes. *Nature*, 254: 524-529

Burwen S.J. and Satir B.H. (1977) Plasma membrane on the mast cell surface and their relationship to secretory activity. *J. Cell Biol.*, 74: 690-697

Corcia A., Pecht I., Hemmerich S., Ran S. and Rivnay B. (1988) Calcium specificity of the antigen-induced channels in rat basophilic leukemia cells. *Biochemistry*, 27: 7499-7506

Daniel J.L., Molish I.R., Rigmaiden M. and Stewart G. (1984) Evidence for a role of myosin phosphorylation in the initiation of the platelet shape change response. *J. Biol. Chem.*,

259: 9826-9831

Durham A.C.H. (1974) A unified theory of the control of actin and myosin in nonmuscle movements. *Cell*, 2: 123-136

Edelman A.M., Blumenthal D.K. and Krebs E.G. (1987) Protein serine-threonine kinase. *Ann. Rev. Biochem.*, 56: 567-614

Ennis M., Truneh A., White J.R. and Pearce F.L. (1980) Calcium pools involved in histamine release from mast cells. *Int. Archs. Allergy Appl. Immunol.*, 62: 467-471

Foreman J.C., Mongar J.L. and Gomperts B.D. (1973) Calcium ionophores and movement of calcium ions following the physiological stimulus to a secretory process. *Nature*, 245: 249-251

Fowler V.M. and Pollard H.B. (1982) Chromaffin granule membrane-F-actin interactions are calcium sensitive. *Nature*, 295: 336-339

Gillespie E., Levine R.J. and Malawista S.E. (1968) Histamine release from rat peritoneal mast cells: inhibition by colchicine and potentiation by deuterium oxide. *J. Pharmacol. Exp. Ther.*, 164: 158-165

Hartwig J.H. and Stossel T.P. (1982) Macrophages: their use in elucidation of the cytoskeletal roles of actin. *Methods Cell Biol.*, 25: 201-225

Highsmith S., Bloebaum P. and Snowdowne K.W. (1986) Sarcoplasmic reticulum interacts with the $Ca(2+)$ indicator precursor Fura-2-AM. *Biochem. Biophys. Res. Commun.*, 138: 1153-1162

Inagaki M., Nishi Y., Nishizawa K., Matsuyama M. and Saito C. (1987) Site-specific phosphorylation induced disassembly of vimentin filament *in vitro*. *Nature*, 328: 649-658

Inagaki M., Gonda Y., Matsuyama M., Nishizawa K., Nishi Y. and Saito C. (1988) Intermediate filament reconstitution *in vitro*. The role of phosphorylation on the assembly-disassembly of desmin. *J. Biol. Chem.*, 263: 5970-5978

Izushi K. and Tasaka K. (1991) Essential role of ATP and possibility of activation of protein kinase C in Ca^{2+}-dependent histamine release from permeabilized rat peritoneal mast cells. *Pharmacology*, 42: 297-308

Jennings L.K., Fox J.E.B., Edwards H.H. and Phillips D.R. (1981) Changes in the cytoskeletal structure of human platelets following thrombin activation. *J. Biol. Chem.*, 256: 6927-6932

Kawamoto S. and Hidaka H. (1984) Ca^{2+}-activated, phospholipid-dependent protein kinase catalyzes the phosphorylation of actin binding proteins. *Biochem. Biophys. Res. Commun.*, 118: 736-742

Kessler S. and Kuhn C. (1975) Scanning electron microscopy of mast cell degranulation. *Lab. Invest.*, 32: 71-77

Kobayashi E., Nakano H., Morimoto M., Tamaoki T. (1989) Calphostin C (UCN-1028C), a novel microbial compound, is a highly potent and specific inhibitor of protein kinase C. *Biochem. Biophys. Res. Commun.*, 159: 548-553

Korn E. D. (1978) Biochemistry of actomyosin-dependent cell motility. *Proc. Natl. Acad. Sci. USA*, 75: 588-599

Lazarides E. (1980) Intermediate filaments as mechanical integrators of cellular space. *Nature*, 283: 249-256

Meyer D.I. and Burger M.M. (1975) The chromaffin granule surface: the presence of actin and the nature of its interaction with membrane. *FEBS Lett.*, 101: 129-133

Mio M. and Tasaka K. (1989) Microfilament-associated, local degranulation of rat peritoneal mast cells. *Int. Archs. Allergy Appl. Immunol.*, 88: 369-371

Nielsen E.H. and Jahn H. (1984) Cytoskeletal studies on Lowicryl K4M embedded and Affi-Gel 731 attached rat peritoneal mast cells. *Virchows Arch.*, 45: 313-323

Nishida E., Kumagai H., Ohtsuki I. and Sakai H. (1979) The interactions between calcium-dependent regulator protein of cyclic nucleotide phosphodiesterase and microtubule proteins. I. Effects of calcium-dependent regulator protein on the calcium sensitivity of microtubule assembly. *J. Biochem.*, 85: 1257-1266

Nishizuka Y. (1984) The role of protein kinase C in cell surface signal transduction and tumour promotion. *Nature*, 308: 693-698

Orr T.S.C., Hall D.E. and Allison A.C. (1972) Role of contractile microfilaments in the release of histamine from mast cells. *Nature*, 236: 350-351

Rölich P. (1975) Membrane-associated actin filaments in the cortical cytoplasm of the rat mast cells. *Exp. Cell. Biol.*, 93: 293-298

Schatzman R.C., Wise B.C. and Kuo J.F. (1981) Phospholipid-selective calcium-dependent protein kinase: inhibition by anti-psychotic drugs. *Biochem. Biophys. Res. Commun.*, 98: 669-676

Sieghart W., Theoharides T., Alper S.L., Douglas W.W. and Greengard P. (1978) Calcium-dependent protein phosphorylation during secretion by exocytosis in the mast cells. *Nature*, 275: 329-330

Tasaka K. and Mio M. (1989) Microfilament-associated degranulation of sensitized guinea-pig lung mast cells. *Agents Actions*, 27: 79-82

Tasaka K., Akagi M. and Miyoshi K. (1986a) Distribution of actin filaments in rat mast cells and its role in histamine release. *Agents Actions*, 18: 49-52

Tasaka K., Mio M. and Okamoto M. (1986b) Intracellular calcium release induced by histamine releasers and its inhibition by some antiallergic drugs. *Ann. Allergy*, 56: 464-469

Tasaka K., Mio M. and Okamoto M. (1987) The role of intracellular Ca^{2+} in the degranulation of skinned mast cells. *Agents Actions.*, 20: 157-160

Tasaka K., Akagi M., Miyoshi K. and Mio M. (1988) Role of microfilaments in the exocytosis of rat peritoneal mast cells. *Int. Archs. Allergy Appl. Immunol.*, 87: 213-221

Tasaka K., Mio M. and Akagi M. (1989) The role of microfilaments in the degranulation and histamine release from mast cells. In 'Bioinformatics', (eds.) Hatase O. and Wang J.H., Elsevier, Amsterdam, pp. 195-203

Tasaka K., Sugimoto Y. and Mio M. (1990) Sequential analysis of histamine release and intracellular Ca^{2+} release from murine mast cells. *Int. Arch. Allergy Appl. Immunol.*, 91: 211-213

Tasaka K., Mio M., Fujisawa K. and Aoki I. (1991a) Role of microtubules on Ca^{2+} release from the endoplasmic reticulum and associated histamine release from rat peritoneal mast cells. *Biochem. Pharmacol.*, 41: 1031-1037

Tasaka K., Mio M., Akagi M., Fujisawa K. and Aoki I. (1991b) Role of the cytoskeleton in Ca^{2+} release from the intracellular Ca store of rat peritoneal mast cells. *Agents Actions*, 33: 44-47

Tasaka K., Mio M. and Izushi K. (1991c) Role of cytoskeletons on Ca^{2+} release from the intracellular Ca store of rat peritoneal mast cells. *Skin Pharmacol.*, 4 (suppl. 1): 43-55

Tellam R. and Friden C. (1982) Cytochalasin D and platelet gelsolin accelerate actin polymer formation: A model for regulation of the extent of actin polymer formation *in vivo*. *Biochemistry*, 21: 3207-3214

Terasaki M., Chen L.B. and Fujiwara K. (1986) Microtubules and the endoplasmic reticulum are highly interdependent structures. *J. Cell Biol.*, 103: 1557-1568

Tokui T., Yamauchi T., Yano T., Nishi Y., Kusagawa M., Yatani R. and Inagaki M. (1990) Ca^{2+}-calmodulin-dependent protein kinase II phosphorylates various types of non-epithelial intermediate filament proteins. *Biochem. Biophys. Res. Commun.*, 169: 896-904

Tsuyama S., Bramblett G.T., Huang K.P. and Flavin M. (1986) Calcium/phospholipid-dependent kinase recognizes sites in microtubule-associated protein 2 which are phosphorylated in living brain and not accessible to other kinases. *J. Biol. Chem.*, 261: 4110-4116

Weber K. and Osborn M. (1981) Microtubule and intermediate filament networks in cells viewed by immunofluorescence microscopy. In 'Cytoskeletal Elements and Plasma Membrane Organization', (eds.) Poste G. and Nicolson G.L., Elsevier, Amsterdam, pp. 1-53

Wells E. and Mann J. (1983) Phosphorylation of a mast cell protein in response to treatment with anti-allergic compounds. Implications for the mode of action of sodium cromoglycate. *Biochem. Pharmacol.*, 32: 827-842

Yazawa M., Yagi K. and Sobue K. (1987) Isolation and characterization of a calmodulin binding fragment of chicken gizzard caldesmon. *J. Biochem.*, 102: 1065-1073

Yoshii N., Mio M. and Tasaka K. (1988) Ca uptake and Ca releasing properties of the endoplasmic reticulum in rat peritoneal mast cells. *Immunopharmacology*, 16: 107-113

Yoshimura N., Mittag T.W. and Podos S.M. (1989) Calcium-dependent phosphorylation of proteins in rabbit ciliary processes. *Invest. Ophthalmol. Vis. Sci.*, 30: 723-730

Chapter 6
Histamine release induced by basic peptides and proteins

Introduction

Histamine release from mast cells is elicited not only by immunological stimuli but also various substances (Uvnäs, 1974; Tasaka, 1986; Foreman, 1988; Akagi and Tasaka 1991). Among them, it is of importance to elucidate the mechanism of histamine release induced by biological substances, such as neuropeptides or proteins, in association with the pathogenesis of allergic reactions. In this chapter, the histamine release induced by substance P, histone and eosinophil major basic protein was investigated.

1. Substance P-induced histamine release from rat peritoneal mast cells and its inhibition by antiallergic agents and calmodulin inhibitors

1.1. Introduction

It has been shown that substance P induces histamine release from mast cells of various tissues in different species (Piotrowski and Foreman, 1985; Pearce et al., 1989). Structure-activity studies of substance P indicated that the amino acid sequences required for histamine release are different from those seen in the activation of other tachykinin receptors in smooth muscles and nervous systems (Fewtrell et al., 1982; Foreman et al., 1983; Repke and Bienert, 1988). It was supposed that basicity around the N-terminal region and hydrophobicity in the C-terminal sequence may be a requisite for histamine release (Repke and Bienert, 1988; Foreman and Jordan, 1983). Since many basic compounds, such as compound 48/80, peptide 401, histone and major basic protein are effective in releasing histamine even in a Ca-free medium (Ennis et al., 1980; Tasaka et al., 1990; Aoki et al., 1991a), it was supposed that the basic compounds may cause histamine release in a similar way. By means of model membrane systems, Shibata et al. (1985) indicated that substance P may interact with acidic phospholipids in the lipid bilayer membrane so as to change its secondary structure from a random conformation to the β-form, and this results in an increase of membrane permeability. In this section, the attention was paid to understand the mechanism of histamine release due to substance P from the different aspects. Furthermore, the effects of some antiallergic agents and calmodulin inhibitors on substance P-induced histamine release were studied.

1.2. Effects of antiallergic compounds on the histamine release from rat peritoneal mast cells induced by substance P

The inhibitory effect of antiallergic drugs on the histamine release from rat peritoneal mast cells induced by substance P $(2\mu M)$ in a Ca-free medium is shown in Table 1. In the control experiment, 37.45 ± 2.58 % of histamine was released from the mast cells. As shown in Table 1, colchicine and cytochalasin D were effective in inhibiting histamine release induced by substance P.

1.3. Effects of antiallergic drugs on the Ca^{2+} release from the intracellular Ca store of rat mast cells elicited by substance P

When quin 2-loaded rat mast cells were exposed to substance P $(2\mu M)$, CGRP $(5\mu M)$ or neurotensin $(10\mu M)$ in a Ca-free medium, the fluorescence intensity derived from quin 2-Ca complex increased promptly, reaching the maximum within 5 sec, and decreasing

Table 1. Inhibitory effects of antiallergic drugs on histamine release from rat peritoneal mast cells induced by substance P $(2\mu M)$

Drugs	concentration (μM)	% histamine release	% inhibition
control		37.45 ± 2.58	
ketotifen	1	33.66 ± 6.46	10.12
	10	32.76 ± 5.47	12.52
	100	$17.94 \pm 2.82**$	52.10
mequitazine	2	36.98 ± 4.34	1.26
	10	$21.96 \pm 1.49*$	41.36
	20	$13.63 \pm 2.89**$	63.60
terfenadine	0.1	35.65 ± 2.33	4.81
	1	29.82 ± 2.74	20.37
	3	$19.34 \pm 3.52**$	48.36
oxatomide	0.1	37.18 ± 2.39	0.72
	1	32.73 ± 2.23	12.60
	10	$17.03 \pm 2.29**$	54.53
DSCG	1	35.23 ± 3.50	5.93
	10	30.94 ± 3.58	17.3
	100	$24.17 \pm 1.60*$	35.46
colchicine	1	29.34 ± 5.28	21.66
	10	26.47 ± 4.60	29.32
	100	$20.56 \pm 3.77**$	45.10
cytochalasin D	0.1	37.23 ± 1.78	0.59
	1	$29.36 \pm 1.15*$	21.60
	10	$19.17 \pm 3.71**$	48.82

Spontaneous histamine release, which was lower than 5 % in each experiment, was subtracted from the gross histamine release and the net value is indicated in the table. Each value represents the mean $\pm$ S.E. M. from 6 separate experiments. $*$ p < 0.05, $**$ p < 0.01. Reproduced from Mio, M., Izushi, K. and Tasaka, K.: Immunopharmacology, 22, 59-66 (1991), with permission.

gradually to the control level after 20 sec (Fig. 1). Since it is known that changes in the fluorescence intensity reflect the amount of intracellular Ca^{2+} (Tasaka et al., 1986), it is clear that substance P was more potent in releasing Ca^{2+} from the intracellular Ca store than the other neuropeptides. As shown in Table 2, when quin 2-loaded mast cells were pretreated with various antiallergic agents before exposure to substance P, the increase in fluorescence intensity was effectively inhibited.

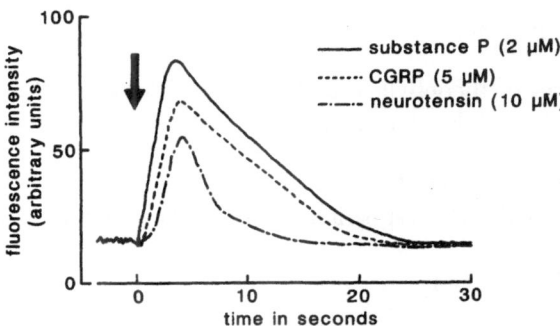

Fig. 1. Changes in intracellular Ca^{2+} concentrations associated with an increase in Ca^{2+} release from the Ca store of rat peritoneal mast cells induced by some neuropeptides. Each line represents the mean value obtained from 5 separate experiments using 10 cells in each measurement. Reproduced from Mio, M., Izushi, K. and Tasaka, K.: Immunopharmacology, 22, 59-66 (1991), with permission.

Table 2. Inhibitory effects of antiallergic agents on fluorescence increase of quin 2-loaded mast cells elicited by substance P $(2\,\mu M)$

Compounds	Concentration (μM)	% Inhibition
ketotifen	1	10.30 ± 4.82
	10	38.31 ± 0.73**
	100	68.92 ± 3.68**
mequitazine	2	27.69 ± 4.16*
	10	69.23 ± 1.26**
	20	84.61 ± 2.68**
terfenadine	1	8.45 ± 3.38
	3	48.27 ± 4.69**
	10	88.79 ± 5.62**
oxatomide	1	28.08 ± 3.27*
	10	68.53 ± 3.32**
	100	88.31 ± 5.14**
DSCG	10	38.36 ± 3.65**
	100	70.09 ± 6.80**

Each value represents the mean $\pm$ S.E.M. determined in 50 cells in one experiment. * $p < 0.05$, ** $p < 0.01$. Reproduced from Mio, M., Izushi, K. and Tasaka, K.: Immunopharmacology, 22, 59-66 (1991), with permission.

1.4. Effects of cytoskeleton-inhibiting agents on the Ca^{2+} release from the intracellular Ca store of rat mast cells induced by substance P

When rat peritoneal mast cells were pretreated with colchicine for 1 hr, the increase in the fluorescence intensity of quin 2-loaded mast cells induced by substance P was suppressed in a dose-dependent fashion, suggesting a participation of microtubules in the Ca^{2+} release from the store (Fig. 2) (Tasaka et al., 1991). However, cytochalasin D treatment was not effective in inhibiting Ca^{2+} release from the intracellular Ca store, even at a concentration of $10\,\mu$M.

1.5. Inhibitory effects of theophylline and db-cAMP on the histamine release and intracellular Ca^{2+} mobilization in rat mast cells induced by substance P

Fig. 3 represents the relationship between the histamine release and intracellular Ca^{2+} release induced by substance P in rat peritoneal mast cells treated with either theophylline

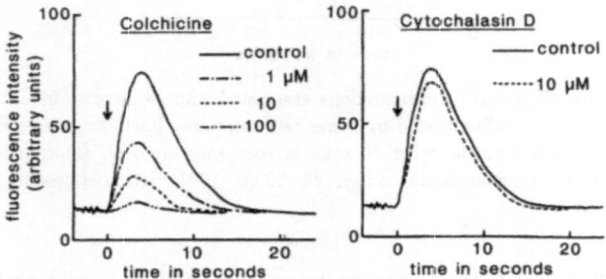

Fig. 2. Effects of colchicine and cytochalasin D on the Ca^{2+} release from the intracellular Ca store of rat peritoneal mast cells induced by substance P ($2\,\mu$M). Each line represents the mean value obtained from 5 separate experiments using 10 cells in each measurement. Reproduced from Mio, M., Izushi, K. and Tasaka, K.: Immunopharmacology, 22, 59-66 (1991), with permission.

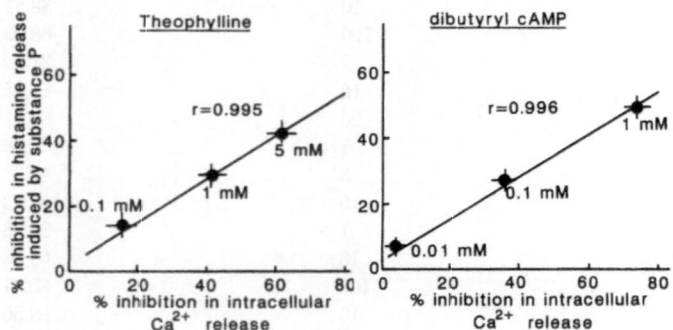

Fig. 3. Inhibitory effects of theophylline and db-cAMP on histamine release and quin 2 signal induced by substance P ($2\,\mu$M) in rat peritoneal mast cells. Each point represents the mean $\pm$ SEM of 5 separate experiments. Reproduced from Mio, M., Izushi, K. and Tasaka, K.: Immunopharmacology, 22, 59-66 (1991), with permission.

or db-cAMP. In both cases, the extent to which they suppressed the histamine release was equivalent to the extent of inhibition of intracellular Ca^{2+} release, and these two parameters increased dose-dependently.

1.6. Inhibitory effects of calmodulin inhibitors on the histamine release from rat peritoneal mast cells induced by substance P

Fig. 4 shows the inhibitory effects of some calmodulin inhibitors on histamine release induced by substance P from rat mast cells. Among these inhibitors, calmidazolium was the most potent. The degree of the inhibitory potency was as follows: calmidazolium > trifluoperazine > W-7 > W-5. When the inhibitory effects (IC50 values) of calmodulin inhibitors on histamine release were compared with those for calmodulin activity represented by calmodulin-dependent phosphodiesterase activity, it became apparent that these two activities were closely related to each other ($r = 0.992$) as shown in Fig. 5.

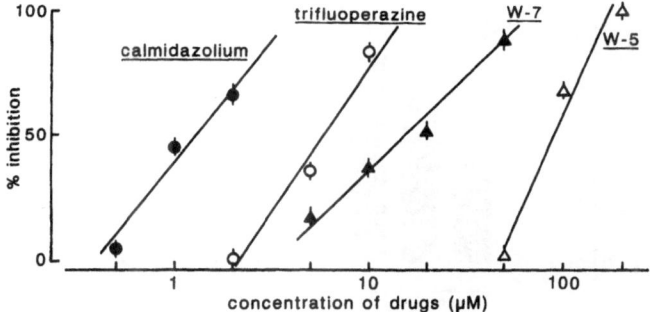

Fig. 4. Inhibitory effects of calmodulin inhibitors on histamine release from rat peritoneal mast cells induced by substance P ($2\,\mu M$). Each point represents the mean ± SEM determined in 5 separate experiments. Reproduced from Mio, M., Izushi, K. and Tasaka, K.: Immunopharmacology, 22, 59-66 (1991), with permission.

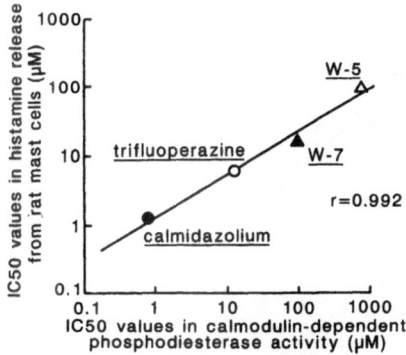

Fig. 5. Inhibitory effects of calmodulin inhibitors on histamine release and calmodulin-dependent phosphodiesterase activity. Reproduced from Mio, M., Izushi, K. and Tasaka, K.: Immunopharmacology, 22, 59-66 (1991), with permission.

1.7. Histamine release from permeabilized mast cells induced by GTPγS, substance P, somatostatin, CGRP and neurotensin

As indicated in Fig. 6, GTPγS effectively liberated histamine from β-escin permeabilized rat mast cells, suggesting that GTPγS directly activates intracellular GTP binding protein (G protein) from the inside of the cell. In contrast, substance P, somatostatin, CGRP and neurotensin led to no release of histamine from the permeabilized cells at all.

1.8. Binding study of ³H-substance P to rat mast cells

When rat mast cells were incubated at 4 ℃ with ³H-substance P, ³H-substance P specifically bound to the cell surface. Fig. 7 represents a Scatchard analysis of the ³H-substance P binding site on the mast cell. As indicated in this figure, it was revealed that rat peritoneal mast cells possess a single binding site with Kd value of 0.585×10^{-7}

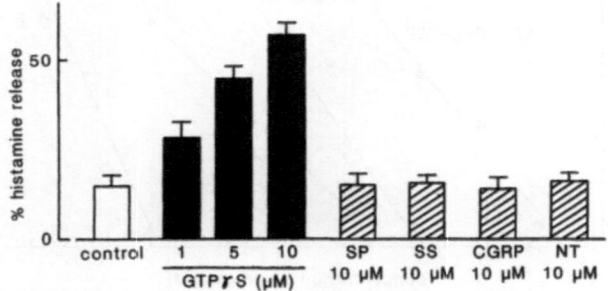

Fig. 6. Histamine release from permeabilized rat mast cells induced by GTPγS, substance P (SP), somatostatin (SS), CGRP and neurotensin (NT). Each column represents the mean ± SEM (n = 6). Reproduced from Mio, M., Izushi, K. and Tasaka, K.: Immunopharmacology, 22, 59-66 (1991), with permission.

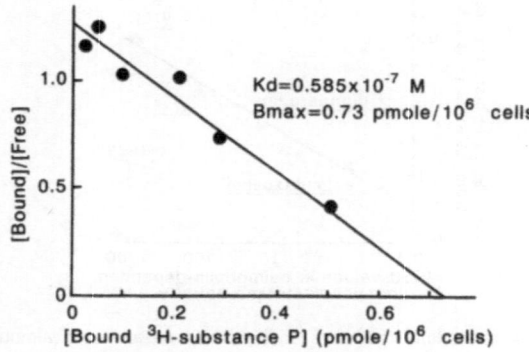

Fig. 7. Scatchard plot analysis of ³H-substance P binding to rat peritoneal mast cells.

M and Bmax of 0.73 pmole/10^6 cells.

1.9. Influence of compound 48/80 on the ³H-substance P binding to mast cells

When compound 48/80, at concentrations ranging from 0.1 to 2 μg/ml, was added to the reaction mixture, the extent of ³H-substance P binding to mast cells was inhibited in a dose-dependent fashion, as indicated in Fig. 8. Almost complete inhibition was elicited at 1 μg/ml of compound 48/80.

1.10. Effects of antiallergic agents on the ³H-substance P binding to mast cells

The effects of antiallergic drugs on the ³H-substance P binding to rat peritoneal mast cells are indicated in Fig. 9. As shown in this figure, basic antiallergic drugs (terfenadine and oxatomide) inhibited the ³H-substance P binding to mast cells, although DSCG, an acidic

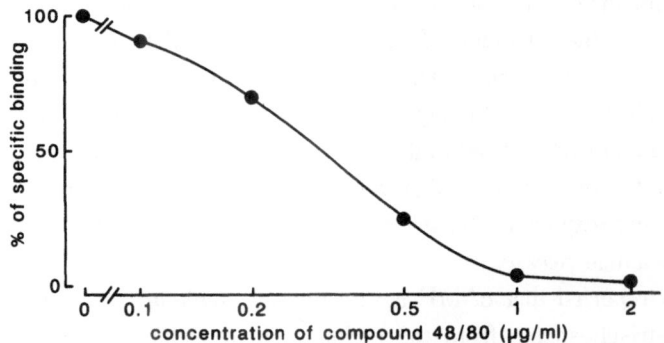

Fig. 8. Inhibitory effect of compound 48/80 on ³H-substance P binding to rat peritoneal mast cells.

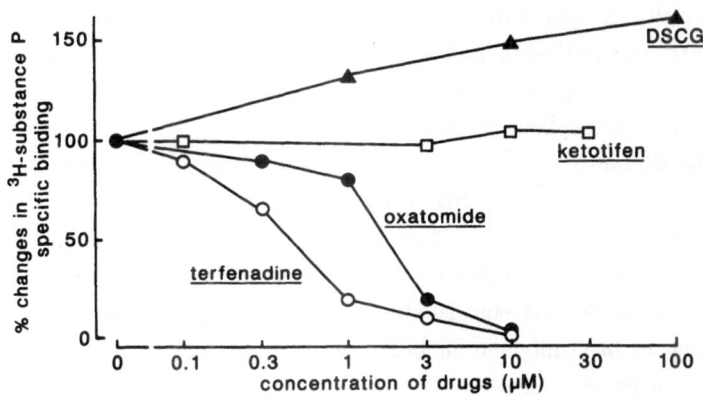

Fig. 9. Influence of antiallergic drugs on ³H-substance P binding to rat peritoneal mast cells.

compound, enhanced ^{3}H-substance P binding to the cell. Ketotifen did not markedly affect the ^{3}H-substance P binding at the concentrations tested.

1.11. Discussion

It has been shown that substance P effectively induces histamine release from not only rat peritoneal mast cells but also the mast cells obtained from some other tissues in various species (Piotrowski and Foreman, 1985; Pearce et al., 1989; Fewtrell et al., 1982). Histamine release from rat peritoneal mast cells can be induced even in a Ca-free medium owing to Ca^{2+} release from the intracellular Ca store (Fewtrell et al., 1982; Tasaka et al., 1986). Also, it was confirmed that not only substance P but also neurotensin and CGRP, are capable of releasing Ca^{2+} from the intracellular Ca store; however, substance P was most potent. The histamine release caused by substance P was inhibited by various antiallergic agents. Since such antiallergic agents inhibited Ca^{2+} release from intracellular store induced by substance P, it was supposed that one of the critical reasons for the effect of antiallergic agents on the histamine release from mast cells may be ascribed to the inhibition of intracellular Ca^{2+} mobilization, as in the case of compound 48/80 (Tasaka et al., 1986, 1987). Colchicine was also effective in inhibiting the intracellular Ca^{2+} release, although cytochalasin D was not. It has been reported that microtubules are intimately related to the Ca^{2+} release from the intracellular store (Tasaka et al., 1991), and it was supposed that the depolymerization of microtubules induced by colchicine treatment may be attributable to the inhibition of Ca^{2+} release and the resulting inhibition of histamine release.

It has been reported that cAMP is effective in inhibiting Ca^{2+} release induced by inositol-1, 4, 5-trisphosphate from the endoplasmic reticulum, which is an important intracellular Ca store (Yoshii et al., 1988). In accordance with this, theophylline and db-cAMP were effective in inhibiting both histamine release and intracellular Ca^{2+} release induced by substance P, as shown in Fig. 3. The inhibitory effects on these two parameters are closely related to each other ($r = 0.995$ and $r = 0.996$ in theophylline and dibutyryl cAMP, respectively), and it was assumed that Ca^{2+} release from the intracellular Ca store caused by substance P might be inhibited by increased cAMP levels induced by either theophylline or db-cAMP treatments.

Calmodulin inhibitors are also effective in inhibiting histamine release from mast cells. Although it has been suggested that calmodulin may be in some way related to the histamine release from mast cells (Chakravarty and Nielsen, 1985; Izushi and Tasaka, 1989), the correlation between the histamine release inhibition and the inhibition of the calmodulin activity is not yet elucidated. As shown in Fig. 5, it was observed that the inhibitory potencies of calmodulin inhibitors on histamine release and calmodulin activity are intimately correlated. Therefore, it was assumed that the activation of calmodulin, due to an increase in intracellular Ca^{2+} concentrations, may precede the histamine releasing process. Since it has been suggested that calmodulin may regulate the functions of microfilaments and microtubules (Yazawa et al., 1987; Mercum et al., 1978) and that both microfilaments and microtubules are intimately related to histamine release (Mio and Tasaka, 1989; Tasaka et al., 1991), it can be supposed that substance P activates such cytoskeletal elements by releasing Ca^{2+} from the intracellular store via an activation of calmodulin. In addition, calmodulin activates phosphodiesterase, which leads to a decrease in intracellular cAMP content. This may be the reason for the participation of calmodulin in the process leading to histamine release from mast cells.

It has been suggested that substance P may directly activate G protein (Mousli et al., 1990). In the present study, GTPγS induced histamine release from permeabilized mast cells, indicating that GTPγS is able to activate G proteins. However, substance P and other neuropeptides did not elicit histamine release from permeabilized mast cells. It was assumed that substance P and other neuropeptides may not be able to activate G protein directly. Studies on G protein activation induced by either mastoparan or substance P were carried out in the presence of phospholipids or detergents (Mousli et al., 1990; Higashijima et al., 1990). However, it has been indicated that both substance P and mastoparan are capable of binding to acidic phospholipid, resulting in the perturbation of phospholipid conformation in the lipid bilayer (Shibata et al., 1985; Higashijima et al., 1983), and that the changes in phospholipid conformation can alter the protein-lipid interaction (Boggs et al., 1982). Therefore, it can be assumed that in the case of mastoparan or substance P, G protein activation may be attributable to the indirect action of phospholipid perturbation rather than the direct action to G protein itself.

It has been reported that among tachykinin peptides, substance P is the most potent in releasing histamine from mast cells (Holzer-Petsche et al., 1985; Foreman, 1988). Although C-terminal region of substance P (a common sequence among tachykinin peptides) is required for the binding of receptors observed in smooth muscles and nervous systems, it was indicated that the C-terminal peptide of substance P is not essential in releasing histamine from mast cells, although N-terminal region is requisite for histamine release (Fewtrell et al., 1982; Foreman and Jordan, 1983). Furthermore, some substance P antagonists, which are effective in antagonizing the action of substance P on smooth muscles and nervous systems, are much more potent than substance P in releasing histamine (Foreman and Jordan, 1983). Moreover, it was found that N-terminal tetrapeptide connected to an alkyl chain, which is not active in smooth muscle contraction, is capable of releasing histamine (Repke et al., 1987). Based on these findings, it was assumed that the binding site for substance P on the mast cell surface may be quite different from those found in smooth muscles (Foreman and Jordan, 1983). Actually, the Kd and Bmax values of substance P to the binding site on the mast cell surface were presented here as shown in Fig. 7. It has been reported that the affinity of substance P to the mast cell binding site was lower while the binding capacity on the mast cells was higher than those reported as substance P receptors (Watson, 1984). Since many basic substances, such as compound 48/80, peptide 401, histone and eosinophil major basic protein, are capable of releasing histamine even in a Ca-free medium (Ennis et al., 1980; Tasaka et al., 1990; Aoki et al., 1991a), it is reasonable to assume that these molecules may share the binding sites on the mast cell membrane. Among substance P analogs, none but $[\text{D-Pro}^4, \text{D-Trp}^{7,9,10}]\text{-SP}_{4-11}$ (SP-A) is able to inhibit histamine release from mast cells induced by substance P (Piotrowski and Foreman, 1985). SP-A is also competent in inhibiting the histamine release induced by compound 48/80 (Piotrowski and Foreman, 1985). This observation indicated that compound 48/80 and substance P may share the same site necessary for the activation of the mast cells (Piotrowski and Foreman, 1985). As shown in Fig. 8, compound 48/80 inhibits ^{3}H-substance P binding to rat mast cells. Therefore, it is almost certain that both substance P and compound 48/80 share the same binding site.

2. Histamine release induced by histone and related morphological changes in mast cells

2.1. Introduction

It is known that various polycations are capable of releasing histamine from mast cells. Sagi-Eisenberg et al. (1985) reported that histone releases histamine from rat mast cells in a Ca-free medium and suggested that protein kinase C may participate in the process leading to histamine release. In this section, mechanism of histamine release due to histone was investigated in relation to the morphological changes in the mast cells.

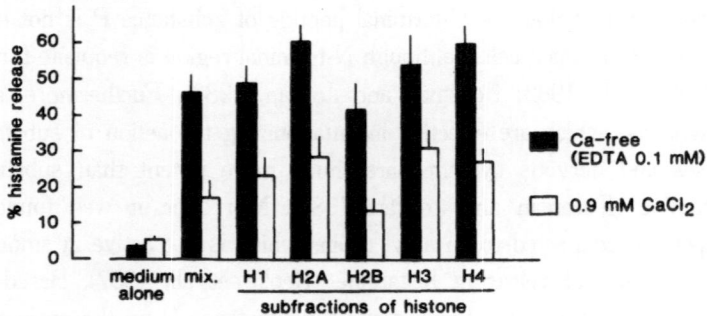

Fig. 10. Histamine release from rat peritoneal mast cells induced by histone mixture and its subfractions. Rat peritoneal mast cells were incubated with 10 μg/ml of histone mixture or each of the subfractions. Each column represents the mean ± SEM (n = 5). Reproduced from Tasaka, K., Mio, M., Akagi, M. and Saito, T.: Agents Actions, 30, 114-117 (1990), with permission.

2.2. Histamine release induced by subfractions of histone protein

The effects of the individual subfractions of histone on the histamine release were studied

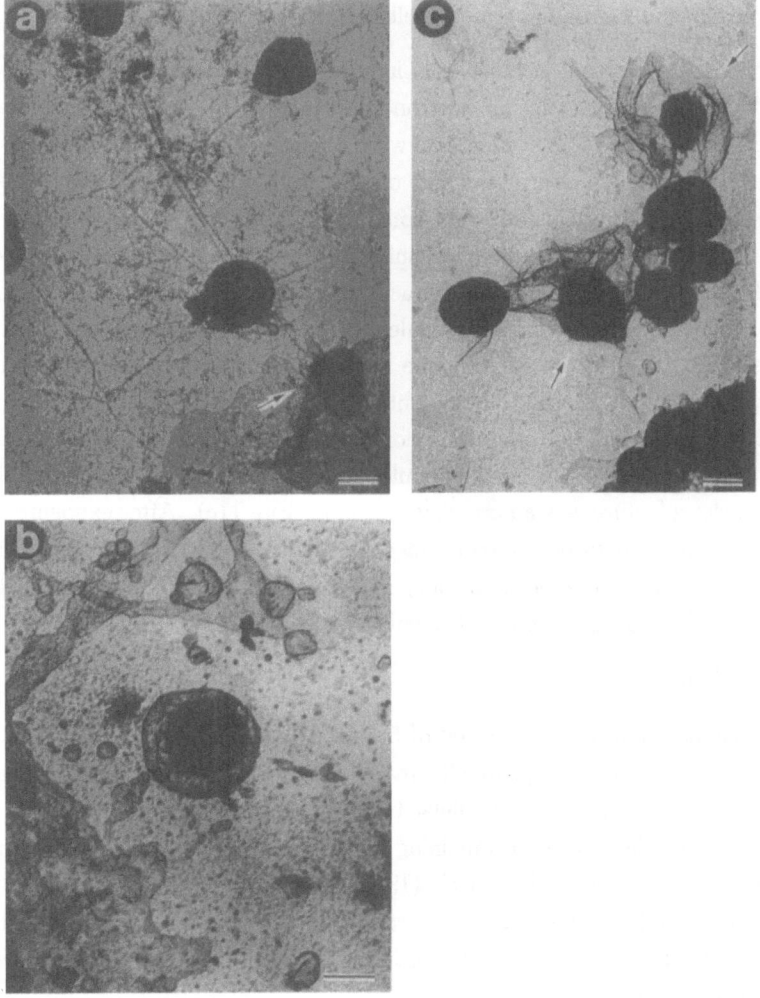

Fig. 11. High voltage-transmission electron microscopic images of rat peritoneal mast cells after histone exposure (10 μg/ml). (a) Many thin filaments were extruded out of the cell in association with granules. In the periphery, the granule (as indicated by an arrow) is about to detach from the cell surface. Many filaments are assembled densely around the bottom of the granule. (b) Membrane enveloped degranulation. An extruded granule enveloped by the granule membrane. (c) Mass degranulation. As indicated by arrows, the granule membranes seemed to be disrupted in these granules. In the lower granule particularly, most of the lower granule has been peeled off. Reproduced from Tasaka, K., Mio, M., Akagi, M. and Saito, T.: Agents Actions, 30, 114-117 (1990), with permission.

in the presence or absence of Ca ion (Fig. 10). All of the histone subfractions, H1, H2A, H2B, H3 and H4, induced histamine release from mast cells in a Ca-free medium and the addition of Ca^{2+} into the medium reduced histamine release as in the case of the histone mixture. The difference in the histamine release between each subfraction was not significant.

2.3. Morphological changes in mast cells stimulated by histone

The morphological changes in mast cells induced by histone mixture were observed by means of high voltage transmission electron microscope (HV-TEM). As indicated in Fig. 11, at least three types of degranulations were observed. After exposure to histone, the granules were observed clearly outside of the cell (Fig. 11a). However, they were interconnected by thin filaments. On some occasions, there was a denser network formation connecting the protruded granules. This can be identified as microfilament-associated degranulation. The other characteristic feature encountered in the degranulation was the appearance of extruded granules which seems to be enveloped by the granule membrane (Fig. 11b). There exists some space and what appear to be several connections between the granule and the membrane. This can be classified as membrane enveloped degranulation. On other occasions, the membranes surrounding the granules seemed to be disrupted, with some granules coming out of their envelopes (Fig. 11c). The third striking feature was a mass degranulation (Fig. 11c). After exposure to histone, granules that appeared to be interconnected by filamentous structure were extruded *en masse* from the cell. On some occasions, approximately 20 granules remained adhering each other on the surface of the cell membrane.

2.4. Discussion

Although histone mixture is composed of 5 subfractions, H1, H2A, H2B, H3 and H4, each subfraction was almost equally effective in releasing histamine from mast cells either in the presence or absence of extracellular Ca^{2+}. In each of these fractions, the histamine release was reduced in a medium containing Ca^{2+}. The results are in good agreement with those reported by Sagi-Eisenberg et al. (1985). It was assumed that Ca^{2+} may compete with histone to attack the negatively charged portions of the surface membrane. However, it should be noted that histone induced Ca^{2+} release from the intracellular Ca store of mast cell, as determined using quin 2.

In the TEM images, it became apparent that the three types of degranulation exhibit characteristic features. After histone exposure, many thin filaments were extruded from the cell in association with granules. This type of degranulation can be called microfilament-associated degranulation, as reported previously (Tasaka and Mio, 1989; Mio and Tasaka, 1989). In such cases, it was postulated that the microfilaments may consist of actin filaments (Tasaka and Mio, 1989; Mio and Tasaka, 1989). Moreover, it has been reported that complicated, dense network formations of cytoskeletons exist inside the cell, interconnecting the granules (Tasaka et al., 1988). Therefore, it can be assumed that when the exocytotic processes are induced simultaneously at several locations, multiple degranulations occur. Since the intergranular filaments were densely connected, these filaments may appear in association with degranulation. Although the constituents of the cytoskeletal systems participating in mass degranulation are not known at present, the microtubules seem to be a plausible candidate. It has already been reported that the extruded granules are seen in a mass close to the cell membrane after local application of compound 48/80 to rat mast cells (Tasaka et al., 1970). In some cases, the extruded granules seemed to be enveloped by membranes even in the extracellular medium. The perigranular space between the granule and the membrane seemed to became wider than that seen in the cytoplasm and several filamentous connections were observed in the perigranular spaces. The widening of the perigranular space may be induced by water entering through the granular membrane and a subsequent cation-exchange reaction and histamine release probably occur in the perigranular space.

3. Guinea pig eosinophil major basic protein as a potent histamine releaser

It is well known that eosinophils accumulate at the sites of allergic inflammation and that proinflammatory proteins such as eosinophil peroxidase, major basic protein (MBP), eosinophil cationic protein (ECP) and eosinophil-derived neurotoxin are released from the activated eosinophils (Ayars et al., 1985; Gleich et al., 1989). MBP is one of these proteins, and may play some important roles in allergic reactions and inflammations, since it is capable of releasing histamine from mast cells and damaging epithelial cells of bronchial tubes (Zheutlin et al., 1984; Gleich et al., 1988). Since the activated mast cells release the eosinophil-activating factors, such as leukotriene B_4, platelet activating factor, ECF-A tetrapeptides and histamine (Pincus et al., 1982; Fechter et al., 1986; Wardlaw et al., 1986), it was assumed that in allergic reactions mast cells and eosinophils activate each other by releasing their own chemical mediators. However, the mechanism of histamine release due to MBP has not been well documented. Moreover, Baker et al. (1988, 1989) reported the cDNA sequence of the MBP and ECP obtained from HL-60 cells. However, the sequence of the MBP derived from experimental animals is not known. Since the guinea pig is one of the most widely used experimental animals in the study of allergic reaction and asthma, it may be of use to determine the cDNA sequence of MBP derived from guinea pig eosinophils. In this section, the mechanism of histamine release induced by guinea pig MBP and the determination of the cDNA sequence of guinea pig eosinophil MBP were investigated.

3.1. Histamine releasing activity of guinea pig major basic protein and its chemical structure

MBP was purified from guinea pig eosinophils according to the method of Gleich et al. (1974) with some modifications (Tasaka et al., 1992a). After Sephadex G-50 column chromatography, the eluted MBP exhibited a single band on SDS-PAGE with a molecular weight of about 11,000. When isoelectric focusing (IEF) was carried out, MBP migrated to the cathode, suggesting that pI value was higher than 10. As indicated in Fig. 12, MBP potently released histamine from rat peritoneal mast cells both in the presence and in the absence of extracellular Ca^{2+} at concentrations higher than $3 \mu g/ml$.

When guinea pig eosinophil MBP was applied to reverse phase HPLC, two distinct peaks were observed (Fig. 13a). By means of SDS-PAGE, the apparent molecular weight of both proteins was 11,000 Dalton. After IEF, the pI values of these two proteins are quite similar to each other. The amino acid compositions of these two peaks are indicated on the right hand side of Fig. 13. Both proteins were rich in arginine residues and the amino acid compositions were quite similar. These two proteins were named GMBP1 and GMBP2, respectively.

When rat peritoneal mast cells were exposed to guinea pig MBP, GMBP1 and GMBP2, each protein effectively liberated histamine from mast cells both in the presence and in the absence of extracellular Ca^{2+} at concentrations higher than $0.3 \mu M$. When the histamine release induced by MBP, GMBP1 and GMBP2 was compared, these proteins

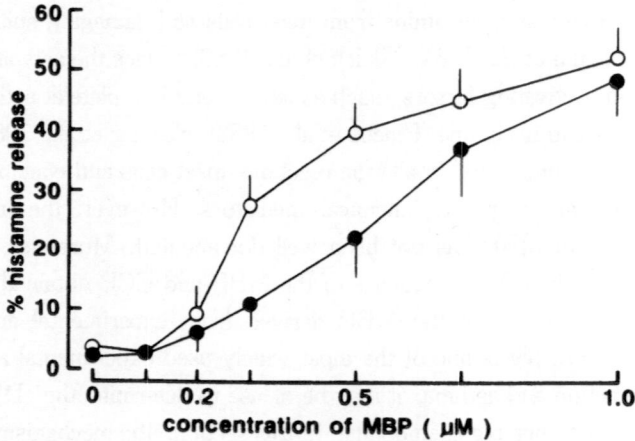

Fig. 12. Histamine release from rat peritoneal mast cells induced by guinea pig MBP in the presence and in the absence of extracellular Ca^{2+}. (●) in the presence of Ca^{2+} (0.9 mM); (○) in the absence of Ca^{2+} (supplemented with 0.1 mM of EGTA). Reproduced from Aoki, I., Shindoh, Y., Nishida, T., Nakai, S., Hong, Y.-M., Mio, M., Saito, T. and Tasaka, K.: FEBS Lett., 279, 330-334 (1991a), with permission.

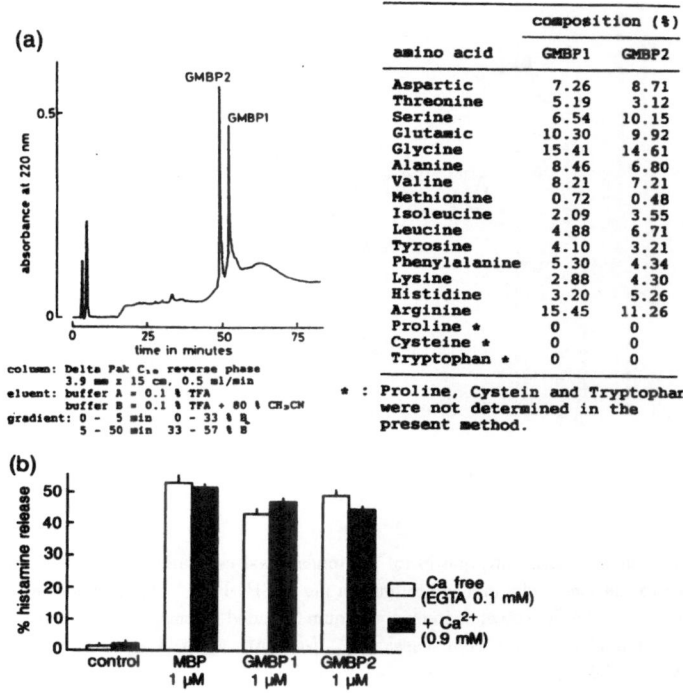

Fig. 13. Isolation of guinea pig MBPs and their histamine releasing activities. (a) Elution profile of guinea pig MBP on reverse phase HPLC and amino acid compositions of GMBP1 and GMBP2. (b) Histamine release induced by guinea pig MBP, GMBP1 and GMBP2 in the presence and in the absence of extracellular Ca²⁺. Each column indicates the mean ± SEM (n = 4). Reproduced from Tasaka, K., Mio, M., Aoki, I. and Saito, T.: Agents Actions, 36/spec. iss., C242-C245 (1992a), with permission.

were similarly effective both in the presence and in the absence of extracellular Ca^{2+} (Fig. 13b).

When rat peritoneal mast cells were exposed to guinea pig MBP at concentrations higher than $0.1\,\mu$M, ^{45}Ca uptake dose-dependently increased as shown in Fig. 14a. As quin 2-loaded rat mast cells were exposed to MBP in a Ca-free medium, the fluorescence intensity derived from quin 2-Ca complex increased rapidly, indicating that Ca^{2+} release takes place from intracellular Ca stores (Fig. 14b).

So far, MBP has been considered to consist of a single protein. Baker et al. (1988) reported that the amino acid sequence of human MBP indicates that MBP is a single protein. In accordance with this, guinea pig MBP exhibited a single band on SDS-PAGE. However, as shown in the reverse phase HPLC, guinea pig MBP is composed of two distinct proteins with similar amino acid compositions. These proteins are similar, not only in amino acid composition, but also in their histamine releasing activity. This

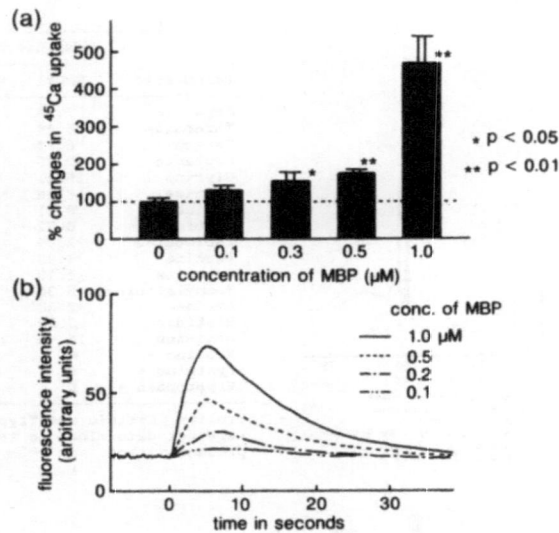

Fig. 14. Changes in Ca²⁺ concentration in rat peritoneal mast cells induced by guinea pig MBP. (a) ^{45}Ca uptake of rat peritoneal mast cells induced by guinea pig MBP. Each column represents the mean ± SEM (n = 4). (b) Changes in the fluorescence intensity of quin 2-loaded rat mast cells induced by guinea pig MBP in a Ca free medium. Reproduced from Tasaka, K., Mio, M., Aoki, I. and Saito, T.: Agents Actions, 36/spec. iss., C242-C245 (1992a), with permission.

may suggest that these two proteins belong to the subtypes of MBP. Furthermore, GMBP1 and GMBP2 are rich in arginine residues and this may be the reason for the high basicity of these proteins. Although there are no reports regarding the MBP subtypes in any other species, it may be assumed that MBP subtypes are likely to exist in other animals including humans.

It was found that GMBP1 and GMBP2 were active in releasing histamine even in a Ca-free medium as was also found with guinea pig MBP. In the absence of extracellular Ca²⁺, MBP caused Ca²⁺ release from intracellular Ca stores. On the other hand, in the presence of extracellular Ca²⁺, MBP elicited Ca uptake into rat mast cells. The influence of extracellular Ca²⁺ on the histamine release due to MBP was similar to that observed in the case of histone (Tasaka et al., 1990). Also, in the cases of compound 48/80 (Ennis et al., 1980) and substance P (Fewtrell et al., 1982), histamine release in response to these polycationic substances takes place even in a Ca-free medium. It is possible that the mechanism of MBP-induced histamine release is similar to that of other polycationic histamine releasing agents.

3.2. Sequencing and cloning of the cDNA of guinea pig MBP

The partial amino acid sequences of GMBP1 and GMBP2 compared with that of human

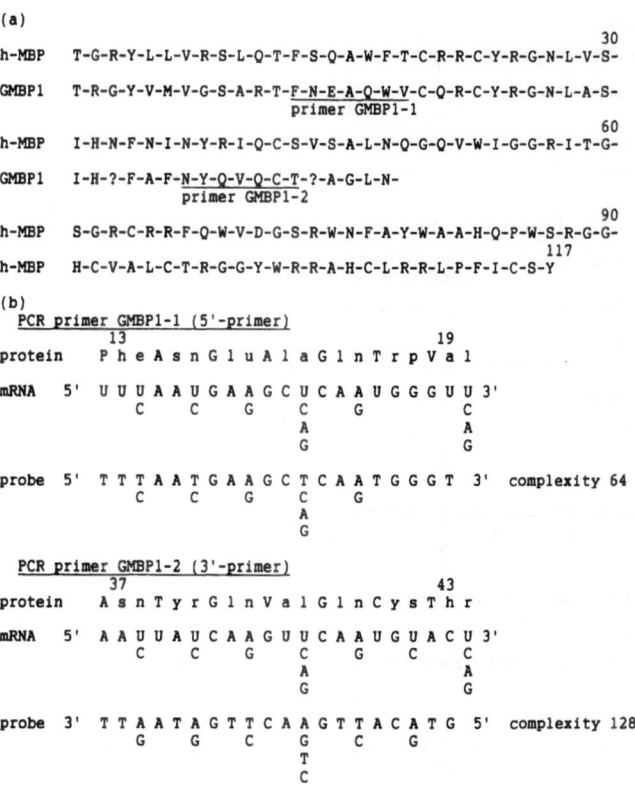

Fig. 15. Partial amino acid sequences of GMBP1 and nucleotide sequence of the synthesized PCR primers. (a) Partial amino acid sequence of GMBP1 in comparison with human MBP (Baker et al., 1988); (b) Nucleotide sequence of the synthesized PCR primers in the underlined positions in Fig. 15a. Reproduced from Aoki, I., Shindoh, Y., Nishida, T., Nakai, S., Hong, Y.-M., Mio, M., Saito, T. and Tasaka, K.: FEBS Lett., 279, 330-334 (1991a), with permission.

MBP are indicated in Fig. 15a and 16a, respectively. According to these partial sequences, polymerase chain reaction (PCR) primers, which corresponds to the underlined positions, were synthesized as indicated in Fig. 15b and 16b. Using these oligonucleotide primers and cDNA obtained from guinea pig eosinophils, PCR was carried out in order to amplify the partial sequences of cDNA of GMBP1 and GMBP2 to obtain the hybridization probes for the screening of the λ gt10 cDNA library. After PCR, the length of amplified DNA of GMBP1 was 92 bp and that of GMBP2 was 185 bp, respectively. The sequences of both oligonulceotides coincided well with that determined from the amino acid sequences, and these oligonucleotides were used for the screening of the cDNA library of eosinophils. After plaque hybridization using ^{32}P-labeled probes, positive clones for GMBP1 and GMBP2 were obtained from 8×10^5 and 2.8×10^6

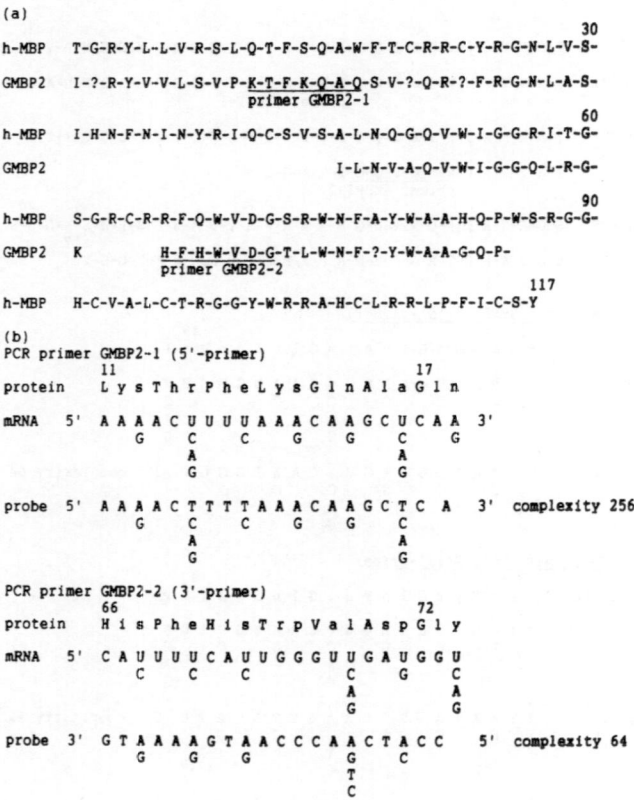

Fig. 16. Partial amino acid sequence of GMBP2 and nucleotide sequence of the synthesized PCR primers. (a) Partial amino acid sequence of GMBP2 in comparison with human MBP (Baker et al., 1988). (b) Nucleotide sequences of the synthesized PCR primers in the underlined positions in Fig. 16a. Reproduced from Aoki, I., Shindoh, Y., Nishida, T., Nakai, S., Hong, Y.-M., Mio, M., Saito, T. and Tasaka, K.: FEBS Lett., 282, 56-60 (1991b), with permission.

independent clones, respectively.

Fig. 17 and 18 represent the determined sequences of the cDNA of GMBP1 and GMBP2, respectively. Comparing the amino acid sequence with human MBP (Baker et al., 1988), and those obtained from GMBP1 and GMBP2 protein, it became apparent that these cDNAs correspond to the full length of the mRNA of MBP. It was indicated that these cDNAs encoded pre-proMBP may be composed of a signal peptide, an acidic domain and mature MBP, as in the case of human MBP (Baker et al., 1988). Although the mature domains of GMBP1 and GMBP2 are highly basic (calculated pI values are 11.7 and 11.3, respectively), pro-proteins of GMBP1 and GMBP2 are slightly acidic (calculated pI values are 5.3 and 5.6, respectively) due to the existence of the acidic

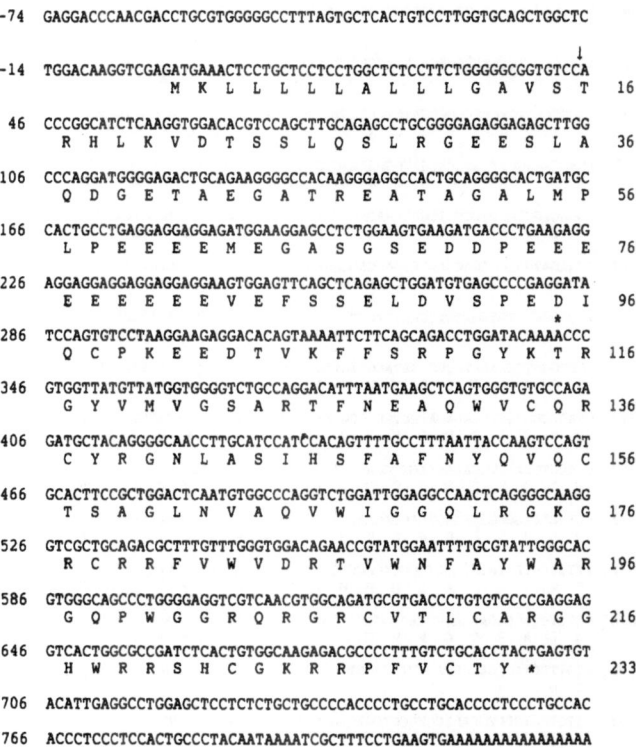

```
 -74   GAGGACCCAACGACCTGCGTGGGGGCCTTTAGTGCTCACTGTCCTTGGTGCAGCTGGCTC

                                                          ↓
 -14   TGGACAAGGTCGAGATGAAACTCCTGCTCCTCCTGGCTCTCCTTCTGGGGGCCGGTGTCCA
                M  K  L  L  L  L  L  A  L  L  L  G  A  V  S  T    16

  46   CCCGGCATCTCAAGGTGGACACGTCCAGCTTGCAGAGCCTGCGGGGAGAGGAGAGCTTGG
        R  H  L  K  V  D  T  S  S  L  Q  S  L  R  G  E  E  S  L  A   36

 106   CCCAGGATGGGGAGACTGCAGAAGGGGCCACAAGGGAGGCCACTGCAGGGGCACTGATGC
        Q  D  G  E  T  A  E  G  A  T  R  E  A  T  A  G  A  L  M  P   56

 166   CACTGCCTGAGGAGGAGGAGATGGAAGGAGCCTCTGGAAGTGAAGATGACCCTGAAGAGG
        L  P  E  E  E  M  E  G  A  S  G  S  E  D  D  P  E  E  E      76

 226   AGGAGGAGGAGGAGGAGGAAGTGGAGTTCAGCTCAGAGCTGGATGTGAGCCCCGAGGATA
        E  E  E  E  E  V  E  F  S  S  E  L  D  V  S  P  E  D  I      96
                                                          *
 286   TCCAGTGTCCTAAGGAAGAGGACACAGTAAAATTCTTCAGCAGACCTGGATACAAAACCC
        Q  C  P  K  E  E  D  T  V  K  F  F  S  R  P  G  Y  K  T  R  116

 346   GTGGTTATGTTATGGTGGGGTCTGCCAGGACATTTAATGAAGCTCAGTGGGTGTGCCAGA
        G  Y  V  M  V  G  S  A  R  T  F  N  E  A  Q  W  V  C  Q  R  136

 406   GATGCTACAGGGGCAACCTTGCATCCATCCACAGTTTTGCCTTTAATTACCAAGTCCAGT
        C  Y  R  G  N  L  A  S  I  H  S  F  A  F  N  Y  Q  V  Q  C  156

 466   GCACTTCCGCTGGACTCAATGTGGCCCAGGTCTGGATTGGAGGCCAACTCAGGGGCAAGG
        T  S  A  G  L  N  V  A  Q  V  W  I  G  G  Q  L  R  G  K  G  176

 526   GTCGCTGCAGACGCTTTGTTTGGGTGGACAGAACCGTATGGAATTTTGCGTATTGGGCAC
        R  C  R  R  F  V  W  V  D  R  T  V  W  N  F  A  Y  W  A  R  196

 586   GTGGGCAGCCCTGGGGAGGTCGTCAACGTGGCAGATGCGTGACCCTGTGTGCCCGAGGAG
        G  Q  P  W  G  G  R  Q  R  G  R  C  V  T  L  C  A  R  G  G  216

 646   GTCACTGGCGCCGATCTCACTGTGGCAAGAGACGCCCCTTTGTCTGCACCTACTGAGTGT
        H  W  R  R  S  H  C  G  K  R  R  P  F  V  C  T  Y  *         233

 706   ACATTGAGGCCTGGAGCTCCTCTCTGCTGCCCCACCCCTGCCTGCACCCCTCCCTGCCAC

 766   ACCCTCCCTCCACTGCCCTACAATAAAATCGCTTTCCTGAAGTGAAAAAAAAAAAAAAAA
```

Fig. 17. Nucleotide sequence of GMBP1 cDNA. Nucleotides -74-825 and amino acid 1-233 are numbered on the left and right hand sides of the figure, respectively. An arrow indicates the putative signal peptide cleavage site. An asterisk indicates the first amino acid in eosinophil granule GMBP1. Reproduced from Aoki, I., Shindoh, Y., Nishida, T., Nakai, S., Hong, Y.-M., Mio, M., Saito, T. and Tasaka, K.: FEBS Lett., **279**, 330-334 (1991a), with permission.

domain of pro-protein. The calculated pI value of the acidic domain was 3.8 in each case.

The calculated molecular weights of GMBP1 and GMBP2 are 13.8 kDa in either case. The apparent molecular weight of guinea pig MBP determined on SDS-PAGE was 11 kDa as reported Gleich et al. (1974). The possible reason for the difference in the molecular weights between the calculated value and that determined on SDS-PAGE can be ascribed to the highly positive charges in MBP, which may influence the migration of SDS-bound proteins on SDS-PAGE to the position corresponding to its molecular weight.

Comparison of the amino acid sequences of human and guinea pig MBPs is indicated in Fig. 19. The homology in the amino acid sequences of pre-pro-proteins between human MBP and GMBP1 was 49.4%, although that of mature MBP was 58%. Similarly, the homology in pre-pro-protein sequences between human MBP and GMBP2

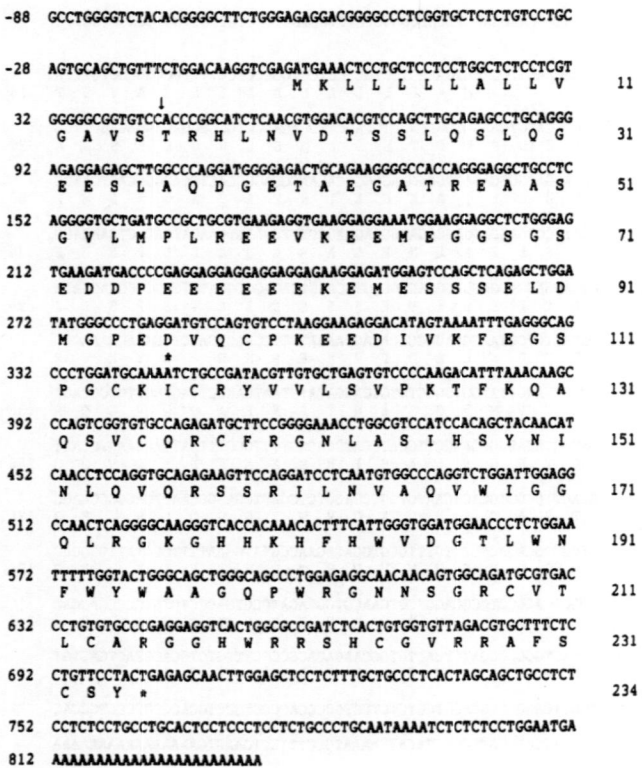

```
-88  GCCTGGGGTCTACACGGGGCTTCTGGGAGAGGACGGGGCCCTCGGTGCTCTCTGTCCTGC

-28  AGTGCAGCTGTTTCTGGACAAGGTCGAGATGAAACTCCTGCTCCTCCTGGCTCTCCTCGT
                                  M  K  L  L  L  L  L  A  L  L  V    11
                              ↓
 32  GGGGGCGGTGTCCACCCGGCATCTCAACGTGGACACGTCCAGCTTGCAGAGCCTGCAGGG
      G  A  V  S  T  R  H  L  N  V  D  T  S  S  L  Q  S  L  Q  G     31

 92  AGAGGAGAGCTTGGCCCAGGATGGGGAGACTGCAGAAGGGGCCACCAGGGAGGCTGCCTC
      E  E  S  L  A  Q  D  G  E  T  A  E  G  A  T  R  E  A  A  S     51

152  AGGGGTGCTGATGCCGCTGCGTGAAGAGGTGAAGGAGGAAATGGAAGGAGGCTCTGGGAG
      G  V  L  M  P  L  R  E  E  V  K  E  E  M  E  G  G  S  G  S     71

212  TGAAGATGACCCGAGGAGGAGGAGGAGGAGAAGGAGATGGAGTCCAGCTCAGAGCTGGA
      E  D  D  P  E  E  E  E  E  K  E  M  E  S  S  E  L  D           91

272  TATGGGCCCTGAGGATGTCCAGTGTCCTAAGGAAGAGGACATAGTAAAATTTGAGGGCAG
      M  G  P  E  D  V  Q  C  P  K  E  E  D  I  V  K  F  E  G  S    111
                                  *
332  CCCTGGATGCAAAATCTGCCGATACGTTGTGCTGAGTGTCCCCAAGACATTTAAACAAGC
      P  G  C  K  I  C  R  Y  V  V  L  S  V  P  K  T  F  K  Q  A    131

392  CCAGTCGGTGTGCCAGAGATGCTTCCGGGGAAACCTGGCGTCCATCCACAGCTACAACAT
      Q  S  V  C  Q  R  C  F  R  G  N  L  A  S  I  H  S  Y  N  I    151

452  CAACCTCCAGGTGCAGAGAAGTTCCAGGATCCTCAATGTGGCCCAGGTCTGGATTGGAGG
      N  L  Q  V  Q  R  S  S  R  I  L  N  V  A  Q  V  W  I  G  G    171

512  CCAACTCAGGGGCAAGGGTCACCACAAACACTTTCATTGGGTGGATGGAACCCTCTGGAA
      Q  L  R  G  K  G  H  H  K  H  F  H  W  V  D  G  T  L  W  N    191

572  TTTTTGGTACTGGGCAGCTGGGCAGCCCTGGAGAGGCAACAACAGTGGCAGATGCGTGAC
      F  W  Y  W  A  A  G  Q  P  W  R  G  N  N  S  G  R  C  V  T    211

632  CCTGTGTGCCCGAGGAGGTCACTGGCGCCGATCTCACTGTGGTGTTAGACGTGCTTTCTC
      L  C  A  R  G  G  H  W  R  R  S  H  C  G  V  R  R  A  F  S    231

692  CTGTTCCTACTGAGAGCAACTTGGAGCTCCTCTTTGCTGCCCTCACTAGCAGCTGCCTCT
      C  S  Y  *                                                   234

752  CCTCTCCTGCCTGCACTCCTCCCTCCTCTGCCCTGCAATAAAATCTCTCTCCTGGAATGA

812  AAAAAAAAAAAAAAAAAAAAAAAA
```

Fig. 18. Nucleotide sequence of GMBP2 cDNA. Nucleotides -88-835 and amino acid 1-234 are numbered on the left and right hand sides of the figure, respectively. An arrow indicates the putative signal peptide cleavage site. An asterisk indicates the first amino acid in eosinophil granule GMBP2. Reproduced from Aoki, I., Shindoh, Y., Nishida, T., Nakai, S., Hong, Y.-M., Mio, M., Saito, T. and Tasaka, K.: FEBS Lett., 282, 56-60 (1991b), with permission.

was 48.7%, although in mature MBPs the homology was 55.4%. The homologies in pre-pro-proteins and mature proteins between GMBP1 and GMBP2 were 74.2% and 67.5%, respectively. In contrast, the nucleotide sequences of the coding regions of GMBP1 and GMBP2 were homologous by 81.1%.

Fig. 20 represents Harr plot analyses on the primary structures between the pre-pro-proteins of GMBP1, GMBP2 and human MBP. As indicated in Fig. 20a, GMBP1 and GMBP2 were homologous in both pro-portion and mature regions. On the other hand, when human MBP was compared with GMBP1 and GMBP2, the basic regions in the mature proteins were similar and it was considered that the basic region is well conserved, although the primary sequences in the acidic regions of pro-portions were not very homologous despite their similar pI values (Fig. 20b, c). In each case, the

a)
```
GMBP1     1' MKLLLLLALLLGAVSTRHLKVDTSSLQSLRGEESLAQDGETAEGATREATAGALMPLPEE
             ********** ******** ********* ******************* * **** **
GMBP2     1" MKLLLLLALLVGAVSTRHLNVDTSSLQSLQGEESLAQDGETAEGATREAASGVLMPLREE

GMBP1    61' --EEMEGASGSEDDPEEEEEEEEEVEFSSELDVSPEDIQCPKEEDTVKFFSRPGYKTRGY
             ************* ****** * * ***** *** ******* *** ** *
GMBP2    61" VKEEMEGGSGSEDDP-EEEEEEKEMESSSELDMGPEDVQCPKEEDIVKFEGSPGCKICRY

GMBP1   119' VMVGSARTFNEAQWVCQRCYRGNLASIHSFAFNYQVQCTSAGLNVAQVWIGGQLRGKGRC
             *   ** ***** ********* * *** ****************
GMBP2   120" VVLSVPKTFKQAQSVCQRCFRGNLASIHSYNINLQVQRSSRILNVAQVWIGGQLRGKGHH

GMBP1   179' RRFVWVDRTVWNFAYWARGQPWGGRQRGRCVTLCARGGHWRRSHCGKRRPFVCTY
             * *** * *** *** **** * ******************** ** * * *
GMBP2   180" KHFHWVDGTLWNFWYWAAGQPWRGNNSGRCVTLCARGGHWRRSHCGVRRAFSCSY
```

b)
```
GMBP1     1' MKLLLLLALLLGAVSTRHLKVDTSSLQSLRGEESLAQDGETAEGATREATAGALMPLPEE
             *** ****** **** ** ** * * *** * * *
h-MBP     1" MKLPLLLALLLFGAVSALHLRSETSTFETPLGAKTLPEDEETPEQEMEETPCREL-----E

GMBP1    61' EEMEGASGSEDDPEEEEEEEEEVEFSSELDVSPEDIQCPKEEDTVKFFSRPGYKTRGYVM
             ** * **** * * * ****** ** * *
h-MBP    56" EEEEWGSGSE-DASKKDGAVESISVPDMVD---KNLTCPEEEDTVKVVGIPGCQTCRYLL

GMBP1   121' VGSARTFNEAQWVCQRCYRGNLASIHSFAFNYQVQCTSAGLNVAQVWIGGQLRGKGRCRR
             * * ** * * ******** *** * ** ** ** ****** * *****
h-MBP   112" VRSLQTFSQAWFTCRRCYRGNLVSIHNFNINYRIQCSVSALNQGQVWIGGRITGSGRCRR

GMBP1   181' FVWVDRTVWNFAYWARGQPWGGRQRGRCVTLCARGGHWRRSHCGKRRPFVCTY
             * *** ******* *** * ** ** *** *** ** * **
h-MBP   172" FQWVDGSRWNFAYWAAHQPW--SRGGHCVALCTRGGYWRRAHCLRRLPFICSY
```

c)
```
GMBP2     1' MKLLLLLALLVGAVSTRHLNVDTSSLQSLQGEESLAQDGETAEGATREAASGVLMPLREE
             *** ****** **** ** ** * * *** * *
h-MBP     1" MKLPLLLLALLFGAVSALHLRSETSTFETPLGAKTLPEDEETPE-QEMEETPC------RE

GMBP2    61' VKEEMEGGSGSEDDPEEEEEEKEMESSSELDMGPEDVQCPKEEDIVKFEGSPGCKICRYV
             ** * ***** ** * ** *** *** ** * *** ***
h-MBP    54" LEEEEWGSGSE---DASKKDGAVESISVPDMVDKNLTCPEEEDTVKVVGIPGCQTCRYL

GMBP2   121' VLSVPKTFKQAQSVCQRCFRGNLASIHSYNINLQVQRSSRILNVAQVWIGGQLRGKGHHK
             ** ** * ** **** *** *** * * ** ****** * *
h-MBP   111" LVRSLQTFSQAWFTCRRCYRGNLVSIHNFNINYRIQCSVSALNQGQVWIGGRITGSGRCR

GMBP2   181' HFHWVDGTLWNFWYWAAGQPWRGNNSGRCVTLCARGGHWRRSHCGVRRAFSCSY
             * **** **** *** ** * ** *** *** ** * * ***
h-MBP   171" RFQWVDGSRWNFAYWAAHQPW--SRGGHCVALCTRGGYWRRAHCLRRLPFICSY
```

Fig. 19. Comparison of the amino acid sequences of pre-proMBP in guinea pig (GMBP1 and GMBP2) and human (h-MBP). (a) GMBP1 and GMBP2, (b) GMBP1 and h-MBP, and (c) GMBP2 and h-MBP. Identical residues are indicated by an asterisk. Gaps have been introduced to achieve maximum sequence homology. Reproduced from Aoki, I., Shindoh, Y., Nishida, T., Nakai, S., Hong, Y.-M., Mio, M., Saito, T. and Tasaka, K.: FEBS Lett., 282, 56-60 (1991b), with permission.

signal peptides closely resembled to each other.

It is well known that MBP is a major constituent of eosinophil granules. Since isolated guinea pig MBP exhibits a single band on SDS-PAGE and IEF (migrated to the cathode position) (Gleich et al., 1974), it had been thought that guinea pig MBP was a single protein. However, when guinea pig MBP was applied to reverse phase HPLC, 2 peaks were recognized (Aoki et al., 1991a; Tasaka et al., 1992a). These two proteins not only exhibited almost identical molecular weight and pI values on SDS-PAGE and IEF (migrated to cathode position), respectively, but also they retained similar histamine releasing activity. Therefore, it is indicated that guinea pig MBP consists of 2 compo-

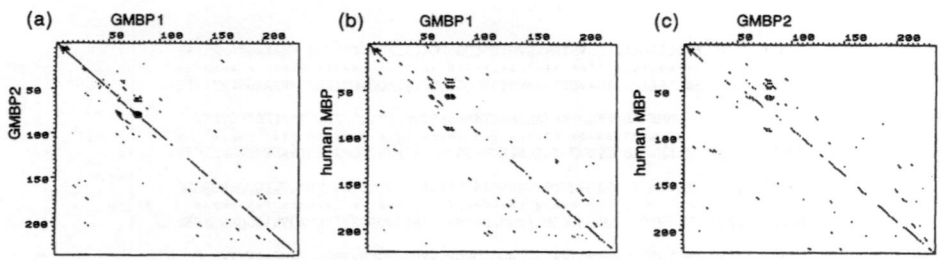

Fig. 20. Harr plot analyses of the primary structures between GMBP1, GMBP2 and human MBP. (a) GMBP1 and GMBP2, (b) GMBP1 and human MBP, and (c) GMBP2 and human MBP. Harr plot was carried out by GENETYX software with a personal computer. The program was conditioned as follows: minimum length $= 5$ and matching $= 3/5$ or more. Each dot represents a common amino acid residue between indicated proteins. Reproduced from Aoki, I., Shindoh, Y., Nishida, T., Nakai, S., Hong, Y.-M., Mio, M., Saito, T. and Tasaka, K.: FEBS Lett., 282, 56-60 (1991b), with permission.

nents, namely GMBP1 and GMBP2 (Aoki et al., 1991a; Tasaka et al., 1992a). Even in cDNA sequences, the calculated molecular weights of GMBP1 and GMBP2 were almost identical (13.8 kDa). This value is exactly the same as that of human MBP (Baker et al., 1988). The calculated pI values of GMBP1 and GMBP2 were also quite similar (11.6 and 11.3, respectively). Since the pH range of the IEF is usually 10 at the highest, it was supposed that both proteins migrated to the cathode position. The molecular weights of GMBP1, GMBP2 and human MBP determined on SDS-PAGE were 11, 11 and 9.3 kDa (Tasaka et al., 1992a; Gleich et al., 1976). The difference in the molecular weights between the calculated values and those determined by SDS-PAGE may be ascribed to the strong basicity of MBPs, which may hinder the migration of proteins to the position corresponding to the molecular weights. The similarity in pI values and molecular weights may well be the reason for the difficulty in separating these two proteins by SDS-PAGE or IEF.

It has been supposed that the cytotoxic activity of MBP may be due to its highly basic charge (Gleich et al., 1974, 1988). Therefore, the acidic domain of pro-protein may be of use to neutralize the basicity of mature GMBPs as in the case of human MBP (Baker et al., 1988). It was assumed that both GMBP1 and GMBP2 would be firstly translated as pre-pro-proteins, which would be converted to pro-proteins and mature proteins in order as in the case of human MBP (Baker et al., 1988).

The amino acid sequence near the cleavage site between pro-portions and mature GMBP2 was slightly different from that seen in GMBP1 (G-S-P-G-C-K*I-C-R-Y-V-V and S-R-P-G-Y-K*T-R-G-Y-V-M, respectively). The common amino acid sequence between these sites is x-x-P-G-x-K*x-x-x-Y-V-x. In the case of human MBP, the cleavage site processing from proMBP to mature MBP is G-I-P-G-C-Q*T-C-R-Y-L-L (Baker et al., 1988). This sequence more closely resembles the sequence in GMBP2 than that seen in GMBP1. The common amino acid sequence at this cleavage site between human MBP and GMBP2 is G-x-P-G-C-x*x-C-R-Y-x-x. Although it is not known what kind of enzyme(s) participates in the processing of proMBP to mature MBP, the above sequence may be a plausible candidate for the recognition site of the endo-protease. In contrast, the amino acid sequences of putative signal peptides and the cleavage sites of these three proteins are quite similar to each other (GMBP1: M-K-L-L-L-L-L-L-A-L-L-L-G-A-V-S*T-R-H-L, GMBP2: M-K-L-L-L-L-L-A-L-L-V-G-A-V-S*T-R-H-L, and human MBP: M-K-L-P-L-L-L-A-L-L-F-G-A-V-S*A-L-H-L) (Baker et al., 1988; Aoki et al., 1991a, b). The lengths of signal peptides of these three proteins are all 15 amino acids. The common sequence in these three proteins is M-K-L-x-L-L-L-A-L-L-x-G-A-V-S*x-x-H-L. It was assumed that a quite similar mechanism would participate in the cleavage process in these proteins. As indicated in the Harr plot analysis, the signal peptides of these three peptides closely resembled each other. Homologous sequences were observed in the basic region of mature MBPs. Although the pI values of the pro-portions of three MBPs are all 3.8, the homology was rather low between human MBP and guinea pig MBPs.

From the present study, it was indicated that GMBP1 and GMBP2 may be two subclasses of guinea pig MBP. Although only one MBP is reported in human eosino-phils, it is possible to assume that there exists a subclass of MBP in other species, including humans.

3.3. Structure-activity relationships of GMBP1 and GMBP2 in releasing histamine

From the studies indicated above, it became apparent that guinea pig MBP is composed of two different proteins (GMBP1 and GMBP2), and that these proteins are similar not only in the extent of histamine releasing activity but also with respect to not requiring extracellular Ca^{2+} for histamine release (Tasaka et al., 1992a). From the cDNA cloning studies, it became clear that the mature MBPs are rich in arginine residues and are highly basic proteins. Therefore, it was assumed that the basic nature of MBP may be responsible for the histamine release from mast cells (Zheutlin et al., 1984). However, the active site(s) prerequisite for histamine release is not yet known. To clarify the structure-activity relationship of MBP in histamine release, partial oligopeptides consist-ing of GMBP1 and GMBP2 were synthesized and the histamine releasing activity of these peptides was compared.

Referring to the amino acid sequences of mature GMBP1 and GMBP2, several portions in the sequences were selected and the oligopeptides corresponding to these regions were synthesized by means of a peptide synthesizer (Fig. 21). The synthesized oligopeptides were purified by means of reverse phase HPLC. For peptides No. 1 and No. 2 the amino acid sequence was the same in both GMBP1 and GMBP2. Peptides No. 3 and 4, peptide No. 5 and 6, and peptide No. 7 and 8 are corresponding peptides

(a)

```
                           1                    2                    3
GMBP1   1  TRGYVMVGSARTFNEAQWVCQRCYRGNLASIHSFAFNYQVQCTSAGLNVAQVWIGGQLRG   60
           **        **  **  *****  *********    *  ***   *  **************
GMBP2   1  ICRYVVLSVPKTFKQAQSVCQRCFRGNLASIHSYNINLQVQRSSRILNVAQVWIGGQLRG   60
                           1                    2                    4
```

```
                           5                          7
GMBP1  61  KGRCRRFVWVDRTVWNFAYWARGQPWGGRQRGRCVTLCARGGHWRRSHCGKRRPFVCTY  119
           **      *  *** *  ***  *** ****  *   ******************** ** * * *
GMBP2  61  KGHHKHFHWVDGTLWNFWYWAAGQPWRGNNSGRCVTLCARGGHWRRSHCGVRRAFSCSY  119
                           6                          8
```

(b)

Peptide No.

1	Gly-Asn-Leu-Ala-Ser-Ile-His-Ser	7.82	common in GMBP1 and 2
2	Leu-Asn-Val-Ala-Gln-Val-Trp-Ile-Gly-Gly	5.96	common in GMBP1 and 2
3	Leu-Arg-Gly-Lys-Gly-Arg-Cys-Arg-Arg-Phe	12.50	GMBP1
4	Leu-Arg-Gly-Lys-Gly-His-His-Lys-His-Phe	11.70	GMBP2
5	Gly-Gly-Arg-Gln-Arg-Gly-Arg-Cys-Val-Thr	12.20	GMBP1
6	Arg-Gly-Asn-Asn-Ser-Gly-Arg-Cys-Val-Thr	10.90	GMBP2
7	Gly-Lys-Arg-Arg-Pro-Phe-Val-Cys-Thr-Tyr	10.42	GMBP1
8	Gly-Val-Arg-Arg-Ala-Phe-Ser-Cys-Ser-Tyr	9.86	GMBP2

Fig. 21. Amino acid sequences of GMBP1, GMBP2 and synthesized oligopeptides. (a) Amino acid sequences of mature GMBP1 and GMBP2. The synthesized oligopeptides are underlined. *: homologous amino acid; R, K, H: basic amino acid; D, E, Y, C: acidic amino acid. As assessed from the number of homologous amino acids, these two proteins are highly homologous. (b) Amino acid sequences and calculated pI values of the synthesized partial oligopeptides. Reproduced from Tasaka, K., Mio, M., Nakai, S. and Aoki, I.: Agents Actions, 36/Spec. Iss., C246-C249 (1992b), with permission.

in these two proteins, respectively. As indicated in Fig. 21b, peptides 3 - 8 are all basic, while peptides No. 1 and 2 are not basic; all amino acid residues are non-polar. Peptide No. 3 was the most basic amongst them. In peptide No. 3, 4, 7 and 8, more than two of the basic amino acids were arranged side by side, and a phenylalanine residue was observed in all of these peptides. In No. 5 and 6, the pI values were high, but basic and neutral amino acids were placed alternatively.

The histamine releasing activity of the oligopepides in the presence of extracellular Ca^{2+} is indicated in Fig. 22a. As shown in this figure, peptide No. 3 was the most potent, and peptides No. 7 and 8 were the next most potent. Other than peptide No. 4, the rest had no effect at all. The histamine releasing activity of these peptides in a Ca free medium is indicated in Fig. 22b. In peptides No. 3, 4, 7 and 8, histamine release was greater than that observed in the presence of Ca^{2+}. When the histamine releasing activity

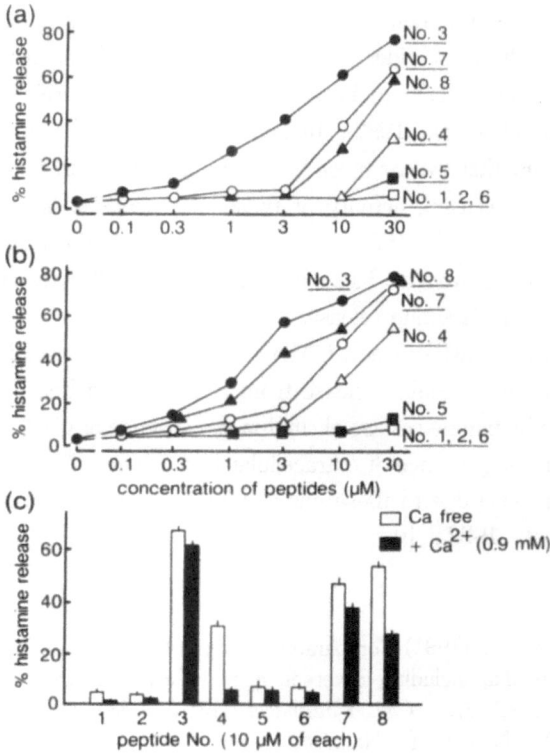

Fig. 22. Histamine release induced by partial oligopeptides of GMBP1 and GMBP2. (a) Histamine release from rat peritoneal mast cells induced by partial oligopeptides of GMBP1 and GMBP2 in the presence of extracellular Ca^{2+} (0.9 mM). The data shown are the mean values from 4 separate experiments. The standard error bars have been omitted for reasons of clarity. (b) Histamine release from rat peritoneal mast cells induced by partial oligopeptides of GMBP1 and GMBP2 in a Ca free medium. The data shown are the mean values from 4 separate experiments. The standard error bars have been omitted for reasons of clarity. (c) Histamine release from rat peritoneal mast cells induced by a fixed concentration (10 μM) of partial oligopeptides of GMBP1 and GMBP2 in the presence and absence of extracellular Ca^{2+}. Each column represents mean $\pm$ SEM (n = 4). Reproduced from Tasaka, K., Mio, M., Nakai, S. and Aoki, I.: Agents Actions, 36/Spec.Iss., C246-C249 (1992b), with permission.

of these oligopeptides was compared in the presence and in the absence of extracellular Ca^{2+} (Fig. 22c), it is obvious that the histamine releasing activity of peptide No. 3 was not dependent upon extracellular Ca^{2+}. The histamine releasing activity of peptide No. 4, 7 and 8 was more pronounced in a Ca free medium.

In the active peptides indicated in Fig. 22, such as No. 3, 4, 7 and 8, more than two of the basic amino acids were arranged side by side. Although the pI values of the inactive peptides (No. 5 and 6) were as high as those of the active peptides, basic and neutral amino acids were arranged alternatively in these two peptides. Neutral or weakly acidic peptides (No. 1 and 2) were not active at all. Moreover, in all of these active peptides, the existence of a phenylalanine residue was recognized. Considering the amino acid sequences of these peptides, it was noticed that both basic amino acids and hydrophcbic amino acids seem to be essential for histamine release, as in the case of substance P (Fewtrell et al., 1982). In addition, in order to elicit effective histamine release it is necessary that two or more basic amino acids occur in sequence. On a molar basis, the histamine releasing activity of these oligopeptides is lower than that of the original MBP (Tasaka et al., 1992a). It is supposed that the original MBP possesses several hydrophobic as well as basic regions in the molecule, and both regions are effectively involved in histamine release.

As in the cases of GMBP1 and GMBP2 (Tasaka et al., 1992a), extracellular Ca^{2+} was not necessary for histamine release from mast cells elicited by these oligopeptides. In most cases, the histamine release elicited in the absence of Ca^{2+} was more pronounced than that elicited in the presence of extracellular Ca^{2+}. The influence of extracellular Ca^{2+} on the histamine release due to these oligopeptides is similar to that observed in mature MBP (Tasaka et al., 1992a, b).

References

Akagi M. and Tasaka K. (1991) Comparative study of the adverse effects of various radiographic contrast media, including ioversol, a new low-osmolarity medium: (I) Histamine release. *Meth. Find. Exp. Clin. Pharmacol.*, 13: 377-384

Aoki I., Shindoh Y., Nishida T., Nakai S., Hong Y.-M., Mio M., Saito T. and Tasaka K. (1991a) Sequencing and cloning of the cDNA of guinea pig eosinophil major basic protein. *FEBS Lett.*, 279: 330-334

Aoki I., Shindoh Y., Nishida T., Nakai S., Hong Y.-M., Mio M., Saito T. and Tasaka K. (1991b) Comparison of the amino acid and nucleotide sequences between human and guinea pig major basic proteins. *FEBS Lett.*, 282: 56-60

Ayars G.H., Altman L.C., Gleich G.J., Loegering D.A. and Baker C.B. (1985) Eosinophil- and eosinophil granule-mediated pneumocyte injury. *J. Allergy Clin. Immunol.*, 76: 595-604

Barker R.L., Gleich G.J. and Pease L.R. (1988) Acidic precursor revealed in human eosinophil granule major basic protein cDNA. *J. Exp. Med.*, 168: 1493-1498

Barker R.L., Loegering D.A., Ten R.M., Hamann K.J., Pease L.R. and Gleich G.J. (1989) Eosinophil cationic protein cDNA. Comparison with other toxic cationic proteins and ribonucleases. *J. Immunol.*, 143: 952-955

Boggs J.M., Moscarello M.A. and Papahadjopoulos D. (1982) Structural organization of myelin — Role of lipid-protein interactions determined in model systems. In 'Lipid-Protein Interactions vol. 2', (eds.) Jost P.C. and Griffith O.H., John Wiley & Sons, New York, pp. 1-51

Chakravarty N. and Nielsen E.M. (1985) Calmodulin in mast cells and its role in histamine secretion. *Agents Actions*, 16: 122-125

Ennis M., Atkinson G. and Pearce F.L. (1980) Inhibition of histamine release induced by compound 48/80 and peptide 401 in the presence and absence of calcium. Implication for the mode of action of anti-allergic compounds. *Agents Actions*, 10: 222-228

Fechter M., Egger D., Auer H. and Popper H. (1986) Experimental eosinophilia and inflammation - The effect of various inflammatory mediators and chemoattractants. *Exp. Pathol.*, 29: 153-158

Fewtrell C.M.S., Foreman J.C., Jordan C.C., Oehme P., Renner H. and Stewart J.M. (1982) The effects of substance P on histamine and 5-hydroxytryptamine release in the rat. *J. Physiol.*, 330: 393-411

Foreman J.C. (1988) The skin as an organ for the study of the pharmacology of neuropeptides. *Skin Pharmacol.*, 1: 77-83

Foreman J. and Jordan C. (1983) Histamine release and vascular changes induced by neuropeptides. *Agents Actions*, 13: 105-116

Foreman J.C., Jordan C.C., Oehme P. and Renner H. (1983) Structure-activity relationships for some substance P-related peptides that cause wheal and flare reactions in human skin. *J. Physiol.*, 335: 449-465

Gleich G.J., Loegering D.A., Kueppers F.H., Bajaj S.P. and Mann K.G. (1974) Physiochemical and biological properties of the major basic protein from guinea pig eosinophil granules. *J. Exp. Med.*, 140: 313-332

Gleich G.J., Loegering D.A., Mann K.G. and Maldonado J.E. (1976) Comparative properties of the Charcot Leyden crystal protein and the major basic protein from human eosinophils. *J. Clin. Invest.*, 57: 633-640

Gleich G.J., Flavahan N.A., Fujisawa T. and Vanhoutte P.M. (1988) The eosinophil as a mediator of damage to respiratory epithelium: A model for bronchial hyperreactivity. *J. Allergy Clin. Immunol.*, 81: 776-781

Gleich G.J., Ottesen E.A., Leiferman K.M. and Ackerman S.J. (1989) Eosinophils and human disease. *Int. Arch. Allergy Appl. Immunol.*, 88: 59-62

Higashijima T., Wakamatsu K., Wakamatsu M., Fujino M., Nakajima T. and Miyazawa T. (1983) Conformational change of mastoparan from wasp venom on binding with phospholipid membrane. *FEBS Lett.*, 152: 227-230

Higashijima T., Burnier J. and Ross E.M. (1990) Regulation of Gi and Go by mastoparan, related amphiphilic peptides, and hydrophobic amines. *J. Biol. Chem.*, 265: 14176-14186

Holzer-Petsche U., Schimek E., Amann R. and Lembeck F. (1985) *In vivo* and *in vitro* actions of mammalian tachykinins. *Naunyn-Schmied. Arch. Pharmacol.*, 330: 130-135

Izushi K. and Tasaka K. (1989) Histamine release from β-escin-permeabilized rat peritoneal mast cells and its inhibition by intracellular Ca^{2+} blockers, calmodulin inhibitors and cAMP. *Immunopharmacology*, 18: 177-186

Mercum J.M., Dedman J.R., Brinkley B.R. and Means A.R. (1978) Control of microtubule assembly-disassembly by calcium-dependent regulator protein. *Proc. Natl. Acad. Sci. USA*, 75: 3771-3775

Mio M. and Tasaka K. (1989) Microfilament-associated, local degranulation of rat peritoneal mast cells. *Int. Arch. Allergy Appl. Immunol.*, 88: 369-371

Mio M., Izushi K. and Tasaka K. (1991) Substance P-induced histamine release from rat peritoneal mast cells and its inhibition by antiallergic agents and calmodulin inhibitors. *Immunopharmacology*, 22: 59-66

Mousli M., Bronner C., Landry Y., Bockaert J. and Rouot B. (1990) Direct activation of

GTP-binding regulatory proteins (G-proteins) by substance P and compound 48/80. *FEBS Lett.*, 259: 260-262

Pearce F.L., Kassessinoff T.A. and Liu W.L. (1989) Characteristics of histamine secretion induced by neuropeptides: implications for the relevance of peptide-mast cell interactions in allergy and inflammation. *Int. Arch. Allergy Appl. Immunol.*, 88: 129-131

Pincus S.H., DiNapoli A.-N. and Schooley W.R. (1982) Superoxide production by eosinophils: Activation by histamine. *J. Invest. Dermatol.*, 79: 53-57

Piotrowski W. and Foreman J.C. (1985) On the action of substance P, somatostatin, and vasoactive intestinal polypeptide on rat peritoneal mast cells and in human skin. *Naunyn-Schmied. Arch. Pharmacol.*, 331: 364-368

Repke H. and Bienert M. (1988) Structural requirements for mast cell triggering by substance P-like peptides. *Agents Actions*, 23: 207-210

Repke H., Piotrowski W., Bienert M. and Foreman J.C. (1987) Histamine release induced by Arg-Pro-Lys-Pro-$(CH_2)_{11}CH_3$ from rat peritoneal mast cells. *J. Pharmacol. Exp. Ther.*, 243: 317-321

Sagi-Eisenberg R., Foreman J.C. and Shelly R. (1985) Histamine release induced by histone and phorbol ester from rat peritoneal mast cells. *Eur. J. Pharmacol.*, 113: 11-17

Shibata H., Mio M. and Tasaka K. (1985) Analysis of the mechanism of histamine release induced by substance P. *Biochim. Biophys. Acta*, 846: 1-7

Tasaka K. (1986) Anti-allergic drugs. *Drugs Today*, 22: 101-133

Tasaka K. and Mio M. (1989) Microfilament-associated degranulation of sensitized guinea-pig lung mast cells. *Agents Actions*, 27: 79-82

Tasaka K., Sugiyama K., Komoto S. and Yamasaki H. (1970) Repeated local degranulation of isolated rat mast cell by microelectrophoresis of basic histamine releasers. *Proc. Japan. Acad.*, 46: 311-316

Tasaka K., Mio M., Okamoto M. (1986) Intracellular calcium release induced by histamine releasers and its inhibition by some antiallergic drugs. *Ann. Allergy*, 56: 464-469

Tasaka K., Akagi M., Mio M., Miyoshi M. and Nakaya N. (1987) Inhibitory effects of oxatomide on intracellular Ca mobilization, Ca uptake and histamine release, using rat peritoneal mast cells. *Int. Arch. Allergy Appl. Immunol.*, 83: 348-353

Tasaka K., Akagi M., Miyoshi K. and Mio M. (1988) Role of microfilaments in the exocytosis of rat peritoneal mast cells. *Int. Archs. Allergy Appl. Immunol.*, 87: 213-221

Tasaka K., Mio M., Akagi M. and Saito T. (1990) Histamine release induced by histone and related morphological changes in mast cells. *Agents Actions*, 30: 114-117

Tasaka K., Mio M., Fujisawa K. and Aoki I. (1991) Role of microtubules on Ca^{2+} release from the endoplasmic reticulum and associated histamine release from rat peritoneal mast cells. *Biochem. Pharmacol.*, 41: 1031-1037

Tasaka K., Mio M., Aoki I. and Saito T. (1992a) Guinea pig eosinophil major basic protein (GMBP) as a potent histamine releaser. (I) Histamine releasing activity of GMBP and its chemical structure. *Agents Actions*, 36/Spec. Iss.: C242-C245

Tasaka K., Mio M., Nakai S. and Aoki I. (1992b) Guinea pig eosinophil major basic protein as a potent histamine releaser. (II) Structure-activity relationships of GMBP1 and GMBP2. *Agents Actions*, 36/Spec. Iss.: C246-C249

Uvnäs B. (1974) Histamine storage and release. *Fed. Proc.*, 33: 2172-2176

Wardlaw A.J., Moqbel R., Cromwell O. and Kay A.B. (1986) Platelet-activating factor. A potent chemotactic and chemokinetic factor for human eosinophils. *J. Clin. Invest.*, 78: 1701-1706

Watson S.P. (1984) Are the proposed substance P receptor sub-types, substance P receptors? *Life Sci.*, 35: 797-808

Yazawa M., Yagi K. and Sobue K. (1987) Isolation and characterization of a calmodulin binding fragment of chicken gizzard caldesmon. *J. Biochem.*, 102: 1065-1073

Yoshii N., Mio M. and Tasaka K. (1988) Ca uptake and Ca releasing properties of the endoplasmic reticulum in rat peritoneal mast cells. *Immunopharmacology*, 16: 107-113

Zheutlin L.M., Ackerman S.J., Gleich G.J. and Thomas L.L. (1984) Stimulation of basophil and rat mast cell histamine release by eosinophil granule-derived cationic proteins. *J. Immunol.*, 133: 2180-2185

Chapter 7

Inhibitory effect of interleukin-2 on the histamine release from rat peritoneal mast cells in association with the production of lipocortin-I

1. Introduction

It has been shown that the development and growth of mast cells are dependent upon interleukin-3 (IL-3), IL-4 and IL-9 (Ihle et al., 1983; Schmitt et al., 1987; Hultner et al., 1990). On the other hand, it has also been reported that histamine release from mast cells or basophils is induced or potentiated by several cytokines, such as IL-1α, IL-1 β, IL-2, IL-3, IL-4 and granulocyte-macrophage colony stimulating factor (GM-CSF) (Massey et al., 1989; Levi-Schaffer et al., 1991; Haak-Frendscho et al., 1988). However, most of these studies are carried out using basophils and mucosal type mast cells, but not with connective tissue type mast cells. Levi-Schaffer et al. (1991) reported that IL-2, IL-3 and IL-4 enhanced both the basal and anti-IgE-mediated histamine release from rat and murine peritoneal mast cells co-cultured with fibroblasts. Moreover, there has been no demonstration about the inhibitory effect of cytokine on the histamine release from mast cells.

On the other hand, it has been shown that corticosteroids are effective in inhibiting histamine release from mast cells (White et al., 1991; Sautebin et al., 1992). Since it has been indicated that corticosteroids induce the production of lipocortins and that lipocortins inhibited histamine release from mast cells, it was suggested that the inhibitory effect of corticosteroids on the histamine release can be ascribed to the lipocortin formation in mast cells (White et al., 1991; Sautebin et al., 1992). However, there has been no report concerning the effect of cytokines on the lipocortin production in mast cells. In this chapter, the inhibitory mechanism of the action of cytokines, especially IL-2 on the histamine release from rat peritoneal mast cells, was investigated.

2. Inhibitory effects of interleukin-2 on histamine release from rat peritoneal mast cells induced by compound 48/80 and concanavalin A

When rat peritoneal mast cells were pretreated with several cytokines at 37 °C for 8 hr, the histamine release induced by compound 48/80 was significantly inhibited only by IL-2, but other cytokines such as IL-1α, IL-3, IL-4 and IL-5 were not effective at all, as shown in Table 1. As indicated in Fig. 1a, IL-2 inhibited the histamine release in a concentration-dependent manner: significant inhibition was observed at concentrations

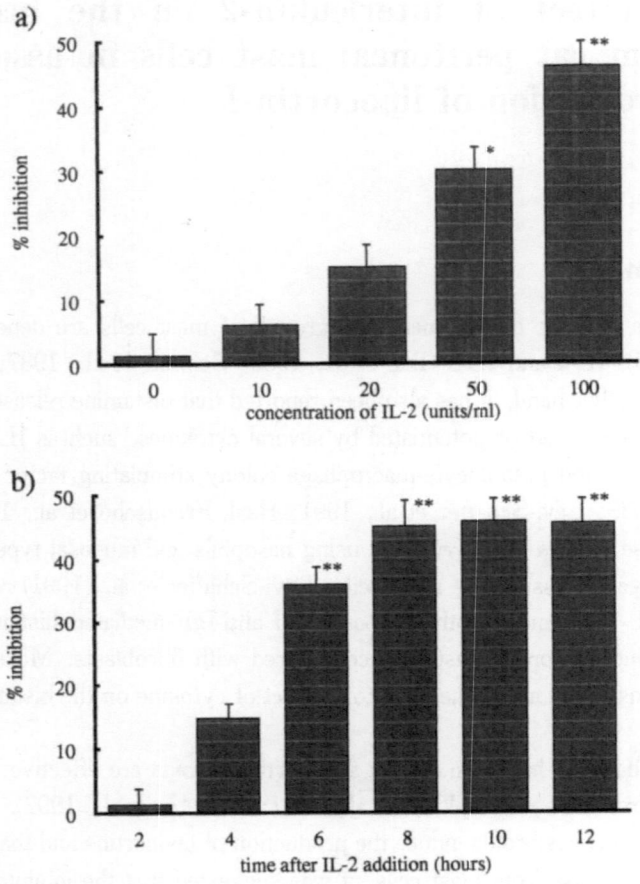

Fig. 1 Inhibitory effect of IL-2 on the histamine release from rat peritoneal mast cells induced by compound 48/80 ($1\,\mu g/ml$). (a) Concentration-response relationship. Rat peritoneal mast cells were incubated with various concentrations of IL-2 for 8 hr. (b) Time course. Rat peritoneal mast cells were incubated with 100 units/ml of IL-2 for various periods of time. Each point represents the mean $\pm$ S.E.M. of 5 separate experiments. * and ** indicate statistical significance in comparison with the control at $p < 0.05$ and $p < 0.01$, respectively.

higher than 50 units/ml. When rat mast cells were incubated with IL-2 for various periods of time, the inhibitory effect of IL-2 on the histamine release became significant at 6 hr and the maximal inhibition achieved at 8 hr (Fig. 1b). When rat mast cells were exposed to concanavalin A at 30 and $100\,\mu g/ml$, histamine release was elicited at 27.2 ± 1.46 and $50.48 \pm 2.39\%$, respectively (n = 5). When the cells were pretreated with 100 units/ml of IL-2 for 8 hr prior to concanavalin A stimulation, the histamine release

Table 1. Effects of various cytokines on the histamine release from rat peritoneal mast cells induced by compound 48/80

cytokines	% histamine release	
	compound 48/80 alone	+ cytokines
IL-1α (100 units/ml)	41.7 $\pm$ 1.7	42.5 $\pm$ 1.3
IL-2 (100 units/ml)	46.5 $\pm$ 3.9	19.0 $\pm$ 1.5**
IL-3 (50 units/ml)	50.9 $\pm$ 5.3	52.4 $\pm$ 2.7
IL-4 (50 ng/ml)	43.7 $\pm$ 2.6	42.3 $\pm$ 2.2
IL-5 (500 units/ml)	41.3 $\pm$ 4.2	45.4 $\pm$ 1.9

Rat peritoneal mast cells were incubated with various cytokines in DMEM at 37°C for 8 hr. Subsequently, the cells were incubated with compound 48/80 (1 μg/ml) for 10 min. ** indicates statistical significance at $p < 0.01$.

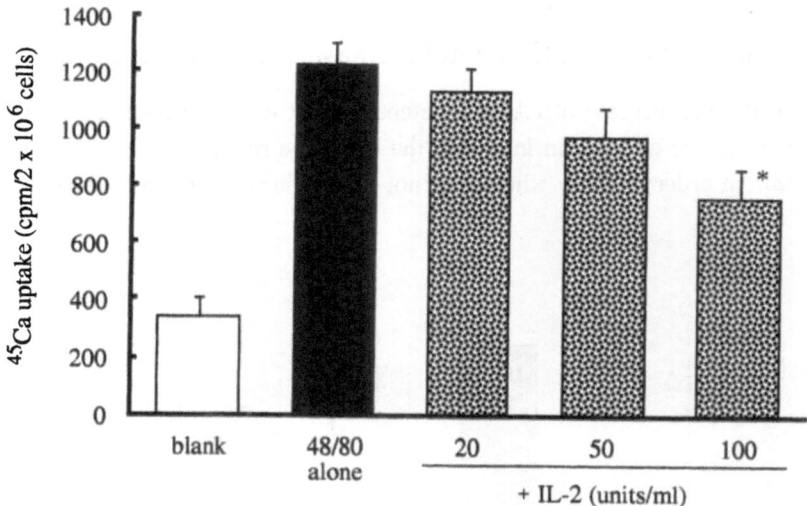

Fig. 2. Inhibitory effect of IL-2 on ^{45}Ca uptake into rat peritoneal mast cells elicited by compound 48/80 (1 μg/ml). Rat mast cells were incubated with various concentrations of IL-2 at 37°C for 8 hr. Thereafter, the cells were stimulated with compound 48/80 (1 μg/ml). Each point represents the mean $\pm$ S.E.M. of 5 separate experiments. * indicates statistical significance in comparison with control at $p < 0.05$.

induced by 30 and 100 μg/ml of concanavalin A was significantly reduced to 20.65 $\pm$ 1.04 ($p < 0.05$) and 26.79 $\pm$ 1.72% ($p < 0.01$), respectively.

3. Effect of IL-2 on ^{45}Ca uptake and IP$_3$ production in rat peritoneal mast cells

It is known that an increase in intracellular Ca^{2+} concentration is a prerequisite for histamine release from mast cells (Tasaka et al., 1986). In addition, it has been shown that the IP_3 content in mast cells increases when the cells are treated with histamine-releasing stimuli and that this induces Ca^{2+} release from intracellular Ca store (Tasaka et al., 1990). In order to study the inhibitory mechanism of IL-2 on the histamine release from mast cells, the effects of IL-2 on the ^{45}Ca uptake and IP_3 production in rat peritoneal mast cells were investigated.

When rat peritoneal mast cells were stimulated by compound 48/80 at $1 \mu g/ml$, ^{45}Ca uptake into mast cells increased to 3.5 times of the spontaneous uptake. However, ^{45}Ca uptake elicited by compound 48/80 was inhibited by IL-2 pretreatment at concentrations higher than 20 units/ml. Significant inhibition was observed at 100 units/ml (Fig. 2). IP_3 content in rat mast cells increased more than 9 times of the control value 5 sec after addition of compound 48/80 $(2 \mu g/ml)$, as reported previously (Tasaka et al., 1991). IL-2 inhibited the compound 48/80-induced increase in IP_3 content in mast cells in a concentration-dependent manner (Fig. 3). Significant inhibition was observed at concentrations higher than 50 units/ml.

4. Effect of IL-2 on cAMP-protein kinase A system in rat mast cells

It is known that an increase of cAMP content in mast cells and resulting activation of protein kinase A are effective in inhibiting the histamine release from mast cells (Izushi et al., 1992). In order to study whether or not IL-2 induced inhibition is related with an

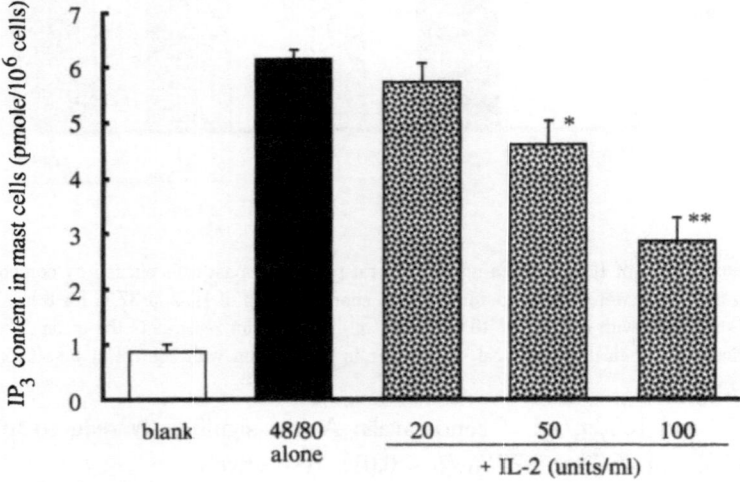

Fig. 3. Influence of IL-2 on the increase in IP_3 contents in rat peritoneal mast cells elicited by compound 48/80 $(1 \mu g/ml)$. Rat peritoneal mast cells were incubated with various concentrations of IL-2 at 37 °C for 8 hr. Thereafter, the cells were stimulated with $1 \mu g/ml$ of compound 48/80 for 5 sec. Each column represents the mean $\pm$ S.E.M. of 4 separate experiments. * and ** indicate statistical significance in comparison with control at $p < 0.05$ and $p < 0.01$, respectively.

activation of cAMP-protein kinase A system, the effect of protein kinase A inhibitors was studied on the histamine release inhibition elicited by IL-2 pretreatment. The inhibitory effect of IL-2 (100 units/ml) on the compound 48/80-induced histamine release was totally unaffected by protein kinase A inhibitors, such as H-8 (0.1 - 30 μg/ml) and KT-5720 (0.01 - 10 μg/ml). Furthermore, the cAMP contents in rat mast cells (2.812 $\pm$ 0.287 pmol/10^6 cells, n = 4) were not affected by 100 units/ml of IL-2 (2.743 $\pm$ 0.184 pmol/10^6 cells, n = 4). When rat mast cells were stimulated with 1 μg/ml of compound 48/80, the cAMP contents were decreased in both non-treated control (1.457 $\pm$ 0.262 pmol/10^6 cells, n = 4) and IL-2 treated cells (1.76 $\pm$ 0.368 pmol/10^6 cells, n = 4).

5. [³H]-Leucine uptake into rat mast cells elicited by IL-2

Since IL-2 requires a long incubation period (more than 4 hr) to be effective in inhibiting the histamine release from mast cells, it was supposed that some proteins may be produced during the incubation period. To make this point clear, [³H]-leucine uptake into rat mast cells was tested along with IL-2 incubation. When rat peritoneal mast cells were incubated with [³H]-leucine in the absence of IL-2, the [³H]-leucine content gradually increased as time elapsed, indicating that a spontaneous protein synthesis takes place (Fig. 4). On the other hand, when rat peritoneal mast cells were incubated with

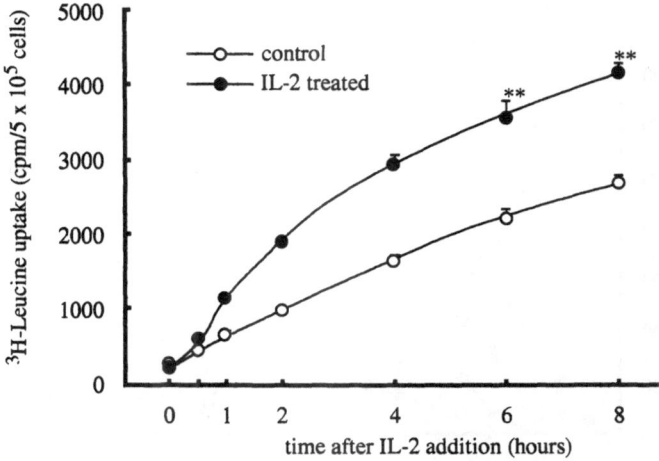

Fig. 4. Sequential changes in [³H]-leucine uptake into rat peritoneal mast cells induced by 100 units/ml of IL-2. Rat peritoneal mast cells were incubated with [³H]-leucine both in the presence and in the absence of 100 units/ml of IL-2 for various periods of time. Thereafter, the cells were disrupted with TCA and the radioactivity incorporated into the mast cell proteins was determined. Each point represents the mean $\pm$ S. E.M. of 4 separate experiments. * and ** indicate statistical significance in comparison with control at $p <$ 0.05 and $p < 0.01$, respectively.

[³H]-leucine in the presence of 100 units/ml of IL-2, significant enhancement of [³H]-leucine uptake was observed. The most remarkable increase in the rate of [³H]-leucine uptake takes place from 0.5 to 2 hr after addition of IL-2.

6. Determination of the protein synthesis elicited by IL-2

In order to examine the protein(s) synthesized in mast cells after addition of IL-2, rat peritoneal mast cells were incubated with 100 units/ml of IL-2 in the presence of [³H]-leucine, and autoradiography of SDS-PAGE gel of mast cell proteins was carried out. The most obvious incorporation was noticed in the protein band having the molecular weight of about 35 kDa. When the SDS-polyacrylamide gel was cut into small pieces after electrophoresis and the radioactivity of each piece was measured, it was indicated that [³H]-leucine uptake into the proteins having molecular weights of 30 - 35 kDa increased more than 4 times the control level (Fig. 5).

7. Western blotting analysis of lipocortin-I induced by IL-2

The newly synthesized protein in mast cells produced after IL-2 exposure seems to possess characteristics similar to lipocortin, since 1) IL-2 is able to inhibit IP$_3$ production in mast cells and 2) molecular weight of newly synthesized protein is about 30 - 35 kDa (White et al., 1991; Machoczek et al., 1989). To make this point clear, Western blotting analysis of mast cell proteins was carried out using an anti-lipocortin-I antibody. As shown in Fig. 6, in the resting state, rat peritoneal mast cells contained only a small

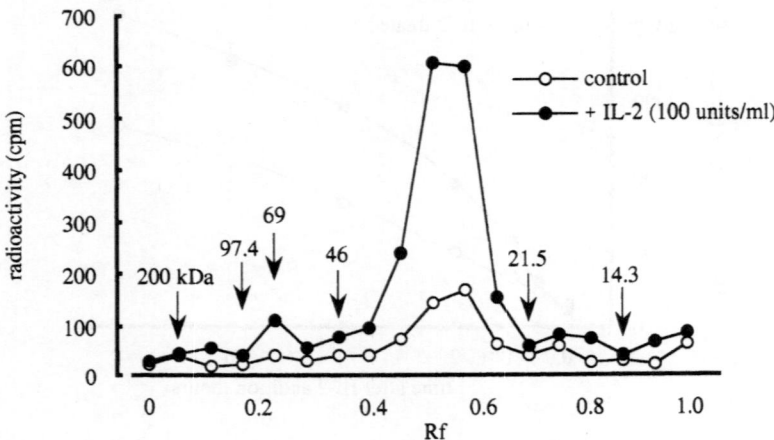

Fig. 5. Quantitative analysis of [³H]-leucine uptake in mast cell proteins induced by IL-2. Rat peritoneal mast cells were incubated with 100 units/ml of IL-2 in the presence of 1 μCi/ml of [³H]-leucine at 37°C for 8 hr. After SDS-PAGE of mast cell proteins, the gel was cut into small pieces and the radioactivity was measured.

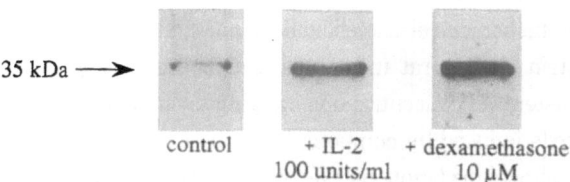

35 kDa ⟶

control   + IL-2   + dexamethasone
100 units/ml   10 μM

Fig. 6. Western blotting analysis of mast cell proteins using anti-lipocortin-I antibody. Rat peritoneal mast cells were incubated with 100 units/ml of IL-2 or 10 μM of dexamethasone at 37°C for 8 hr. Thereafter, Western blotting analysis of mast cell proteins was carried out using an anti-lipocortin-I antibody.

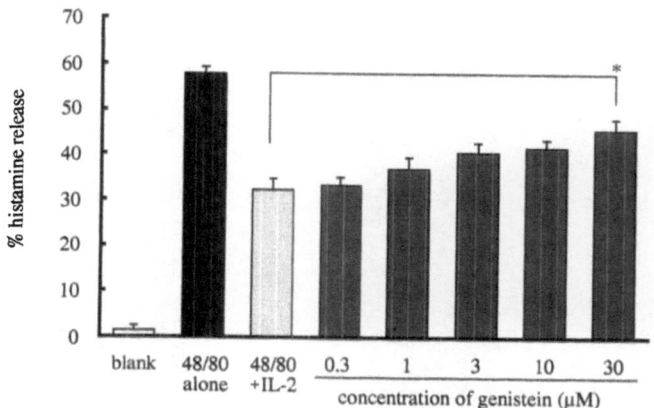

Fig. 7. Effect of the tyrosine kinase inhibitor, genistein, on the histamine release inhibition due to IL-2. Rat peritoneal mast cells were exposed to genistein 30 min prior to IL-2 addition. Thereafter, the mast cells were incubated with 100 units/ml of IL-2 at 37°C for 8 hr. After that, the cells were stimulated by 1 μg/ml of compound 48/80 at 37°C for 10 min. Each column represents the mean ± S.E.M. of 5 separate experiments. * indicates statistical significance in comparison with control at $p < 0.05$.

amount of lipocortin-I with the molecular weight of 35 kDa. When the cells were incubated with 100 units/ml of IL-2 for 8 hr, the content of lipocortin-I markedly increased, indicating that IL-2 caused a significant increase of lipocortin-I formation. Similar increase in lipocortin-I production was also observed when mast cells were incubated with 10 μM of dexamethasone at 37°C for 8 hr.

8. Participation of tyrosine kinase in histamine release inhibition due to IL-2

It has been shown that the IL-2 receptor possesses tyrosine kinase activity (Waldmann, 1991). To make clear whether or not tyrosine kinase plays an important role in inhibiting histamine release, the influence of the tyrosine kinase inhibitor, genistein, was investigated on the histamine release inhibition caused by IL-2. As shown in Fig. 7, the inhibitory effect of IL-2 on the histamine release elicited by compound 48/80 was dose-dependently abrogated by genistein. When rat mast cells that had not been exposed to IL-2 were pretreated with genistein at concentrations ranging from 0.3 to 30 μM, the histamine release from mast cells induced by compound 48/80 (1 μg/ml) was not affected (data not shown). When the phosphorylation profile of mast cell protein was determined, it was indicated that the phosphorylation of 18 kDa protein was markedly increased 1 min after IL-2 addition (Fig. 8). When the radioactivity of phosphorylated protein bands was measured, the phosphorylation of 18 kDa protein reached the maximum, 1 min after

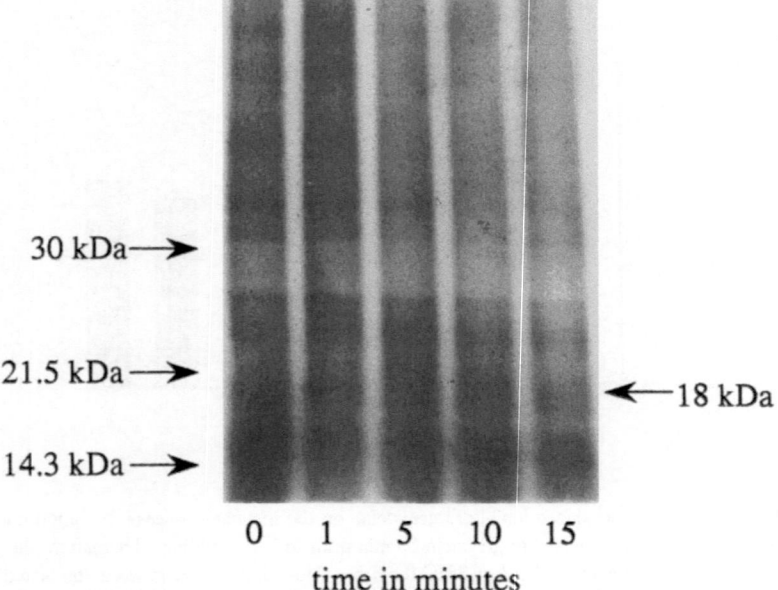

Fig. 8. Phosphorylation profile of rat mast cell proteins after addition of IL-2 (100 units/ml).

Table 2. Sequential changes in the phosphorylation of 18 kDa protein in rat mast cells induced by IL-2 (100 units/ml).

Time in minutes	radioactivity (% of control)
0	100.0
1	137.9
5	116.1
10	108.6
15	103.2

Rat peritoneal mast cells were incubated with 100 units/ml of IL-2 at 37°C for various periods of time in the presence of 500 μCi/ml of [^{32}P]-orthophosphate. Thereafter, SDS-PAGE of mast cell proteins was carried out. Radioactivity of 18 kDa protein band was determined by means of Image Analyzer BAS-2000.

IL-2 addition and decreased to the control level at 15 min (Table 2). Such sequential changes were consistently observed.

9. Discussion

It has been suggested that there exists IL-2 receptor on human mast cells and basophils (Maggiano et al., 1990). Levi-Schaffer et al. (1991) reported that IL-2 enhanced histamine release from rat peritoneal mast cells co-cultured with fibroblasts. However, there has been no report about the inhibitory effect of cytokines on the histamine release from mast cells. In this study, it became apparent that IL-2 inhibited histamine release from rat peritoneal mast cells. This result seems to be contradictory to that reported by Levi-Schaffer et al. (1991). Although Levi-Schaffer et al. cultured rat peritoneal mast cells with fibroblasts, we examined the direct action of IL-2 on mast cells. Therefore, it was supposed that in the results of Levi-Schaffer et al. (1991) the direct action of IL-2 may be modified by the indirect action of IL-2 which was elicited by co-culture with fibroblasts. In the present experiment, other cytokines did not affect histamine release from rat peritoneal mast cells induced by compound 48/80. It has been reported that IL-3 and IL-4 enhance histamine release from mucosal type mast cells and basophils (Massey et al., 1989; Haak-Frendscho et al., 1988). Moreover, when bone marrow cells are co-cultured with IL-3, the cells differentiated to bone marrow-derived mast cells, which can be classified as mucosal type mast cells (Ihle et al., 1983). Therefore, it was assumed that connective tissue type mast cells, such as rat peritoneal mast cells, may not have the receptors for IL-3 and IL-4, and the differentiation pathway of connective tissue mast cells may be different from that of mucosal type mast cells.

As shown in the present results, IL-2 inhibited ^{45}Ca uptake (Fig. 2) and IP_3 production (Fig. 3) in rat peritoneal mast cells stimulated by compound 48/80 without affecting intracellular cAMP levels. Protein kinase A inhibitors did not affect the histamine release inhibition due to IL-2. These results clearly indicate that IL-2 prevented the histamine release from mast cells by inhibiting the increase in intracellular Ca^{2+} concentrations without affecting cAMP-protein kinase A system. Although it has been reported that an activation of cAMP-protein kinase A system is effective in inhibiting an increase in intracellular Ca^{2+} concentration (Tasaka et al., 1986; Izushi et al., 1992), the present results indicate that something other than cAMP may be involved in the histamine release inhibition due to IL-2. Since an increase in intracellular Ca^{2+} concentration is a prerequisite for the histamine release from mast cells, the histamine release inhibition due to IL-2 can be ascribed to the inhibitions of IP_3 production and Ca uptake. Since it requires several hours to exhibit the inhibitory activity of IL-2 on the histamine release from mast cells and IL-2 enhanced [^3H]-leucine uptake into mast cells, it was supposed that the histamine release inhibition may be associated with protein synthesis in mast cells. Actually, IL-2 stimulated the synthesis of protein, which has a molecular weight of about

35 kDa and was determined as lipocortin-I based on Western blotting analysis. Lipocortin-II was not detected in mast cell proteins (data not shown). In accordance with this, it has been shown that glucocorticoids are effective in inhibiting histamine release from mast cells and that this inhibition may be intimately related to the production of lipocortins (White et al., 1991; Sautebin et al., 1992). The production of lipocortin-I in the presence of dexamethasone was also confirmed in this study as shown in Fig. 6. In both cases, lipocortin production became evident at 4 hr after additions, and reached the maximum at 8 hr. The time course of the protein synthesis almost corresponds with that of histamine release inhibition induced by IL-2, suggesting that lipocortin-I production may be intimately related to histamine release inhibition.

It is well known that lipocortin inhibits not only phospholipase A_2 but also phospholipase C, especially polyphosphoinositide-specific phospholipase C, and this subsequently results in an inhibition of IP_3 production (Machoczek et al., 1989; Weber and Ferber, 1990). As shown in Fig. 3, IL-2 inhibited IP_3 production, which is an essential trigger for Ca^{2+} release from intracellular Ca store, and, also, the subsequent histamine release from mast cells (Tasaka et al., 1990). Therefore, it was assumed that the histamine release inhibition due to IL-2 could be ascribed to the production of lipocortin-I and the resulting inhibitions of phospholipase C and IP_3 production. In addition, since it has been reported that inositol 1, 3, 4, 5-tetrakisphosphate (IP_4), which is a phosphorylated product of IP_3, stimulates the opening of the Ca channel (Pittet et al., 1989), the inhibitory effect of IL-2 on ^{45}Ca uptake may also be ascribed to the inhibition of phospholipase C and resulting inhibition of IP_3 and IP_4 productions.

It has been indicated that the IL-2 receptor possesses tyrosine kinase activity (Waldmann, 1991). As shown in Fig. 7, the histamine release inhibition due to IL-2 was reversed by genistein, an inhibitor of tyrosine kinase, in a dose-dependent manner. A phosphorylation of 18 kDa protein in mast cells was also elicited by IL-2 treatment. These results may indicate that tyrosine kinase probably participates in histamine release inhibition due to IL-2. Although the function of 18 kDa protein in mast cells is not clear at the present time, it may be reasonable to assume that IL-2 may increase tyrosine kinase activity of the IL-2 receptor so as to phosphorylate 18 kDa protein and, as a consequence of this, an increase of the lipocortin-I production may take place. The time course of the phosphorylation and dephosphorylation of the 18 kDa protein is much more rapid than that seen in lipocortin-I formation. This may suggest that the phosphorylation of 18 kDa protein may be involved in the signal transduction from the IL-2 receptor to the nucleus in the process leading to lipocortin-I production. The newly synthesized lipocortin-I seems to be effective in inhibiting the phospholipase C activity and this may be effective in inhibiting histamine release from mast cells.

In the case of the glucocorticoid-induced lipocortin production, it has been indicated that the intracellular glucocorticoid receptor participates in the gene expression of lipocortins (Flower, 1988). On the other hand, it has been shown that IL-2 induces several gene expressions and/or differentiations in various types of cells (Emilie et al., 1988; Espinoza-Delgado, 1990; Smyth et al., 1990; Sabath, 1990). At present time, it is still uncertain whether or not IL-2 induces the gene expression of lipocortins or whether the pathway of IL-2-induced lipocortin-I production shares the same way of glucocorticoid. However, the fact that both glucocorticoid and IL-2 induced lipocortin-I production in mast cells may suggest that a similar mechanism may exist between the IL-2-induced and glucocorticoid-induced gene expression of lipocortins.

Based on these observations, it was concluded that IL-2 induces lipocortin-I production in rat peritoneal mast cells. The newly synthesized lipocortin-I inhibits an activation of phospholipase C and, as a consequence of this, an increase in intracellular Ca^{2+} concentration may be blocked. This may further explain the histamine release inhibition from mast cells.

References

Emilie D., Karray S., Merle-Beral H., Debre P. and Galanaud P. (1988) Induction of differentiation in human leukemic B cells by interleukin 2 alone: Differential effect on the expression of μ and J chain genes. *Eur. J. Immunol.*, 18: 1479-1483

Espinoza-Delgado I., Longo D.L., Gusella G.L. and Varesio L. (1990) IL-2 enhances c-fms expression in human monocytes. *J. Immunol.*, 145: 1137-1143

Flower R.J. (1988) Lipocortin and the mechanism of action of the glucocorticoids. *Br. J. Pharmacol.*, 94: 987-1015

Haak-Frendscho M., Arai N. and Arai K. (1988) Human recombinant granulocyte-macrophage colony-stimulating factor and interleukin-3 cause basophil histamine release. *J. Clin. Invest.*, 82: 17-20

Hultner L., Druez C., Moeller J., Uyttenhove C., Schmitt E., Rude E., Dormer P. and Van Snick J. (1990) Mast cell growth-enhancing activity (MEA) is structurally related and functionally identical to the novel mouse T cell growth factor P40/TCGFIII (interleukin 9). *Eur. J. Immunol.*, 20: 1413-1416

Ihle J.N., Keller J., Oroszlan S., Henderson L.E., Copeland T.D., Fitch F, Prystowsky M. B., Goldwasser E., Schrader J.W., Palaszynski E., Dy M. and Lebel B. (1983) Biologic properties of homogeneous interleukin 3. I. Demonstration of WEHI-3 growth factor activity, mast cell growth activity, P cell-stimulating factor activity, colony-stimulating factor activity, and histamine-producing cell-stimulating factor activity. *J. Immunol.*, 131: 282-287

Izushi K., Shirasaka T., Chokki M. and Tasaka K. (1992) Phosphorylation of smg p21B in rat peritoneal mast cells in association with histamine release inhibition by dibutyryl-cAMP. *FEBS Lett.*, 314: 241-245

Levi-Schaffer F., Segal V. and Shalit M. (1991) Effect of interleukins on connective tissue type mast cells co-cultured with fibroblasts. *Immunology*, 72: 174-180

Machoczek K., Fischer M. and Soling H.-D. (1989) Lipocortin I and lipocortin II inhibit phosphoinositide- and polyphosphoinositide-specific phospholipase C. The effect results from interaction with the substrates. *FEBS Lett.*, 251: 207-212

Maggiano M., Cotta F., Castellino F., Ricci R., Valitutti S., Larocca L.M. and Musiani P. (1990) Interleukin-2 receptor expression in human mast cells and basophils. *Int. Arch. Allergy Appl. Immunol.*, 91: 8-14

Massey A.W., Randall C.T., Kagey J.A., MacDonald M.S., Gillis S., Allison C.A. and Lichtenstein M.L. (1989) Recombinant human IL-1α and -1β potentiate IgE-mediated histamine release from human basophils. *J. Immunol.*, 143: 1875-1880

Pittet D., Lew D.P., Mayr G.W., Monod A. and Schlegel W. (1989) Chemoattractant receptor promotion of Ca^{2+} influx across the plasma membrane of HL-60 cells. A role for cytosolic free calcium elevations and inositol 1, 3, 4, 5-tetrakisphosphate production. *J. Biol. Chem.*, 264: 7251-7261

Sabath D.E., Broome H.E. and Prystowsky M.B. (1990) Glyceraldehyde-3-phosphate dehydrogenase mRNA is a major interleukin 2-induced transcript in a cloned T-helper lymphocyte. *Gene*, 91: 185-191

Sautebin L., Carnuccio R., Ialenti A. and Di Rosa M. (1992) Lipocortin and vasocortin: Two species of anti-inflammetory proteins mimicking the effects of glucocorticoids. *Pharmacol. Res.*, 25: 1-12

Schmitt E., Fassbender B., Beyreuther K., Spaeth E., Schwarzkopf R. and Ruede E. (1987) Characterization of a T cell-derived lymphokine that acts synergistically with IL3 on the growth of murine mast cells and is identical with IL4. *Immunobiology*, 174: 406-419

Smyth M.J., Ortaldo J.R., Bere W., Yagita H., Okumura K. and Young H.A. (1990) IL-2 and IL-6 synergize to augment the pore-forming protein gene expression and cytotoxic potential of human peripheral blood T cells. *J. Immunol.*, 145: 1159-1166

Tasaka K., Mio M. and Okamoto M. (1986) Intracellular calcium ion release induced by histamine releasers and its inhibition by some antiallergic drugs. *Ann. Allergy*, 56: 464-469

Tasaka K., Sugimoto Y. and Mio M. (1990) Sequential analysis of histamine release and intracellular Ca^{2+} release from murine mast cells. *Int. Arch. Allergy Appl. Immunol.*, 91: 211-213

Tasaka K., Mio M., Fujisawa K. and Aoki I. (1991) Role of microtubules on Ca^{2+} release from the endoplasmic reticulum and associated histamine release from rat peritoneal mast cells. *Biochem. Pharmacol.*, 41: 1031-1037

Waldmann A. T. (1991) The interleukin-2 receptor. *J. Biol. Chem.*, 286: 2681-2684

Weber G. and Ferber E. (1990) Selective inhibition of phospholipase A_2 by different lipocortins. *J. Biol. Chem.*, 371: 725-731

White M.V., Igarashi Y., Lundgren J.D., Shelhamer J. and Kaliner M. (1991) Hydrocortisone inhibits rat basophilic leukemia cell mediator release induced by neutrophil-derived histamine releasing activity as well as by anti-IgE. *J. Immunol.*, 147: 667-673

Chapter 8
Histamine-induced leukocytosis

1. Introduction

From the early years of the introduction of H_2 antagonists for clinical uses, many reports have been published that H_2 antagonists induce various kinds of cytopenia, such as neutropenia, agranulocytosis, thrombocytopenia, pancytopenia and lymphopenia. At the present time, it has become evident that histamine exerts a stimulative effect on the differentiation and proliferation of bone marrow stem cells, especially in neutrophil progenitors, via an H_2 receptor stimulation (Tasaka, 1991). However, as early as 8 years before the discovery of the H_2 receptor and histamine H_2 antagonist by Black et al. (1972), Tasaka and Code (1964) reported that chronic injection of histamine induces leukocytosis in various kinds of experimental animals. This was the first report which clearly pointed out the hematopoietic action of histamine *in vivo*. A growing body of evidence indicates that histamine is not only a physiological hematopoietic substance but also that it interacts with other hematopoietic cytokines, such as granulocytic colony stimulating factor (G-CSF) and interleukin-1α (IL-1), at physiological concentration range. The stimulative effect of histamine on several leukemia cells has also been reported (Nonaka et al., 1992). In this chapter, the mechanism of histamine-induced leukocytosis is reviewed.

2. Histamine-induced leukocytosis in experimental animals

It has been reported that after an intramuscular injection of histamine (1-6 mg/kg) in a beeswax-sesame oil mixture, to dogs, rabbits and guinea pigs, the number of leukocytes increased gradually, reached a peak in 6-8 hours, and then declined gradually, usually reaching the control levels within 24 hours after the injection (Tasaka and Code, 1964; Tasaka et al., 1992a) (Fig. 1). In this experiment, histamine was slowly released from the admixture with beeswax, maintaining a blood histamine concentration within 10-110 nM (3 mg/kg i.m. in dog) for prolonged periods. The mixture composed of beeswax and sesame oil alone did not affect leukocyte counts at all. When the plasma histamine concentration and leukocyte counts were measured simultaneously, it became apparent that a close relationship exists between peripheral leukocyte counts and blood histamine concentrations (Fig. 2).

In differential leukocyte counts, an increase in the number of neutrophils accounted for most of the leukocytosis in all species tested (Fig. 3). Erythrocyte counts and

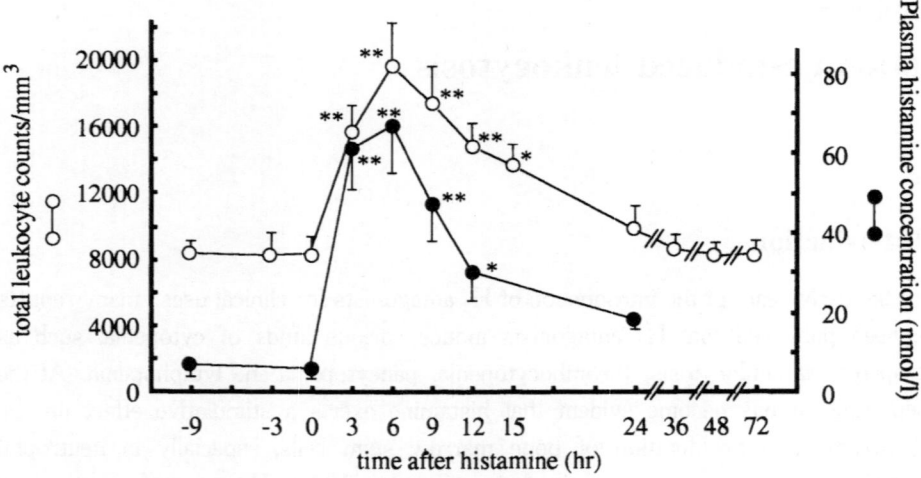

Fig. 1. Sequential changes in the number of total leukocytes and plasma histamine concentration after single administration of histamine in beeswax (3 mg/kg, i.m.) in dogs (n = 5). On day 0, histamine-beeswax mixture was injected. Significant differences from the preinjection level at *p < 0.05 and **p < 0.01, respectively. Reproduced from Tasaka, K., Nakaya, N. and Code, C.F.: Meth. Find. Exp. Clin. Pharmacol., 14, 667-675 (1992a), with permission.

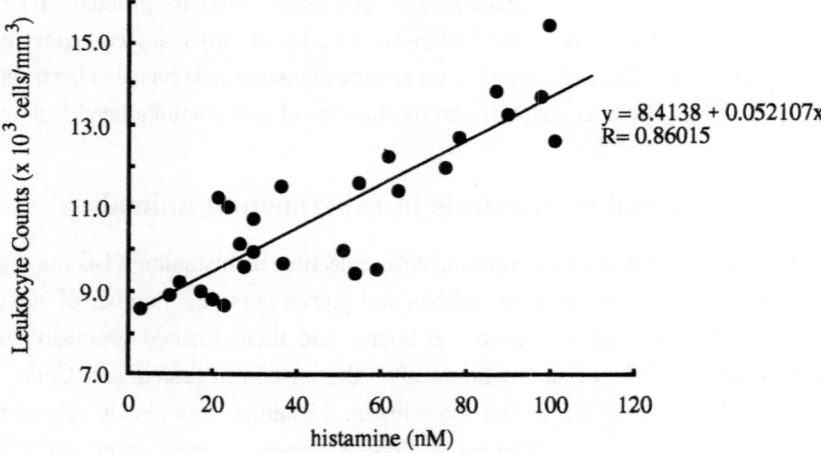

Fig. 2. Correlation between peripheral leukocyte counts and plasma histamine concentration after injection of histamine in beeswax (3 mg/kg, i.m.) in dogs (n = 5). Reproduced from Tasaka, K., Nakaya, N. and Code, C.F.: Meth. Find. Exp. Clin. Pharmacol., 14, 667-675 (1992a), with permission.

hematocrit values were not affected by histamine in any species. The absolute number of lymphocytes decreased somewhat; however, changes in the number of monocytes after histamine injection were small or nil. These findings are similar to the hematological

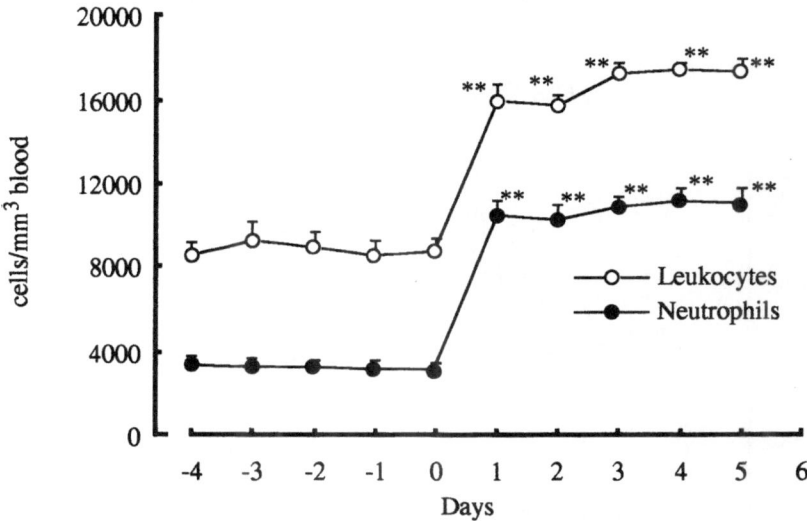

Fig. 3. Changes in the number of total leukocytes and neutrophils before and after daily injections of histamine (3 mg/kg, i.m.) in dogs (n = 3). Leukocyte counts were measured every day at 3:00 p.m., 6 hr after histamine injection. Reproduced from Tasaka, K., Nakaya, N. and Code, C.F.: Meth. Find. Exp. Clin. Pharmacol., 14, 667-675 (1992a), with permission.

picture seen frequently in patients suffering from acute inflammations. The plasma osmolarity of dogs was not altered after histamine injection (301-307 mOsm/l) and erythrocyte counts were actually unchanged in the presence of histamine, clearly indicating that histamine leukocytosis is not due to simple hemoconcentration. In regard to chronic administration of histamine, it is well known that duodenal ulceration occasionally occurs in dogs receiving large doses of histamine (Hay et al., 1942). However, no ulcers were found during the course of this experiment, either in the stomach or in the duodenum. It was also confirmed by macroscopic as well as histological examinations that no signs of inflammation existed around the injection site. These observations further support the idea that leukocytosis is decisively due to histamine. When the injection of histamine was repeated on consecutive days, the extent of leukocytosis subsided in some cases. Such a decline in histamine effect was observed in all species tested. However, simultaneous administration of aminoguanidine restored leukocytosis or, on some occasions, caused the leukocytosis to become more pronounced (Fig. 4) (Tasaka et al., 1992a). It is well known that an elevation of histaminase (diamine oxidase) levels takes place during pregnancy, and intravenous administration of aminoguanidine revealed a high concentration of plasma histamine (Lingberg and Törnqvist, 1966). Based on these findings, a sustained high concentration of histamine in plasma seems to cause an increase in plasma histaminase levels.

When pyrilamine, an H_1 antagonist, at a dose equivalent to that of histamine in molar

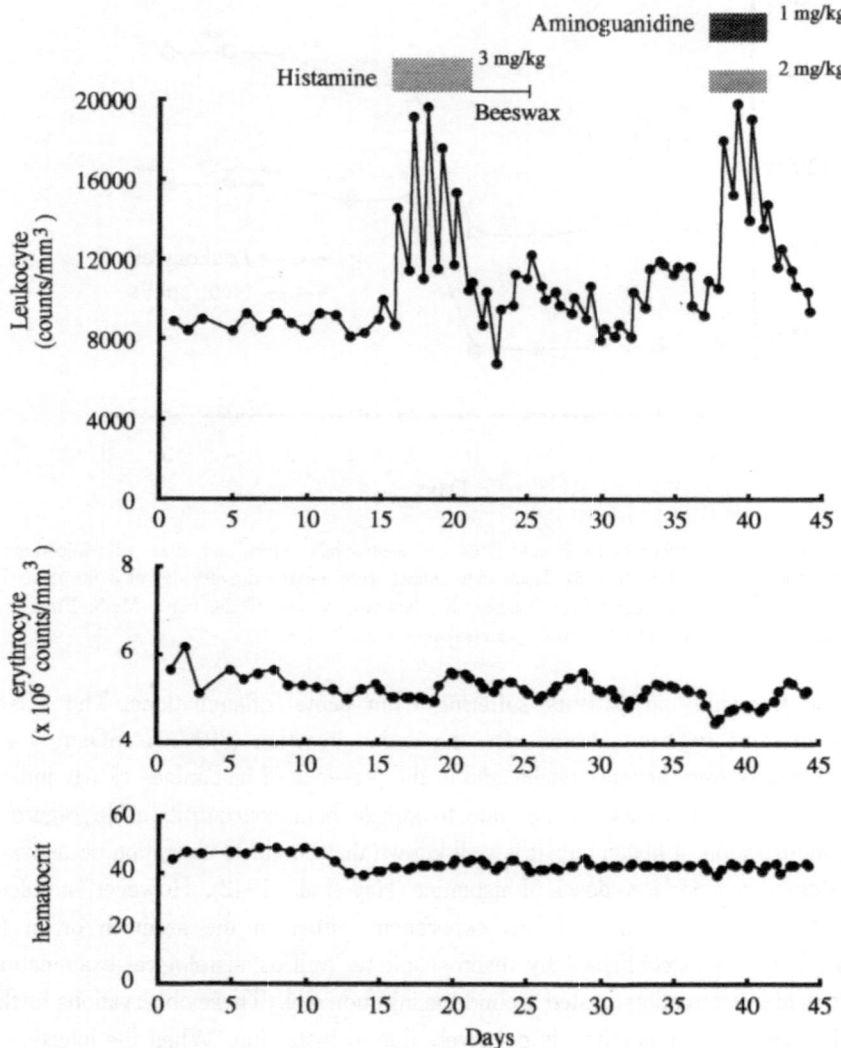

Fig. 4. Changes in the number of leukocytes and erythrocytes in dogs in association with histamine and aminoguanidine. After 3 mg/kg of histamine in beeswax, a marked leukocytosis was detected. However, when the injection was switched to beeswax alone, leukocyte counts rapidly decreased. In the second series, 2 mg/kg of histamine with 1 mg/kg of aminoguanidine again increased leukocyte counts to almost the level seen in the first series. This shows one of the typical experiments carried out simultaneously in 3 dogs. Erythrocyte counts and hematocrits were not altered in association with histamine administration. Reproduced from Tasaka, K., Nakaya, N. and Code, C.F.: Meth. Find. Clin. Pharmacol., 14, 667-675 (1992a), with permission.

base, was given simultaneously with histamine, the histamine-induced leukocytosis was not influenced at all (Tasaka and Code, 1964; Tasaka et al., 1992a). However, when

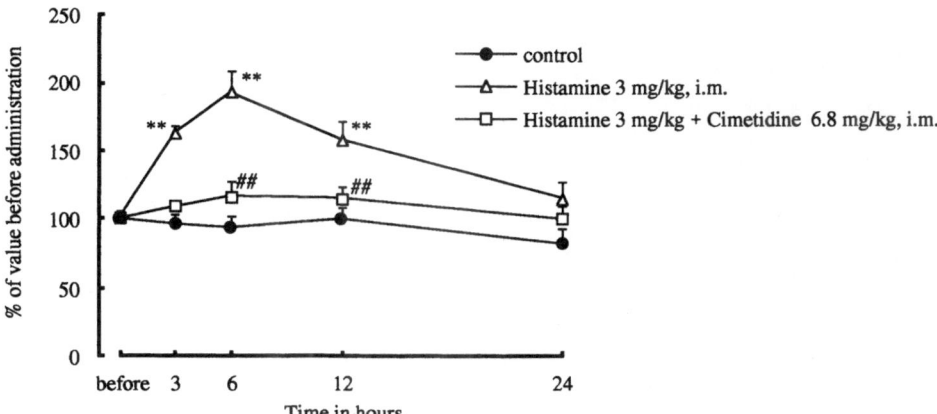

Fig. 5. Effect of cimetidine on histamine-induced leukocytosis (dogs). Leukocyte counts were largely suppressed in the presence of cimetidine. Preinjection level of total leukocyte counts was taken as 100%. Significant increase from the preinjection level at ** p < 0.01. Significant difference from the value produced by histamine-beeswax at ## p < 0.01. Reproduced from Tasaka, K., Nakaya, N. and Code, C.F.: Meth. Find. Exp. Clin. Pharmacol., 14, 667-675 (1992a), with permission.

histamine was injected simultaneously with metiamide or cimetidine, the increase in the leukocyte count seen after histamine injection was suppressed to a level close to the control values (Fig. 5) (Code, 1976; Tasaka et al., 1992a). This seems to suggest that the histamine receptors associated with leukocytosis are H_2 receptors.

There have been many reports describing how histamine stimulates the chemotaxis of neutrophils (Radermecker and Maldague, 1981) and eosinophils (Clark et al., 1975). This action seems to be associated with the movement of leukocytes from the vascular (marginal) pool to the tissues. However, when histamine in beeswax was injected consecutively, the leukocytes were often 2 or 3 times as numerous in the blood as under control conditions (Tasaka et al., 1992a). It seemed unlikely that all of the additional cells, over such a long period, came only from the marginal pool. In order to clarify the origin of the increased cells, bone marrow cells were labelled by a single intravenous injection of ^{3}H-thymidine. Subsequently, the number of labelled cells appearing in the blood of animals given histamine in beeswax was compared with the number in the blood of control animals (Tasaka et al., 1992b).

Neither the number of labelled mononucleocytes nor the time of their appearance in the blood were affected by the administration of histamine in beeswax. Labelled neutrophils, however, appeared earlier and were more numerous in peripheral blood after injection of histamine. They reached a maximal number earlier when histamine leu-

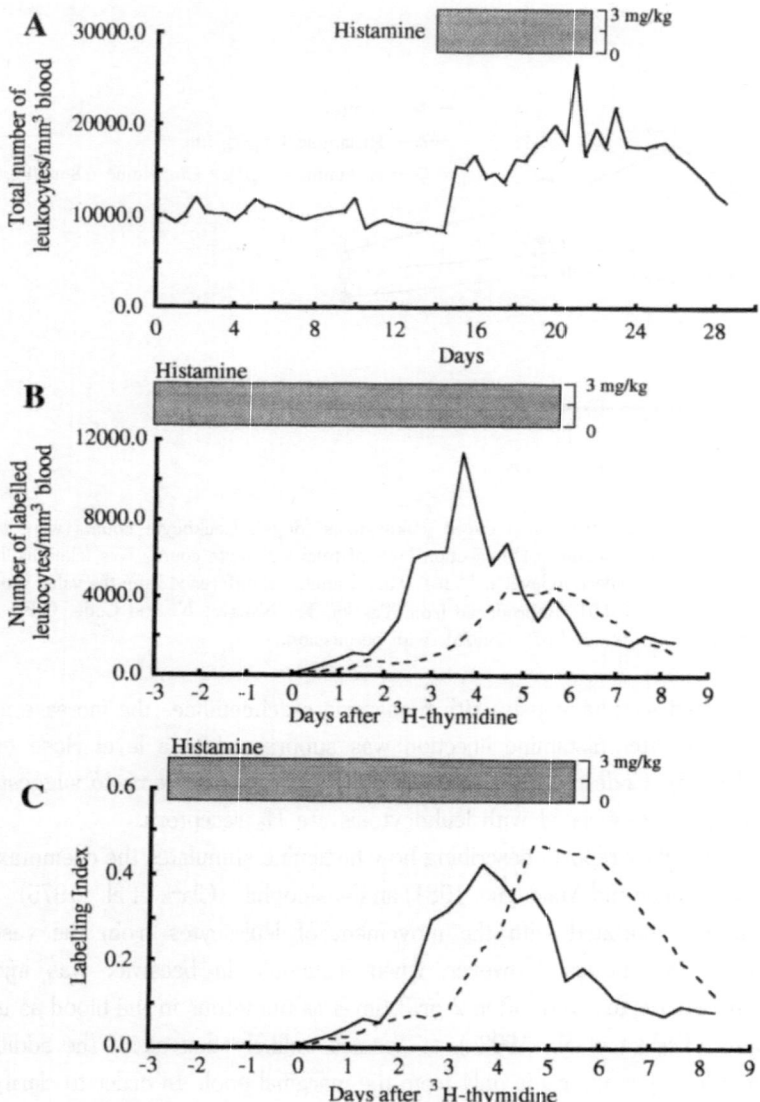

Fig. 6. Appearance of labelled leukocytes in 1 dog during histamine leukocytosis. A) Histamine leukocytosis; B and C) appearance of labelled cells during histamine leukocytosis compared to mean appearance in control tests (interrupted lines). Three similar experiments were carried out. Results shown are one typical experiment. (Labelling index: labelled cells/total leukocytes) Reproduced from Tasaka, K., Shorter, R.G. and Code, C.F.: Meth. Find. Exp. Clin. Pharmacol., 14, 799-804 (1992b), with permission.

kocytosis was present than when it was not (Fig. 6) (Tasaka et al., 1992b). In addition to appearing sooner and in greater number, the labelled granulocytes disappeared earlier

Table 1. Maximal numbers of labelled leukocytes in circulating blood after intravenous injection of ^{3}H-thymidine (1.0 mCi/kg) in dogs.

Group	Conditions	Dog	Labelled leukocytes max. n./mm^3
1	Control: No histamine in beeswax	1	5,619
		2	5,858
		3	4,650
		4	3,234
		Mean	4,840
2	Histamine in beeswax given before and after thymidine	1	11,389
		2	7,400
		3	15,482
		Mean	10,829
3	Histamine in beeswax started 3rd day after thymidine	1	14,650
		2	6,150
		3	13,200
		Mean	11,333
4	Histamine in beeswax started 4th day after thymidine	1	10,428
		2	8,383
		3	15,200
		Mean	11,337

from the blood of the animals with histamine leukocytosis. As shown in Fig. 6, the labelling index shifted markedly to the left and reached the maximum between days 3 and 4; on day 5 a sudden decrease occurred.

When ^{3}H-thymidine was given to dogs having histamine-induced leukocytosis, labelled neutrophils appeared earlier in the blood than in control animals; and when histamine leukocytosis was induced on the 3rd day after injection of ^{3}H-thymidine, labelled neutrophils appeared prematurely. Thus, histamine released labelled leukocytes from the bone marrow. Once the peak or near-peak number of labelled leukocytes had reached the blood, that is on the 4th day after injection of ^{3}H-thymidine, histamine did not affect the percentage of labelled cells in the blood, although it did increase their number. The maximal number of labelled cells appearing in the blood was about the same whether histamine in beeswax was given before and/or after administration of ^{3}H-thymidine (Table 1) (Tasaka et al., 1992b). These results may indicate that leukocytosis induced by chronic action of histamine is due to 1) stimulated proliferation and differentiation of neutrophil precursor cells in the bone marrow and 2) the release of mature leukocytes from the bone marrow (Tasaka et al., 1992b), and that 3) histamine-induced leukocytosis is characterized by a marked increase in neutrophils. Although a tremendous number of leukocytes appeared in the peripheral blood, no immature cell was found. This indicates that histamine potently stimulated the production of leukocytes in the bone marrow but it certainly did not squeeze the immature leukocytes out of the bone marrow to the peripheral blood.

3. Effect on neutrophil precursors

In the 1970s, it was reported that the administration of metiamide (Forrest et al., 1975; Feldman and Isenberg, 1976) or cimetidine (Klotz and Kay, 1978; Johnson et al., 1977) for long periods causes severe agranulocytosis, and much attention was paid to the effects of these drugs on the bone marrow. In 1976, Byron showed that H_2 receptors exist in the pluripotent stem cells (CFUs) of murine bone marrow and H_2 blockers, such as metiamide or cimetidine, inhibited the cell cycle (specifically the change from the G_0 to S phase) in these stem cells (Byron, 1980). In addition, 4-methylhistamine stimulated corresponding changes in the cell cycle. It was reported that the administration of cimetidine caused not only granulocytopenia but also thrombocytopenia (Idvall, 1979) or, in some cases, pancytopenia. Thus, it was assumed that the drug has suppressive effects on CFUs, since these cells share the common progenitors of all blood cells, including megakaryocytes (de Galoscy and van Ypersele de Strihou 1979).

In 1987, Tasaka and Nakaya reported that histamine-reactive, immature cells exist in murine normal bone marrow cells. When whole bone marrow cells of mice were cultured with histamine for 3 days, histamine promoted the differentiation and proliferation of the immature granulocytes; the number of neutrophils and metamyelocytes in the medium increased dose-dependently. Cimetidine potently inhibited the differentiation of bone marrow cells induced by histamine, as indicated in Table 2 (Nakaya and Tasaka, 1988a). When cimetidine (10^{-7} M) alone was added into the culture medium, no change in cell

Table 2. Changes in murine bone marrow cells after culture with histamine.

cell population	before	Histamine (M)				
		0	10^{-8}	10^{-7}	10^{-6}	$10^{-7}+Ci10^{-7a)}$
neutrophil	2.788 ± 0.202	1.859 ± 0.478	$3.752 \pm 0.342*$	$3.738 \pm 0.342**$	$5.499 \pm 0.345**$	2.201 ± 0.252
eosinophil	0.111 ± 0.012	0.066 ± 0.028	0.207 ± 0.050	$0.206 \pm 0.038*$	$0.252 \pm 0.088*$	0.222 ± 0.068
metamylocyte	1.012 ± 0.151	0.433 ± 0.192	1.458 ± 0.382	$1.240 \pm 0.130**$	$2.232 \pm 0.419**$	0.708 ± 0.163
myelocyte	0.541 ± 0.119	0.400 ± 0.102	0.993 ± 0.229	0.502 ± 0.050	1.089 ± 0.296	0.500 ± 0.063
promyelocyte	0.726 ± 0.155	0.627 ± 0.146	0.841 ± 0.196	0.841 ± 0.196	0.526 ± 0.161	0.630 ± 0.072
myelobalst	0.686 ± 0.162	0.561 ± 0.180	0.686 ± 0.045	0.838 ± 0.140	1.011 ± 0.089	0.600 ± 0.125
lymphocyte-like cell	3.500 ± 0.254	0.066 ± 0.029	0.168 ± 0.039	0.118 ± 0.022	0.585 ± 0.247	0.072 ± 0.025
megakaryocytic cell	0.120 ± 0.005	0.141 ± 0.032	0.149 ± 0.079	0.050 ± 0.026	0.098 ± 0.031	0.147 ± 0.069
nucleated erythroid cell	0.632 ± 0.089	0.070 ± 0.032	0.159 ± 0.083	0.054 ± 0.027	0.378 ± 0.181	0.163 ± 0.099

($\times 10^5$ cells)

a): Bone marrow cells were cultured in combination with histamine (10^{-7}M) and cimetidine (10^{-7}M). n = 5; * $p < 0.05$, ** $p < 0.01$. Reproduced from Nakaya, N. and Tasaka, K.: Life Sci., 42, 999-1010 (1988a), with permission.

Table 3. Changes in populations of ^{3}H-thymidine labelled cells in the presence and in the absence of histamine.

cell population	% of labelled cells	
	control	histamine 10^{-6}M
neutrophil eosinophil metamyelocyte	10.9 ± 1.1	10.2 ± 0.5
myelocyte	52.1 ± 3.9	$66.5 \pm 4.6^*$
promyelocyte	37.9 ± 4.4	$55.4 \pm 2.5^{**}$
myeloblast	23.6 ± 3.6	$41.2 \pm 4.4^{**}$
lymphocyte-like cell	0.4 ± 0.2	0.8 ± 0.2
monocyte-like cell	0.1 ± 0.1	0 ± 0
megakaryocytic cell	0 ± 0	0 ± 0
nucleated erythroid cell	0 ± 0	0.1 ± 0.1

* and ** represent significant difference from control groups with $p < 0.05$ and $p < 0.01$, respectively. n = 6. Reproduced from Nakaya, N. and Tasaka, K.: Life Sci., 42, 999-1010 (1988a), with permission.

Table 4. Autoradiographic observation of ^{3}H-histamine incorporation into murine bone marrow cells.

cell population	grain counts (mean $\pm$ S.E.M.)	% of labelled cells[a]
neutrophil	0.37 ± 0.07	0
eosinophil	7.00 ± 0.76	72.0
metamyelocyte	0.55 ± 0.08	0
myelocyte	0.81 ± 0.14	0
promyelocyte	6.46 ± 1.19	43.9
myeloblast	2.95 ± 0.84	22.7
lymphocyte-like cell	0.76 ± 0.12	1.1
megakaryocyte	0 ± 0	0
nucleated erythroid cell	0.20 ± 0.06	0

a) Labelled cells were the cells on which more than ten grains were found. n = 6. Reproduced from Nakaya, N. and Tasaka, K.: Life Sci., 42, 999-1010 (1988a), with permission.

population was observed. Similar results were obtained in both experiments when ranitidine (10^{-7} M) was employed (in combination with histamine (10^{-7} M) and with single applications) instead of cimetidine. On the other hand, H_1 antagonists, such as pyrilamine (10^{-7} M) or diphenhydramine (10^{-7} M), did not alter the effect of histamine (10^{-7} M). These findings indicate that histamine stimulates the differentiation of promyelocytes and myeloblasts through H_2 receptor stimulation, which increases the number of descendant cells such as metamyelocytes and neutrophils.

When bone marrow cells were incubated for 6 hr with ^{3}H-thymidine, the amount of the latter incorporated into the DNA increased in the presence of histamine (10^{-6} M). From the autoradiography of the bone marrow cells, it became apparent the cells which incorporated ^{3}H-thymidine were largely of the granulocytic series, especially promyelocytes and myeloblasts (Table 3) (Nakaya and Tasaka, 1988b). Simultaneously incubated at equimolar concentrations with histamine, neither histamine antagonists (pyrilamine, diphenhydramine, cimetidine or ranitidine) nor histamine agonists (4-methylhistamine or 2-methylhistamine) altered the effect of histamine.

As shown in Table 4, when murine bone marrow cells were incubated with histamine (5×10^{-7} M) containing ^{3}H-histamine, there was a high incidence of labelled cells within the eosinophil, promyelocyte and myeloblast groups (Nakaya and Tasaka, 1988a). Similar incorporation of ^{3}H-histamine is also reported by Gespach et al. (1985a). This high incidence of histamine incorporation into eosinophils is in agreement with the observation that eosinophil is an active histamine scavenger (Catini et al., 1984). Neither histamine antagonists (cimetidine, ranitidine, diphenhydramine or pyrilamine) nor histamine agonists (2-methylhistamine or 4-methylhistamine) affect the ^{3}H-histamine uptake, indicating that neither H_1 nor H_2 receptors are involved in the histamine uptake. The incorporation of ^{3}H-histamine was inhibited by α-2, 4-DNP, cycloheximide and actinomycin D (Tasaka and Nakaya, 1987; Nakaya and Tasaka, 1988b). This may indicate that some protein synthesis is involved in the process leading to histamine uptake, which may act as a transport system. Once inside the cell, histamine is incorporated into the nuclei of myeloblasts and promyelocytes. In order to investigate the interaction between incorporated histamine and nucleus, the binding capacity of histamine to either DNA or to chromatin proteins was measured (Table 5). When histamine binding to DNA extracted from progenitor cells and to that of calf thymus was compared, no difference was noticed in the dissociation constant. However, the comparison of non-

Table 5. Dissociation constants in the binding of histamine to DNA and to chromatin-related proteins extracted from the nuclei of myeloblasts and promyelocytes.

Source	DNA and associated proteins	Kd (M)
progenitor cells	DNA	2.48×10^{-9}
	histone	N.B.
	NHCP	2.54×10^{-10}
	chromatin	5.22×10^{-9}
calf thymus	DNA	3.10×10^{-9}
	histone	N.B.
BSA		N.B.

N.B.: No binding with ^{3}H-histamine was detected. Reproduced from Nakaya, N. and Tasaka, K. Histamine incorporation into murine myeloblasts and promyelocytes. Formation of a histamine transport system. Biochem. Pharmacol., 37: 4523-4530 (1988b), with permission.

histone chromatin protein (NHCP) and DNA, both extracted from neutrophil progenitors, showed that the former has an approximately 10 times higher affinity for histamine than the latter. A negligible amount of histamine was bound to calf thymus histone, histone extracted from progenitor cells or bovine serum albumin (Nakaya and Tasaka, 1988b). Given these results, it can be assumed that the histamine incorporated into these cells is transferred into the nucleus and that it may act to stimulate DNA replication, which may be an important stage in promoting the differentiation of granulocytic precursor cells.

Table 6. Changes in cell populations after cell culture of myeloblasts and promyelocytes with histamine for 24 hr.

cell population	Histamine (M)				
	before	0	10^{-8}	10^{-7}	10^{-6}
neutrophil	0.079 ± 0.010	0.127 ± 0.020	$0.937 \pm 0.104**$	$1.307 \pm 0.118**$	$1.590 \pm 0.099**$
eosinophil	0.317 ± 0.059	0.504 ± 0.093	$0.869 \pm 0.096**$	$0.885 \pm 0.124**$	$1.209 \pm 0.136**$
metamyelocyte	1.583 ± 0.243	1.869 ± 0.095	2.485 ± 0.302	$3.197 \pm 0.262**$	$3.270 \pm 0.416**$
myelocyte	1.266 ± 0.225	1.221 ± 0.237	$2.138 \pm 0.100**$	$1.931 \pm 0.085*$	1.164 ± 0.020
promyelocyte	2.295 ± 0.431	1.903 ± 0.091	$1.164 \pm 0.139*$	$0.746 \pm 0.084**$	$1.008 \pm 0.089*$
myeloblast	5.302 ± 0.449	3.032 ± 0.376	$0.675 \pm 0.082**$	$0.764 \pm 0.025**$	$0.941 \pm 0.100**$

(x 10^5 cells/culture)

n = 6. * p < 0.05. ** p < 0.01. Reproduced from Nakaya, N. and Tasaka, K. Life Sci., 42: 999-1010, with permission.

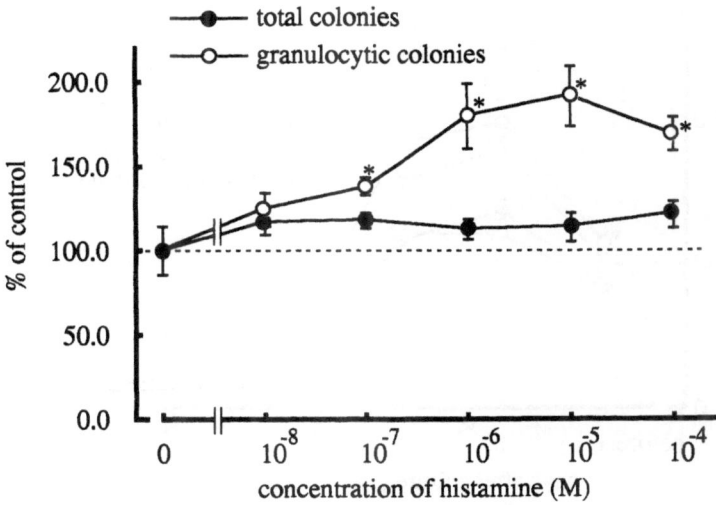

Fig. 7. Changes in the number of CFU-c in the murine bone marrow cells. The number of CFU-c counted immediately before the assay was taken as 100 % in the ordinate. * represents significant difference at the 0.05 level. n = 6. Reproduced from Nakaya, N. and Tasaka, K., Life Sci., 42: 999-1010 (1988a), with permission.

Since histamine acts preferentially to stimulate the differentiation of myeloblasts and promyelocytes, its effect on these cells was studied. Myeloblasts and promyelocytes were fractionated from murine bone marrow cells by means of Percoll density gradient centrifugation (Nakaya and Tasaka, 1988a). According to this procedure, myeloblasts and promyelocytes constituted 70-80 % of the total cells in the obtained fractions. Changes in the cell population of these fractions induced by 24 hr-treatment with histamine are shown in Table 6. A significant decrease in myeloblasts and promyelocytes was observed; conversely, the number of myelocytes, metamyelocytes and matured neutrophils increased.

On the other hand, when the effect of histamine on the number of colony-forming units in culture (CFU-c) was determined after 3 days of incubation, histamine caused a dose-dependent increase in granulocyte colonies, as shown in Fig. 7. A moderate increase was observed at 10^{-8}M, while maximum effect was achieved at 10^{-5}M. However, the total number of CFU-c was only slightly increased. This finding indicates that histamine does not increase macrophage colonies, but rather, acts selectively on granulocyte colonies (Nakaya and Tasaka, 1988a). When histamine (10^{-7}M) alone was added to the medium, the number of CFU-c reached a maximum 3 days after incubation, thereafter, CFU-c decreased slightly up to 5 days (Fig. 8), and a rapid and marked decrease in CFU-c followed. Although simultaneous addition of pyrilamine (10^{-7}M) did not alter the effect of histamine (10^{-7}M), the addition of cimetidine or ranitidine (10^{-7}M) diminished it almost completely. Although neither H_2 blocker (10^{-7}M) alone significantly

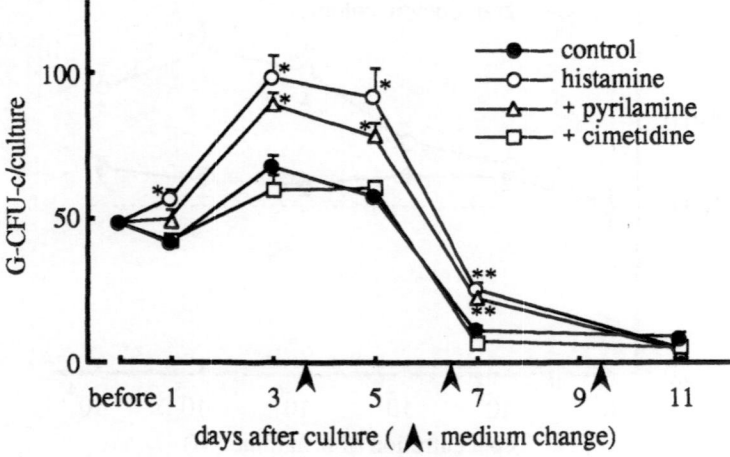

Fig. 8. Effects of histamine (10^{-7} M) and antihistamines (10^{-7} M) on the number of CFU-c. Almost all CFU-c developed into granulocytic myeloid cells. * and ** represent significant differences at $p < 0.05$ and $p < 0.01$, respectively ($n = 6$). Each culture medium was changed every 72 hr. Reproduced from Nakaya, N. and Tasaka, K., Life Sci., 42: 999-1010 (1988a) with permission.

affected the number of CFU-c (n = 6), the addition of 4-methylhistamine at 10^{-7}M increased this number by 51.2 ± 2.5 % (p < 0.05) from the control level when determined after 3 days of incubation (n = 6). The effect of 4-methylhistamine corresponds to 42 % of that induced by histamine. When 2-methylhistamine (10^{-7}M) was added into the culture medium, it decreased the CFU-c number moderately but not significantly, by 10.6 ± 2.1 % (n = 6) (Nakaya and Tasaka, 1988a).

From the findings indicated above, it became apparent that histamine, at physiological concentration range, stimulates the differentiation and proliferation of murine neutrophil progenitor cells in bone marrow, and the mode of action seems to differ depending on the stages of differentiation. Since H_2 blocking agents caused remarkable inhibition on the CFU-c stage, and since 4-methylhistamine mimics the effect of histamine, it seems certain that histamine exerts a stimulatory effect on the cells more immature than CFU-c, differentiating the cells to be matured neutrophils via the H_2 receptor. In the different stage, histamine may be accumulated in myeloblasts and promyelocytes so as to induce the differentiation and proliferation of these cells, probably through interaction with the nucleus.

4. Histamine and cell proliferation

There are many phenomena which reveal that the action of histamine leads to cell proliferation. 4-Methylhistamine also facilitates the alteration of the cell cycle in pluripotent stem cells. The induction of suppressor T cells is one of the well-known effects of histamine in this area. The inhibitory effects of histamine on the immune system are expressed as the inhibition of blastogenesis as well as the proliferation of lymphocytes stimulated by antigen after the induction of suppressor T cells (Rocklin and Haberek-Davidson, 1981). This effect is elicited by histamine at concentrations of 10^{-8}-10^{-3}M and mediated by H_2 receptors. In this case, histamine indirectly inhibits the proliferation of cells through the production of the suppressor factor derived from suppressor T cells (Rocklin et al., 1979). Histamine also inhibits the proliferation of epidermal cells (Harper et al., 1974). This action is also mediated by H_2 receptors and intracellular cAMP contents are elevated markedly after the addition of histamine (Iizuka et al., 1978). By contrast, Adachi et al. (1980) reported that histamine stimulated the proliferation of epidermal cells, elevating intracellular cGMP levels through H_1 receptors. Since the elevation of intracellular cAMP levels leads to the inhibition of human skin growth (Voorhees et al., 1972), the action of cAMP in relation to cell growth is complicated and there are probably differences between species. There are many reports about the abnormal histamine contents of tumor tissues (Maslinski et al., 1984; Burtin et al., 1981). Burtin (1986) investigated the relationship between histamine content and the tumor growth. He thought that the elevation of histamine content in tumor tissues may act as a protective factor, and actually showed that histamine inhibits tumor growth using

W/W^V mast cell deficient mice. Histamine-induced stimulation of angiogenesis (Marks et al., 1986) and synthesis of collagen (Hatamochi et al., 1985) were also observed.

5. Histamine-induced differentiation of HL-60 cells: Analysis of the mechanism of action of histamine

As described above, in the differentiation of murine neutrophil progenitors, the site of action of histamine is distributed among various differentiation stages ranging from colony-forming units in culture (CFU-c) to metamyelocytes (Nakaya and Tasaka, 1988a, 1988b). However, in the bone marrow, neutrophil progenitors are diversely differentiated over a wide range, making it difficult to determine the stage at which histamine acts to differentiate the immature cells and to elucidate the mechanism of this action. To eliminate these problems, it is favorable to use the cells remaining in the same differentiation stages. In 1982, Gespach et al. found that histamine H$_2$ agonists increased the intracellular cAMP content in HL-60 cells (the human promyelocytic leukemia cell line) and that the H$_2$ receptors mediate the uptake of histamine into HL-60 (Gespach et al., 1985a). Gespach et al. (1985b, 1986b) also reported the occurrence of H$_2$ receptors in U-937, a human histocytic lymphoma cell line. Sawutz et al. (1984) found that HL-60 cells treated with dimaprit differentiated into mature neutrophils, and that intracellular cAMP levels increased markedly. The result clearly indicates that the effect of histamine on murine bone marrow cells is similar to that observed in HL-60 cells. Furthermore, since HL-60 cells are on the promyelocytic stage and consist of a homogeneous cell population, we employed HL-60 cells to analyze the mechanism of action by which histamine induces the differentiation to neutrophils (Nonaka et al., 1992).

Before addition of histamine, 92.3 ± 3.8 % (n = 4) of the total HL-60 cells were promyelocytic cells. When histamine was added to the incubation medium of HL-60 cells,

Table 7. Effect of histamine on the differentiation of HL-60 cells during 6 days of culture.

Days of culture	Concentration of histamine (μM)			
	0	0.1	1	10
0	7.7 ± 1.3			
1	3.3 ± 0.3	0.3 ± 4.7	8.3 ± 0.7	12.0 ± 1.5
2	5.7 ± 1.5	6.0 ± 0.1	7.7 ± 0.9	8.0 ± 3.0
3	7.0 ± 0.9	4.7 ± 0.9	9.0 ± 0.1	$14.0 \pm 1.2^*$
4	15.3 ± 4.6	20.7 ± 2.2	$23.7 \pm 1.2^*$	$24.3 \pm 1.2^*$
5	19.0 ± 3.6	21.3 ± 2.2	$32.0 \pm 1.2^*$	$28.0 \pm 0.1^*$
6	18.0 ± 4.4	23.0 ± 1.5	$30.3 \pm 2.7^*$	$41.7 \pm 2.0^{**}$

Each value is the % of differentiated cells, mean $\pm$ S.E.M. (n = 4). * and ** indicate statistical significance in comparison with non-stimulated group on each day at $p < 0.05$ and $p < 0.01$, respectively. Reproduced from Nonaka, T., Mio, M., Doi, M. and Tasaka, K.: Biochem. Pharmacol., 44, 1115-1121 (1992), with permission.

Table 8. Effect of histamine on the myeloperoxidase activity in HL-60 cells after 6 days of culture.

Concentration of histamine (μM)	Myeloperoxidase activity ($\times 10^{-6}$ mole tetraguaiacol/min/10^6 cells)
0	16.90 ± 0.97
0.1	15.48 ± 1.17
1	11.40 ± 0.62**
10	10.55 ± 0.77**

The enzyme activity was measured after 6 days of culture and represented as the formed amount of tetraguaiacol/min/10^6 cells (Nonaka et al., 1992). Each value represents the mean ± SEM (n = 4). ** indicates statistical significance at p < 0.01. Reproduced from Nonaka, T., Mio, M., Doi, M. and Tasaka, K.: Biochem. Pharmacol., 44, 1115-1121 (1992), with permission.

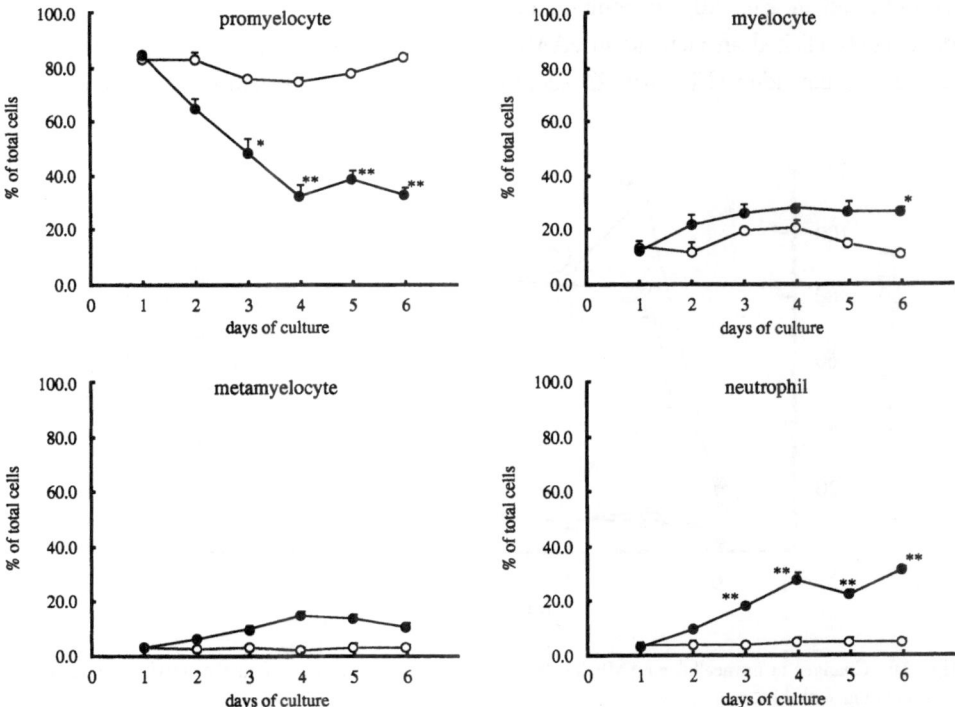

Fig. 9. Morphological changes in HL-60 cells treated with histamine. The cells were incubated with histamine for 6 days. Each point represents the mean ± S.E.M. (n = 3). * and ** indicate significantly different from the control at p < 0.05 and p < 0.01, respectively. ○: control, ●: histamine (1μM). Reproduced from Nonaka, T., Mio, M., Doi, M. and Tasaka, K.: Biochem. Pharmacol., 44, 1115-1121 (1992), with permission.

a significant differentiation of the cells was elicited in a dose-dependent manner during 6 days of culture (Table 7). In this case, the number of differentiated cells was expressed as the total cells differentiating at various stages ranging from myelocytes to neutrophils.

The populational changes in HL-60 cells induced by 1 μM of histamine are indicated in Fig. 9. As the number of promyelocytes decreased, those of metamyelocytes and neutrophils increased, indicating that histamine induced the differentiation of HL-60 cells. The stimulative effect of histamine on the differentiation of HL-60 cells became remarkable after three days of culture. In connection with this, the myeloperoxidase activity of HL-60 cells was significantly reduced by histamine treatment in a dose-dependent manner (Table 8). Since myeloperoxidase activity of HL-60 cells decreased in accordance with cell differentiation from promyelocytes to neutrophils (Bainton et al., 1971), this also supports the hypothesis that histamine effectively induces differentiation of the neutrophil progenitors.

In order to analyze the mechanism of the intracellular signal transduction in HL-60 cells after histamine-stimulation, changes in cAMP content of the cells were measured. As indicated in Fig. 10, an addition of histamine (1 μM) to the incubation medium immediately elicited an increase in cAMP content of HL-60 cells, reaching a plateau 30 min after stimulation (Fig. 10). In addition, the activity of protein kinase A (A kinase)

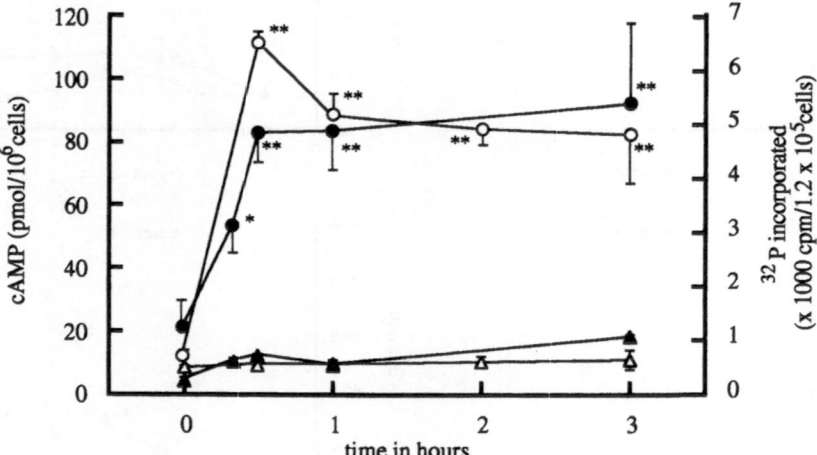

Fig. 10. Changes in intracellular cAMP levels and cAMP-dependent protein kinase activity in HL-60 cells in association with histamine treatment. $\bigcirc$, $\triangle$: cAMP content; the cells were incubated for 3 hr with or without 1μM of histamine, respectively. $\bullet$, $\blacktriangle$: cAMP-dependent protein kinase activity; the cells were incubated for 3 hr with or without 1 μM of histamine, respectively; after a disruption of the cells, the protein kinase activity was determined in the presence of exogenous cAMP (10 μM). Each point represents the mean ± S.E.M. (n = 3). * and ** indicate significantly different from the control at p < 0.05 and p < 0.01, respectively. Reproduced from Nonaka, T., Mio, M., Doi, M. and Tasaka, K.: Biochem. Pharmacol., **44**, 1115-1121 (1992), with permission.

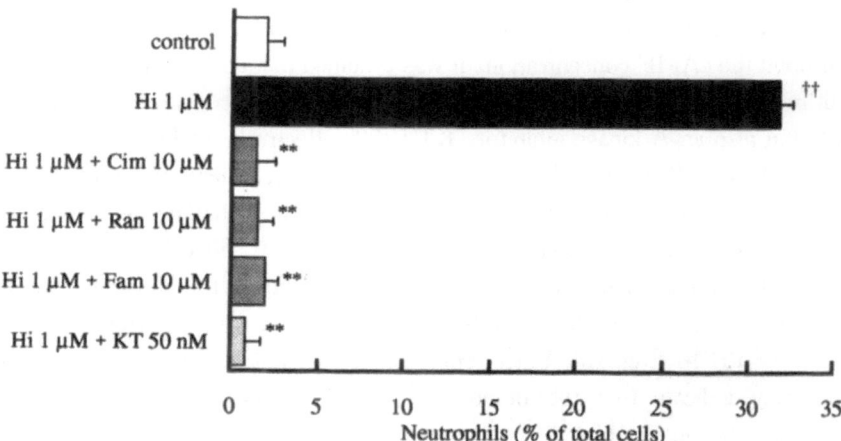

Fig. 11. Effect of H₂ antagonists and KT-5720 on the differentiation of HL-60 cells to neutrophils in the presence of histamine. The cells were cultured with 1µM of histamine both in the presence and in the absence of test compounds for 6 days. Each column represents the mean ± S.E.M. (n = 3). †† indicates significantly different from the control group at p < 0.01. ** indicates significantly different from the histamine-treated group at p < 0.01. Hi: histamine, Cim: cimetidine, Ran: ranitidine, Fam: famotidine, KT: KT-5720. Reproduced from Nonaka, T., Mio, M., Doi, M. and Tasaka, K.: Biochem. Pharmacol., 44, 1115-1121 (1992), with permission.

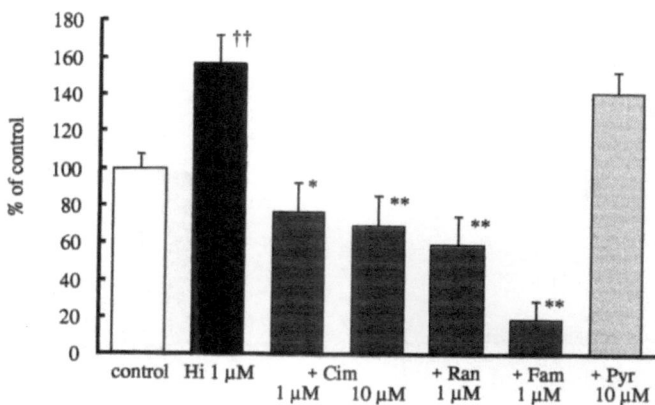

Fig. 12. Effects of histamine antagonists on phagocytic activity of HL-60 cells engulfing the yeast provoked by histamine (4 days of culture). The cells were cultured with 1 µM of histamine both in the presence and in the absencnce of test compounds. Each column represents the mean ± S.E.M. (n = 3). †† indicates significantly different from the control group at p < 0.01. * and ** indicate significantly different from the histamine-treated group at p < 0.05 and p < 0.01, respectively. Hi: histamine, Cim: cimetidine, Ran: ranitidine, Fam: famotidine, Pyr: pyrilamine. Reproduced from Nonaka, T., Mio, M., Doi, M. and Tasaka, K.: Biochem. Pharmacol., 44, 1115-1121 (1992), with permission.

increased almost in parallel with the increase in cAMP level; in both cases a rapid increase takes place after addition of histamine. Since the activity of A kinase is highly dependent upon the cAMP concentration, it was assumed that an early increase in cAMP may result in the subsequent activation of A kinase. In accordance with this, not only H_2 antagonists but also an A kinase inhibitor, KT-5720, effectively inhibited the differentiation of HL-60 cells (Fig. 11). Yeast phagocytic activity, which is an indicative of neutrophil differentiation, was significantly increased by histamine at 10^{-6}M. The histamine-induced increase in yeast-phagocytic activity of HL-60 cells was remarkably inhibited by H_2 antagonists at equi-molar concentrations to histamine, although pyrilamine, an H_1 antagonist, did not affect histamine-induced phagocytosis, even at a concentration of 10^{-5}M (Fig. 12). Since promyelocytes do not exhibit phagocytic activity, this result may indicate that the number of phagocytic cells, such as myelocytes, metamyelocytes and neutrophils, increased in response to histamine.

When added to the incubation medium of HL-60 cells, KT-5720, a specific A kinase inhibitor, alone slightly enhanced ^{3}H-thymidine uptake into HL-60 cells. This may suggest that KT-5720 enhanced cellular proliferation by inhibiting a spontaneous differentiation of HL-60 cells. In connection with this, KT-5720 dose-dependently inhibited the A kinase activity of resting HL-60 cells as well as those stimulated by histamine (Fig. 13). This may suggest that the inhibitory effect of KT-5720 is actually exerted by an inhibition of A kinase activity.

Since it was assumed that the activation of A kinase may play some important role

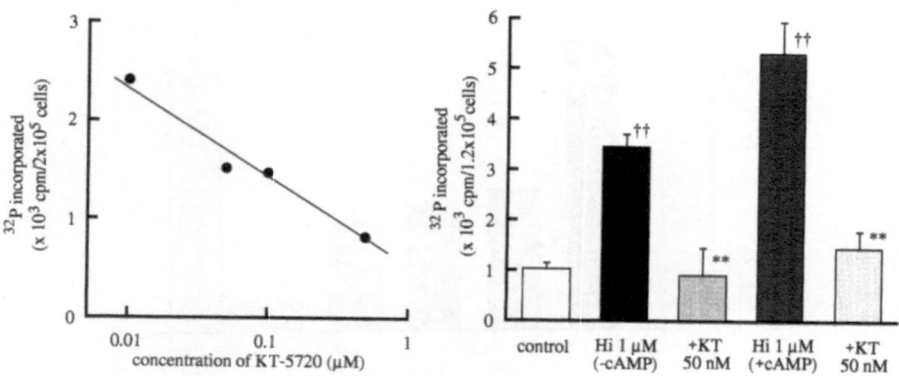

Fig. 13. Effect of KT-5720 on the cAMP-dependent protein kinase activity in resting HL-60 cells (*left*) and histamine-treated HL-60 cells (*right*). The cells were incubated with KT-5720 both in the presence and in the absence of histamine (1 μM) for 3 hr. Thereafter, the protein kinase activity was determined (Nonaka et al., 1992). Each point or column represents the mean ± S.E.M. (n = 3). †† indicates significantly different from the control group at p < 0.01 and ** indicates significantly different from the histamine-treated group. Reproduced from Nonaka, T., Mio, M., Doi, M. and Tasaka, K.: Biochem. Pharmacol., 44, 1115-1121 (1992), with permission.

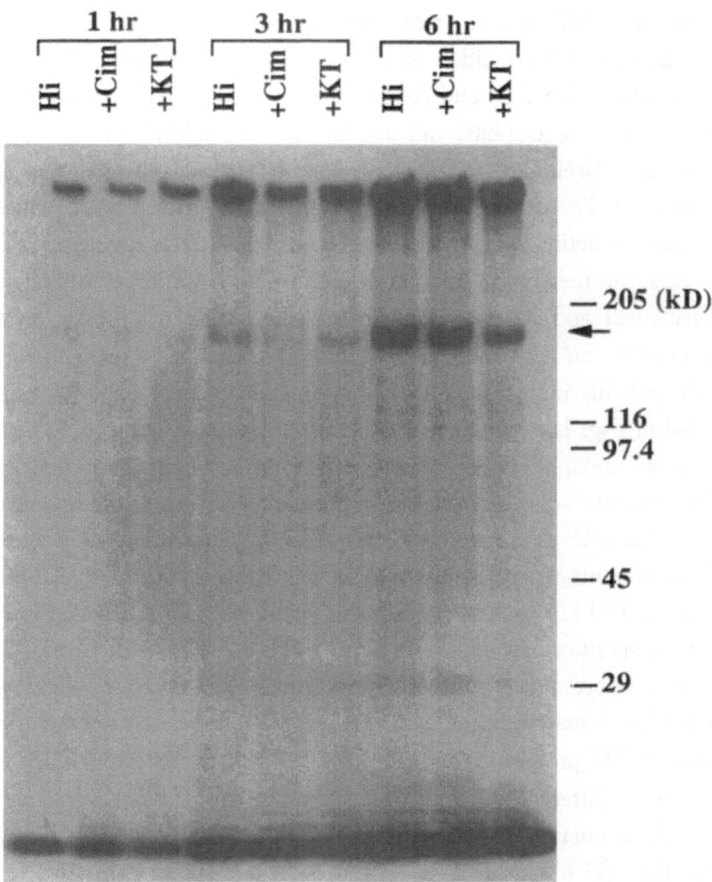

Fig. 14. Effects of cimetidine and KT-5720 on the protein phosphorylation in HL-60 cells elicited by histamine. The cells were stimulated with histamine (1 μM) both in the presence and in the absence of test compounds for 6 hr. Cim: cimetidine (10 μM), KT: KT-5720 (50 nM). Arrow indicates 160 kD protein band. Reproduced from Nonaka, T., Mio, M., Doi, M. and Tasaka, K.: Biochem. Pharmacol., 44, 1115-1121 (1992), with permission.

in the cell differentiation induced by histamine, histamine-induced protein phosphorylation was examined. As shown in Fig. 14, histamine at 1 μM strongly phosphorylated the 160 kD protein band. The 160 kD protein phosphorylation due to histamine was remarkably inhibited in the presence of either cimetidine or KT-5720; conversely, this may suggest that both H_2 receptor stimulation and A kinase activation are somehow crucial to the phosphorylation of 160 kD protein.

It has been shown that the compounds which cause an increase in cAMP contents, such as dibutyryl-cAMP, prostaglandin E_2 and theophylline, markedly promote the differentiation of neutrophil progenitors (Chaplinski and Niedel, 1986); it was supposed that an increase in cAMP and A kinase activity due to histamine stimulation may be essential to trigger the differentiation of HL-60 cells. In contrast, it has been reported that dimethyl sulfoxide (DMSO) and retinoic acid, which potently induce the differentiation of HL-60 cells to neutrophils, did not increase the cAMP content (Collins et al., 1978; Imaizumi and Breitman, 1988). However, it is also reported that DMSO and retinoic acid may act directly on the microcircumstances of A kinase, which is located very near the site of action of cAMP, thus translocating the regulatory subunit of A kinase from cytosol to membrane fraction, which would be a proper trigger for A kinase activation (Chaplinski and Niedel, 1986; Plet et al., 1982). However, in the case of histamine stimulation, an activation of A kinase is intimately related to an increase in cAMP content and this may trigger the differentiation of neutrophil progenitors.

As indicated in Fig. 14, histamine induced the phosphorylation of 160 kD protein in HL-60 cells, while phosphorylation was inhibited by H_2 antagonists and the A kinase inhibitor. This strongly suggests that phosphorylation of this protein is exerted by an activation of A kinase in association with H_2 receptor stimulation. It was assumed that this 160 kD protein may play some roles in the differentiation of HL-60 cells. The phosphorylation of 160 kD protein slightly and spontaneously proceeded in resting cells, suggesting that a spontaneous differentiation of HL-60 cells took place. On the other hand, it has been reported that phorbol esters, which are capable of eliciting differentiation of HL-60 cells to macrophages and monocytes, induced the phosphorylation of 64 kD protein and 68 kD protein via an activation of protein kinase C (Kiss et al., 1987; Evans et al., 1987). Different phosphorylated proteins and protein kinases in HL-60 cells may lead the differentiation in different directions. Based on these findings, it was concluded that the 160 kD protein in HL-60 cells may play some critical role in the differentiation toward neutrophils.

6. Effect of histamine on IL-1 production in bone marrow stromal cells

Various cytokines are known to stimulate the proliferation and the differentiation of bone marrow stem cells. As one of such cytokines, interleukin-1 (IL-1) is known to exert a stimulative effect on the hematopoietic systems through an increase in the release of colony-stimulating factors (CSFs) from stromal cells (Fibbe et al., 1988). It has also been reported that IL-1 is capable of stimulating the differentiation of pluripotent stem cells in cooperation with CSFs (Mochizuki et al., 1987). Using a long-term culture of murine bone marrow cells, Dexter and Testa (1976) observed that the bone marrow stromal cells play some important roles in the proliferation and differentiation of hematopoietic cells in culture. It has been reported that the IL-1 synthesized in bone marrow stromal cells may stimulate the bone marrow stromal cells so as to produce CSFs (Fibbe et al., 1988). However, little is known about the effect of histamine on bone marrow stromal cells and the production of IL-1. In order to clarify this problem, the effects of histamine and its antagonists on bone marrow stromal cells were investigated (Tasaka et al., 1993).

After murine bone marrow cells were incubated in RPMI-1640 supplemented with 10 % FCS containing various concentrations of histamine at 37°C for 24 hr, CFU-c assay was carried out. In this experiment, murine bone marrow cells containing adherent stromal cells were incubated with test compounds for 24 hr. Thereafter, non-adherent cells were suspended in a CFU-c assay medium [RPMI-1640 medium supplemented with 0.6 % methylcellulose, 20 % horse serum and 5 ng/ml of recombinant human granulocyte colony-stimulating factor (rG-CSF)] and cultured for 7 days at 37°C. After that, the colonies consisting of more than 50 cells each were counted microscopically. The activity stimulating colony formation was regarded as CFU-c stimulation activity. As indicated in Fig. 15, histamine increased the number of colonies consisting mostly of granulocytes (G-CFU-c) in a dose-dependent fashion at concentrations equal to or higher than 10^{-7}M without affecting the total CFU-c count. A significant effect was observed at concentrations equal to or higher than 10^{-6}M. 4-Methylhistamine, an H_2 agonist, was also effective in increasing the G-CFU-c count (Fig. 15). On the other hand, 2-methylhistamine, an H_1 agonist, did not affect either G-CFU-c or the total colony counts. However, when either histamine or 4-methylhistamine was added directly, without pretreatment with the bone marrow cells, to the CFU-c assay medium containing

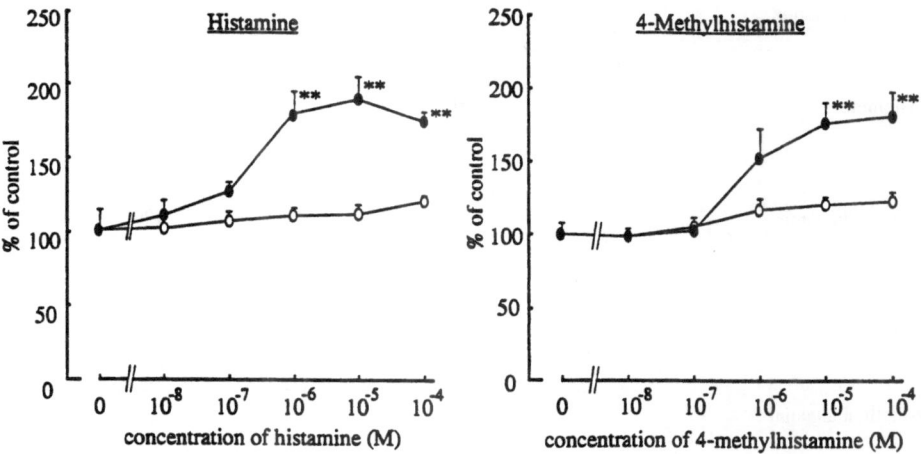

Fig. 15. Effects of histamine and 4-methylhistamine on the G-CFU-c counts in murine bone marrow cells. After treatment of murine bone marrow cells with test compounds at 37°C for 24 hr, CFU-c assay was carried out. ○ and ● represent the total colony count and the granulocyte colony count, respectively. Each point represents the mean ± S.E.M. of the data obtained from 5 separate experiments. ** indicates statistical significance in comparison with control at $p < 0.01$. Reproduced from Tasaka, K., Mio, M., Shimazawa, M. and Nakaya, N.: Mol. Pharmacol., 43, 365-371 (1993), with permission.

non-adherent cells, the G-CFU-c count did not increase at all. This seems to indicate that histamine stimulates H_2 receptor of bone marrow stromal cells so as to induce a production of hematopoietic substance(s) having a G-CFU-c stimulating activity. A similar effect was observed when theophylline and db-cAMP were used in place of histamine, indicating that an increase in cAMP level in bone marrow stromal cells may be involved in the process which triggers the production of such a hematopoietic substance.

Since it was supposed that adherent stromal cells may be involved in increasing the G-CFU-c counts in association with histamine pretreatment, murine stromal cells were cultured in the presence of histamine and its related compounds at 37°C for 24 hr according to the method of Dexter and Testa (1976) and thymocyte comitogenic activity in the supernatant of the culture medium was determined. Thymocyte comitogenic activity, which is known as the most representative IL-1 activity, was determined by the method of Rosenwasser and Dinarello (1981). It has been reported that the stromal cells are composed of mononuclear phagocytic cells, epithelial cells, giant fat cells, endothelial cells, reticular cells, fibroblasts and adipocytes (Dexter et al., 1977). The same types of cells were observed in this study. As indicated in Table 9, the thymocyte comitogenic

Table 9. Effects of histamine agonists on the production of thymocyte comitogenic activity in the supernatant of the culture medium of murine bone marrow stromal cells.

compounds	concentration (μM)	thymocyte comitogenic activity (units/ml)
control		0.390 ± 0.046
histamine	0.01	$0.767 \pm 0.055^*$
	0.1	$0.806 \pm 0.030^{**}$
	1	$0.884 \pm 0.068^{**}$
	10	$1.302 \pm 0.195^{**}$
4-methylhistamine	0.01	0.429 ± 0.039
	0.1	$0.715 \pm 0.065^*$
	1	$0.926 \pm 0.051^{**}$
	10	$1.157 \pm 0.046^{**}$
dimaprit	0.1	0.507 ± 0.065
	1	0.572 ± 0.104
	10	$1.482 \pm 0.098^{**}$
2-methylhistamine	1	0.325 ± 0.065
	10	0.312 ± 0.081

Murine bone marrow stromal cells were cultured in the presence of test compounds at 37°C for 24 hr. After that, thymocyte comitogenic activity of the supernatant was determined. Half maximal activity of murine IL-1α was taken as 1 unit. Each value represents the mean $\pm$ S.E.M. of the data obtained from 5 separate experiments. * and ** indicate statistical significance in comparison with control at $p < 0.05$ and $p < 0.01$, respectively. Reproduced from Tasaka, K., Mio, M., Shimazawa, M. and Nakaya, N.: Mol. Pharmacol., 43, 365-371 (1993), with permission.

Table 10. Effects of histamine antagonists on the histamine-induced production of thymocyte comitogenic activity in the culture medium of murine bone marrow stromal cells.

drugs	concentration (μM)	thymocyte comitogenic activity (units/ml)	
		drug alone	+ histamine (1 μM)
control		0.438 ± 0.021	0.863 ± 0.087**
cimetidine	0.1	0.427 ± 0.032	0.707 ± 0.076
	1	0.435 ± 0.015	0.551 ± 0.051†
	10	0.434 ± 0.022	0.470 ± 0.061††
ranitidine	0.1	0.441 ± 0.034	0.769 ± 0.073
	1	0.429 ± 0.019	0.530 ± 0.126†
	10	0.431 ± 0.042	0.432 ± 0.088††
famotidine	0.01	0.440 ± 0.038	0.697 ± 0.055
	0.1	0.436 ± 0.027	0.383 ± 0.073††
pyrilamine	10	0.433 ± 0.045	0.860 ± 0.082

Murine bone marrow stromal cells were cultured in the presence of test compounds at 37 °C for 24 hr. After that, thymocyte comitogenic activity of the supernatant was determined. Half maximal activity of murine IL-1α was taken as 1 unit. Each value represents the mean $\pm$ S.E.M. of the data obtained from 5 separate experiments. ** indicates statistical significance in comparison with blank at $p < 0.01$. † and †† indicate statistical significance in comparison with histamine (1 μM)-treated group at $p < 0.05$ and $p < 0.01$, respectively. Reproduced from Tasaka, K., Mio, M., Shimazawa, M. and Nakaya, N.: Mol. Pharmacol., 43, 365-371 (1993), with permission.

activity in the culture medium of murine stromal cells significantly increased after treatment with histamine at concentrations equal to or higher than 0.01 μM in a concentration-dependent manner. Although H$_2$ agonists, 4-methylhistamine and dimaprit, also increased the thymocyte comitogenic activity in the culture medium of stromal cells, H$_1$ agonist, 2-methylhistamine, did not induce such an activity. Table 10 represents the effect of histamine antagonists on the histamine-induced production of thymocyte comitogenic activity in murine stromal cells. In this case, histamine antagonists were added simultaneously with histamine to the culture medium of stromal cells. As shown in this table, histamine (1 μM)-induced production of thymocyte comitogenic activity was significantly inhibited by H$_2$ antagonists, such as famotidine, cimetidine and ranitidine, at concentrations equal to or higher than 0.1 μM, 1 μM and 1 μM, respectively. However, pyrilamine, an H$_1$ antagonist, did not affect the histamine-induced production of thymocyte comitogenic activity even at a concentration of 10 μM.

In order to confirm whether or not the substance produced by histamine in the culture medium of stromal cells actually increases the colony formation in bone marrow cells, the colony promoting activity in the culture medium of stromal cells treated with histamine was determined. As indicated in Table 11, histamine at 1 μM significantly increased colony promoting activity in the culture medium of stromal cells. Histamine-induced production of colony promoting activity was completely inhibited by simultaneous addition of the H$_2$ antagonists, cimetidine, ranitidine and famotidine (10 μM each) (Table 11).

Table 11. Histamine-induced production of colony promoting activity in murine bone marrow stromal cells and its inhibition by H_2 antagonists.

drugs	concentration (μM)	colony promoting activity(% of control)	
		drug alone	+ histamine(1 μM)
control		100.0 ± 14.0	193.6 ± 24.4**
cimetidine	10	89.8 ± 12.3	84.6 ± 5.9††
ranitidine	10	97.9 ± 8.6	101.3 ± 14.1†
famotidine	10	90.7 ± 10.6	87.2 ± 5.2††

Murine bone marrow stromal cells were incubated with 1 μM of histamine both in the presence and in the absence of H_2 antagonists. The supernatant was added to the CFU-c assay medium and CFU-c assay was carried out. Each value represents the mean ± S.E.M. of the data obtained from 5 separate experiments. ** indicates statistical significance in comparison with blank at $p < 0.01$. † and †† indicate statistical significance in comparison with histamine (1 μM)-treated group at $p < 0.05$ and $p < 0.01$, respectively. Reproduced from Tasaka, K., Mio, M., Shimazawa, M. and Nakaya, N.: Mol. Pharmacol., 43, 365-371 (1993), with permission.

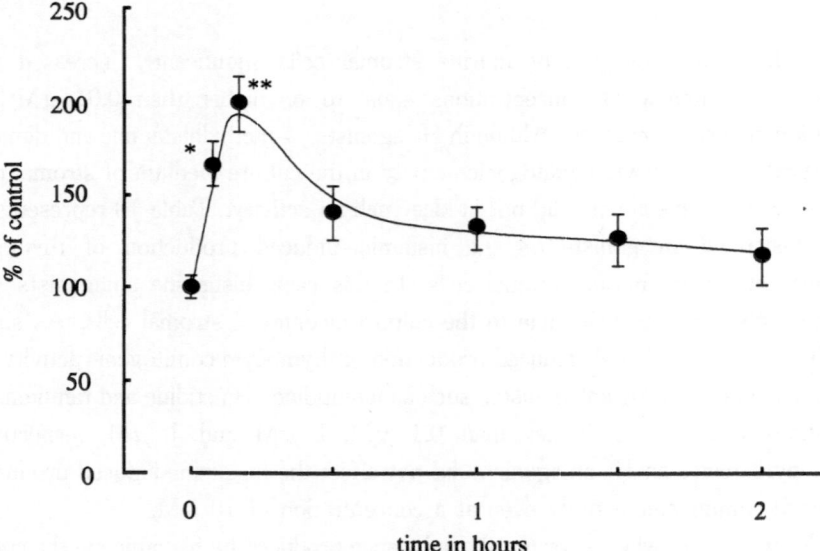

Fig. 16. Effect of histamine on the cAMP level of murine bone marrow stromal cells. 1 μM histamine was added to murine bone marrow stromal cells and the changes in cAMP content were determined by means of radioimmunoassay. Each point represents the mean ± S.E.M. of the data obtained from 4 separate experiments. The level of cAMP in control cells was 6.324 ± 0.435 pmol/10^6 cells (n = 4). * and ** indicate statistical significance in comparison with control at $p < 0.05$ and $p < 0.01$, respectively. Reproduced from Tasaka, K., Mio, M., Shimazawa, M. and Nakaya, N.: Mol. Pharmacol., 43, 365-371 (1993), with permission.

As indicated in Fig. 16, when bone marrow stromal cells were stimulated by histamine (1 μM), cAMP contents increased rapidly, reaching a maximum (almost twice the control) 10 min after stimulation, and, thereafter, decreased gradually. Moreover, when stromal cells were stimulated by db-cAMP at 37°C for 24 hr, thymocyte comitogenic activity in the culture medium increased significantly and dose-dependently at concentrations equal to or higher than 50 μM. In accordance with these observations, the histamine-induced production of thymocyte comitogenic activity was significantly inhibited in the presence of KT-5720, an inhibitor of protein kinase A (Table 12).

From the findings indicated above, it became apparent that when bone marrow stromal cells were stimulated by histamine, both thymocyte comitogenic activity and colony promoting activity in the culture medium significantly increased. Such effects of histamine were exerted by a stimulation of the H_2 receptor and the resulting increase in the cAMP level in stromal cells.

In order to clarify the molecular characteristics of the substance(s) produced by stromal cells by histamine stimulation, the culture medium, after incubating the bone marrow stromal cells with histamine both in the presence and in the absence of H_2 antagonists, was fractionated by means of gel filtration and colony promoting activity in each fraction was determined. Fig. 17 represents the gel filtration profiles of colony promoting activity produced by murine stromal cells after histamine treatment. As indicated in this figure, the colony promoting activity determined in the fractions corresponding to about 15-20 kDa significantly increased following the treatment with 1 μM histamine. In addition, the histamine-induced increase of colony promoting activity in these fractions was completely suppressed by H_2 antagonists, ranitidine and famotidine (10 μM each). This result may indicate that the molecular weight of the active substance

Table 12. Effects of KT-5720 on the histamine-induced production of thymocyte comitogenic activity in the culture medium of murine bone marrow stromal cells.

concentration of KT-5720 (μM)	thymocyte comitogenic activity (units/ml)	
	drug alone	+ histamine (1 μM)
0	0.419 ± 0.022	0.859 ± 0.066**
0.03	0.425 ± 0.034	0.824 ± 0.059
0.1	0.413 ± 0.031	0.595 ± 0.123
0.3	0.372 ± 0.075	0.356 ± 0.066††
1.0	0.328 ± 0.084	0.251 ± 0.024††

Murine bone marrow stromal cells were cultured in the presence of test compounds at 37°C for 24 hr. After that, thymocyte comitogenic activity of the supernatant was determined. Half maximal activity of murine IL-1α was taken as 1 unit. Each value represents the mean ± S.E.M. of the data obtained from 5 separate experiments. ** indicates statistical significance in comparison with blank at $p < 0.01$. †† indicates statistical significance in comparison with histamine (1 μM)-treated group at $p < 0.01$. Reproduced from Tasaka, K., Mio, M., Shimazawa, M. and Nakaya, N.: Mol. Pharmacol., 43, 365-371 (1993), with permission.

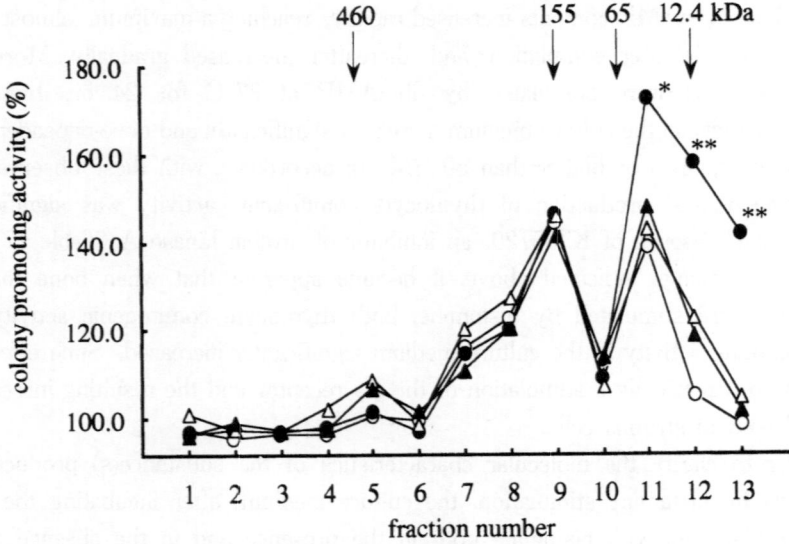

Fig. 17. Gel filtration profiles of colony promoting activity produced by murine bone marrow stromal cells. Murine bone marrow stromal cells were incubated both in the presence and in the absence of histamine and/ or H_2 antagonists at 37 °C for 24 hr. The supernatant was applied on Sephacryl S-200 and eluted by saline. The colony promoting activity of each fraction was determined using murine bone marrow cells. Spontaneous increase in the colony counts in murine bone marrow cells was taken as 100 %. ○: control, ●: histamine (1 μM), △: histamine (1 μM) + ranitidine (10 μM), ▲: histamine (1 μM) + famotidine (10 μM). Each point represents the mean value obtained from 4 separate experiments. Error bars were omitted for clarity. * and ** indicate statistical significance in comparison with control at $p < 0.05$ and $p < 0.01$, respectively. Reproduced from Tasaka, K., Mio, M., Shimazawa, M. and Nakaya, N.: Mol. Pharmacol., 43, 365-371 (1993), with permission.

produced by histamine-stimulation is 15-20 kDa.

From the observations indicated above, the active substance, produced by bone marrow stromal cells after stimulation by histamine, possesses both thymocyte comitogenic activity and colony promoting activity and its molecular weight is 15-20 kDa. Since such properties are very similar to those of IL-1, Western blot analysis was carried out using anti-IL-1α antibody. When murine bone marrow stromal cells were stimulated by histamine at concentrations equal to or higher than 10^{-8}M, production of 32 kDa protein in a concentration-dependent manner was detected by anti-IL-1α, as shown in Fig. 18. The molecular weight of this protein seems to correspond to that of pro-IL-1α (Dinarello, 1988). This may suggest that the production of pro-IL-1α was increased by a histamine stimulation. IL-1α released in the culture medium was not detected since the concentration was much lower than that detectable by means of IL-1α antibody. These results clearly indicates that murine bone marrow stromal cells produced mainly IL-1α while the production of IL-1β was negligible.

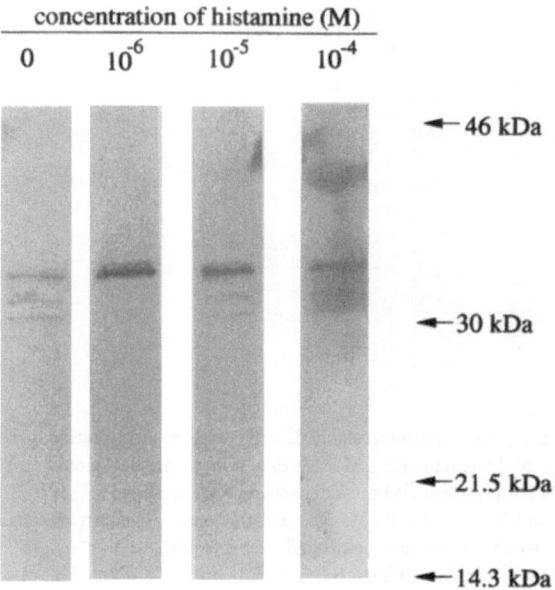

concentration of histamine (M)

0   10^{-6} 10^{-5} 10^{-4}

◄— 46 kDa

◄—30 kDa

◄—21.5 kDa

◄—14.3 kDa

Fig. 18. Western blot analysis of intracellular IL1-α production in murine bone marrow cells cultured in the presence of histamine. Murine bone marrow stromal cells were cultured with various concentrations of histamine at 37 °C for 24 hr. Thereafter, the cells were harvested and intracellular IL-1α was determined by Western blot analysis using anti-IL-1α antibody. The molecular weight of the detected band corresponded to that of pro-IL-1α (32 kDa). Reproduced from Tasaka, K., Mio, M., Shimazawa, M. and Nakaya, N.: Mol. Pharmacol., 43, 365-371 (1993), with permission.

The effect of H_2 antagonists on the histamine-induced production of pro-IL-1α in murine bone marrow stromal cells was investigated. As shown in Fig. 19, H_2 antagonists, such as cimetidine, ranitidine and famotidine, inhibited the production of pro-IL-1α almost to the control level, indicating that the effect of histamine on the production of pro-IL-1α may be exerted via H_2 receptor stimulation. However, these drugs did not affect the spontaneous production of pro-IL-1α at the concentrations employed.

In Fig. 20, the effect of several protein kinase inhibitors on the histamine-induced production of pro-IL-1α is indicated. As shown in this figure, inhibitors of protein kinase A, such as KT-5720 and HA1004, inhibited the histamine-induced production of pro-IL-1α to the control level, suggesting that protein kinase A may also play some important role in the histamine-induced production of pro-IL-1α. These compounds did not affect the spontaneous production of pro-IL-1α at the concentrations employed.

It is known that macrophages, endothelial cells, epithelial cells and fibroblasts are capable of producing IL-1 (Dinarello, 1988). So far, no report has indicated whether or not histamine interacts with these cells to produce IL-1 via H_2 receptor stimulation.

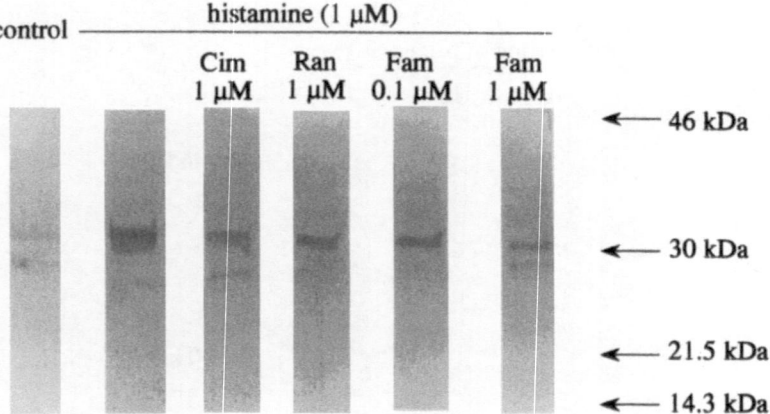

Fig. 19. Effects of H$_2$ antagonists on intracellular IL-1α production in murine bone marrow stromal cells cultured in the presence of histamine (1 μM). Murine bone marrow stromal cells were cultured with histamine (1 μM) both in the presence and in the absence of H$_2$ antagonists at 37 °C for 24 hr. Thereafter, the cells were harvested and intracellular IL-1α was determined by Western blot analysis using anti-IL-1α antibody. The molecular weight of the detected band corresponded to that of pro-IL-1α (32 kDa). Cim: cimetidine, Ran: ranitidine, Fam: famotidine. Reproduced from Tasaka, K., Mio, M., Shimazawa, M. and Nakaya, N.: Mol. Pharmacol., 43, 365-371 (1993), with permission.

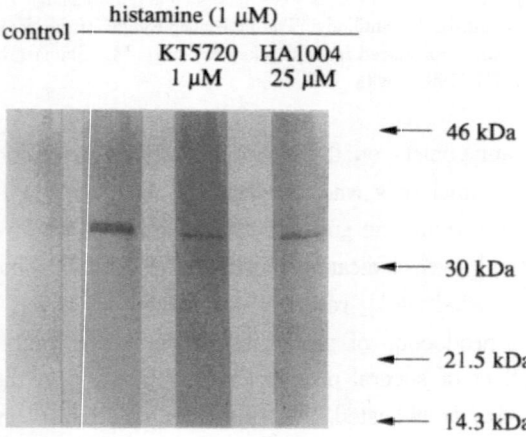

Fig. 20. Effects of protein kinase inhibitors on intracellular IL-1α production in murine bone marrow stromal cells elicited by histamine (1 μM). Murine bone marrow stromal cells were cultured with histamine (1 μM) both in the presence and in the absence of protein kinase inhibitors at 37 °C for 24 hr. Thereafter, the cells were harvested and intracellular IL-1α was determined by Western blot analysis using anti-IL-1α antibody. The molecular weight of the detected band corresponded to that of pro-IL-1α (32 kDa). Reproduced from Tasaka, K., Mio, M., Shimazawa, M. and Nakaya, N.: Mol. Pharmacol., 43, 365-371 (1993), with permission.

However, it has been reported that histamine facilitates the release of free fatty acids from fat cells, and this effect seems to be mediated by H_2-receptors (Akiyama et al, 1990). It is known that adipocytes are important in maintaining the stem cells in the bone marrow (Lanotte et al., 1982). Therefore, giant fat cells and adipocytes in the bone marrow may be useful for IL-1α release in response to histamine.

It has been reported that pro-IL-1 is translated as a 31 kD protein from mRNA and the precursor protein is processed by serine proteases, such as elastase and plasmin, into 17.5, 11 and 4 kD proteins (Dinarello, 1988). It has also been reported that a phosphorylation of pro-IL-1 molecule by protein kinase A markedly increases the susceptibility to protease (Kobayasi et al., 1988). From these observations, it was assumed that an increase in cAMP levels via H_2 receptor stimulation may be intimately related to the digestion of pro-IL-1α and the resulting release of IL-1α. In addition, the present findings indicate that an activation of cAMP-protein kinase A system due to H_2 receptor stimulation is also related to the production of pro-IL-1α in bone marrow stromal cells.

From these observations, it was concluded that histamine induces the production of IL-1α in murine bone marrow stromal cells through an activation of the H_2 receptor in association with the activation of the cAMP-protein kinase A system. As reported previously, IL-1 is effective in producing various CSFs from bone marrow stromal cells, fibroblasts and endothelial cells (Fibbe et al., 1988; Zsebo et al., 1988; Zucali et al., 1986). It has also been reported that IL-1 is capable of activating pluripotent stem cells so as to induce proliferation and differentiation (Mochizuki et al., 1987). Therefore, it is possible to assume that the G-CFU-c stimulation exerted by histamine may be derived from the production of IL-1α and, as a consequence of this, that G-CSF production from bone marrow stromal cells may be induced by the released IL-1α. Furthermore, it has been reported that IL-1 is capable of inducing IL-1 production as a positive feedback mechanism (Dinarello et al., 1987). It is well known that both histamine and IL-1 are released in the case of infections, injuries and allergic reactions. Moreover, the production of IL-1 markedly decreased in patients with aplastic anemia and it has been suggested that IL-1 plays physiologically important roles as a hematopoietic substance (Nakao et al., 1989). In the bone marrow, stromal cells possess important roles in proliferation and differentiation of stem cells. Therefore, it is possible to assume that IL-1α released from stromal cells in the bone marrow due to histamine may stimulate hematopoietic progenitor cells not only by the production of CSFs but also by direct action on pluripotent stem cells. In addition, histamine also directly acts on the myeloblasts and promyelocytes so as to induce the differentiation through the H_2 receptors (Nakaya and Tasaka, 1988a).

7. Histamine-induced externalization of G-CSF receptor

The production of neutrophils in normal bone marrow is influenced by many factors such as a variety of cytokines and the microenvironments surrounding the progenitor cells. So far, many biological substances which are capable of stimulating the proliferation and differentiation of neutrophil progenitor cells have been reported (Rapoport et al., 1992). G-CSF possesses the ability to form the colony which predominantly consists of granulocyte progenitor cells in semisolid agar (Nomura et al., 1986). It is known that G-CSF induces the differentiation and proliferation of neutrophil progenitors and consequently, it increases the number of differentiated cells. On the other hand, as indicated above, when murine bone marrow cells were cultured with histamine at concentrations higher than 10^{-8}M, histamine markedly stimulated the differentiation and proliferation of neutrophil precursor cells through H_2 receptors (Tasaka and Nakaya, 1987; Nakaya and Tasaka, 1988a). In many biological responses mediated by H_2 receptors, a histamine-sensitive adenylate cyclase was stimulated, and such a response is mimicked by cAMP analogue, indicating that cAMP is the second messenger (Schwartz et al., 1991). Since G-CSF and histamine have almost the same effect on neutrophil differentiation, it was assumed that there exists some common mechanism(s) between histamine action and G-CSF action on bone marrow cells. In connection with this, it has been reported that exogenous cAMP stimulates bone marrow cells to form colonies *in vitro* (Fleming and McNeill, 1976; Nonaka et al., 1992). The following experiments were carried out to elucidate the mechanism of histamine reinforcement in the G-CSF-induced differentiation of neutrophil precursors (Tasaka et al., 1994).

Fig. 21a shows the changes in intracellular cAMP contents in myeloblasts and

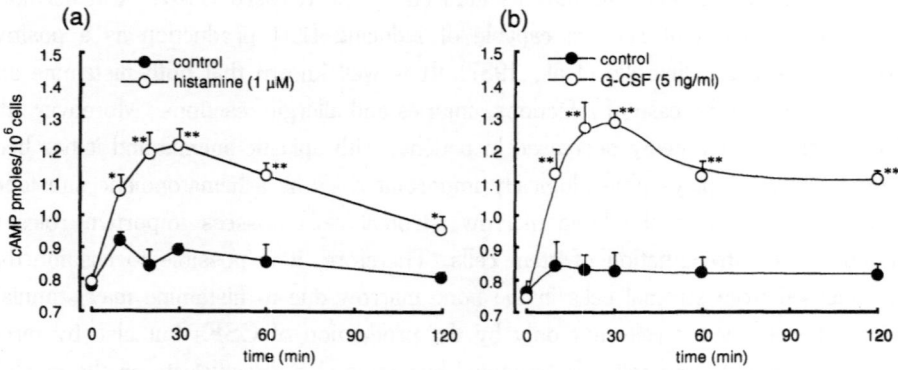

Fig. 21. Effects of rG-CSF and histamine on intracellular cAMP contents in murine myeloblasts and promyelocytes. Cells were treated with (a) histamine (1 μM) or (b) rG-CSF (5 ng/ml) at 37°C. Reaction was terminated by the addition of 5 % TCA and thereafter, cAMP contents were determined by means of radioimmunoassay kit. Each value is the mean ± S.E.M. from 5 separate experiments. (* P < 0.05, ** P < 0.01). Reproduced from Tasaka, K., Doi, M., Nakaya, N. and Mio, M.: Mol. Pharmacol., 45, 837-845 (1994), with permission.

promyelocytes treated in the presence and in the absence of histamine. In the resting state, cyclic AMP level in these cells was 780 ± 12 fmole/10^6 cells. However, a significant increase of intracellular cAMP was noticed 30 min after exposure to histamine (1 μM). This increase reached the maximum level of 1.22 pmoles/10^6 cells 1 hr later. However, no changes in cAMP level were found in the control cells. When rG-CSF (5 ng/ml) was employed instead of histamine, a marked elevation of intracellular cAMP level was also noticed. The increase of cAMP reached the maximum (1.28 pmoles/10^6 cells) within 30 min and thereafter decreased rather rapidly (Fig. 21b).

Promyelocytes and myeloblasts obtained from murine bone marrow were incubated with 1 μM of histamine for 2 to 24 hr, and then the cells were washed and cultured in a fresh culture medium for 24 hr. As shown in Fig. 22a, histamine induced no stimulative effect on the cell differentiation within 2 hr of treatment. However, significant

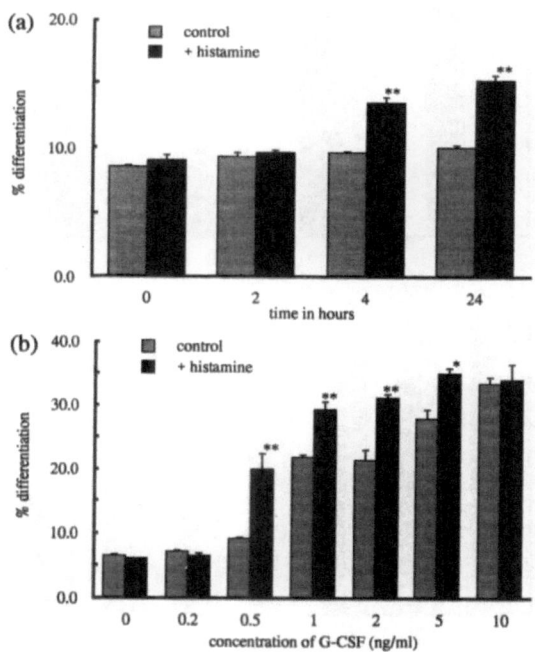

Fig. 22. Changes in the population of matured neutrophils. Murine myeloblasts and promyelocytes were incubated with (a) histamine or (b) rG-CSF in combination with or without histamine (1 μM) pretreatment. The percent of neutrophil was used as an indicator of cell differentiation. In (a), the control value was determined in a medium without histamine, and histamine incubation was continued for various periods of time. In (b), when the cells were cultured in a medium containing rG-CSF alone, the percentage of neutrophil was indicated as the control. * and ** represent the significant difference between the control and histamine-pretreated groups at $P < 0.05$ and $P < 0.01$, respectively. Each value represents the mean $\pm$ S.E. M. in 5 separate experiments. Reproduced from Tasaka, K., Doi, M., Nakaya, N. and Mio, M.: Mol. Pharmacol., 45, 837-845 (1994), with permission.

differentiation was noticed when the cells were treated with histamine longer than 4 hr. In another experiment, histamine (1 μM) was similarly pretreated for 2 hr and thereafter, progenitor cells were washed and incubated for 24 hr in the same medium containing various concentrations of rG-CSF. As shown in Fig. 22b, the differentiating effect of rG-CSF was noticed at concentrations higher than 1 ng/ml and the differentiation increased dose-dependently. On the other hand, in the histamine-pretreated group, rG-CSF caused significant differentiation at a concentration of 0.5 ng/ml. As the concentration of rG-CSF increased, the extent of differentiation in the histamine pretreated group became more evident than those treated with rG-CSF alone. The maximum

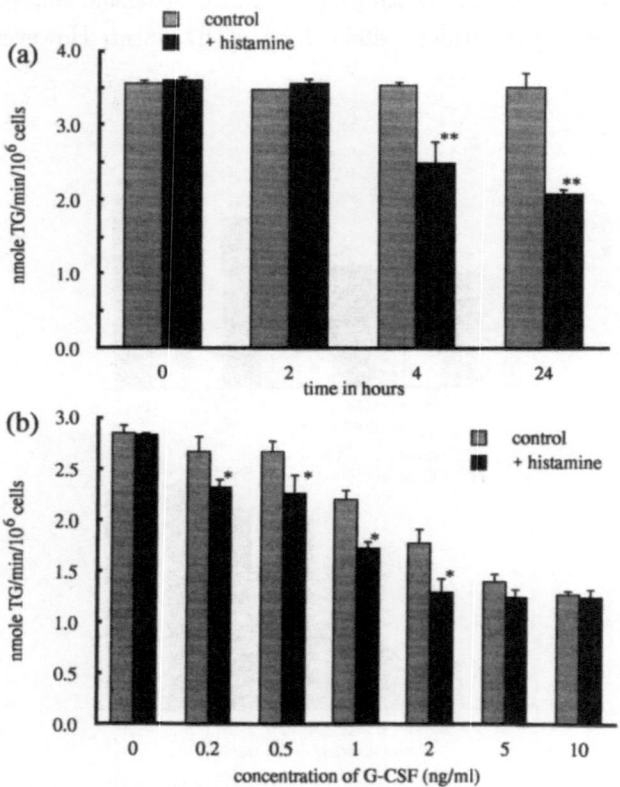

Fig. 23. Changes in myeloperoxidase activity in murine myeloblasts and promyelocytes. (a) Sequential changes in the enzyme activity of the medium in both the presence and the absence of histamine (1 μM). The difference between two groups became significant when incubation was continued for longer than 4 hr. ** indicates P < 0.01. (b) Comparison of the reduction in enzyme activity between the control group (only rG-CSF treated) and histamine pretreated (+ rG-CSF) group. * represents the significant difference between the two groups at P < 0.05. Each value represents the mean ± SEM in 5 separate experiments. Reproduced from Tasaka, K., Doi, M., Nakaya, N. and Mio, M.: Mol. Pharmacol., 45, 837-845 (1994), with permission.

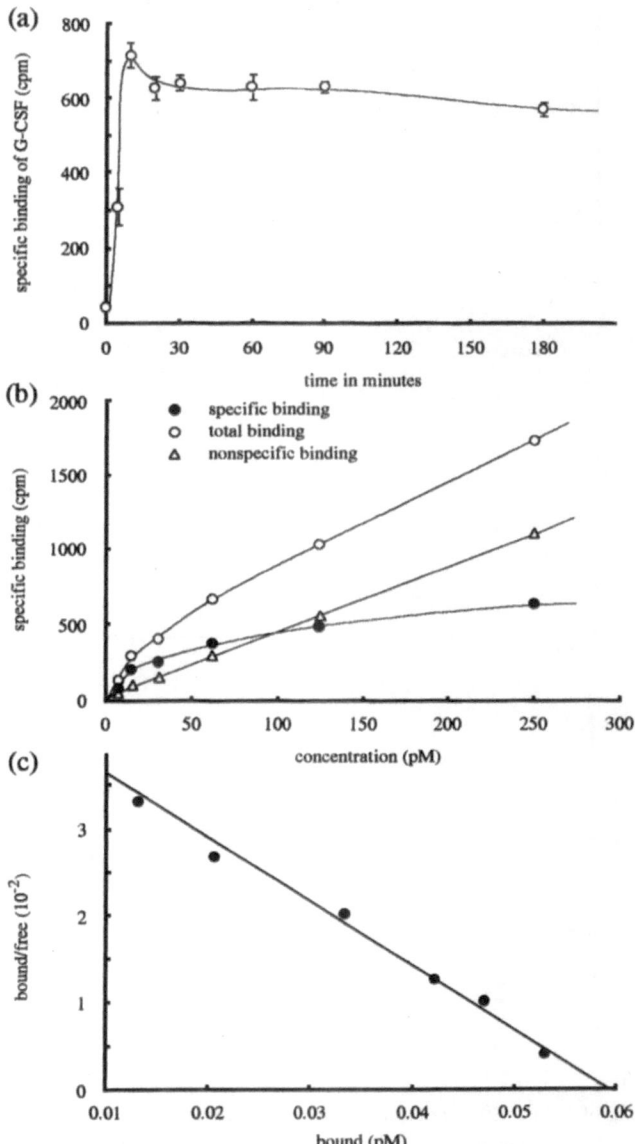

Fig. 24. Specific binding of ^{125}I-rG-CSF to murine neutrophil progenitors. (a) Sequential changes in specific binding of ^{125}I-rG-CSF to myeloblasts and promyelocytes. Each point indicates the mean ± S.E.M. of 5 separate experiments. (b) Saturable binding of ^{125}I-rG-CSF to murine myeloblasts and promyelocytes. Each point represents the mean of 5 separate experiments. (c) Scatchard analysis. Reproduced from Tasaka, K., Doi, M., Nakaya, N. and Mio, M.: Mol. Pharmacol., 45, 837-845 (1994), with permission.

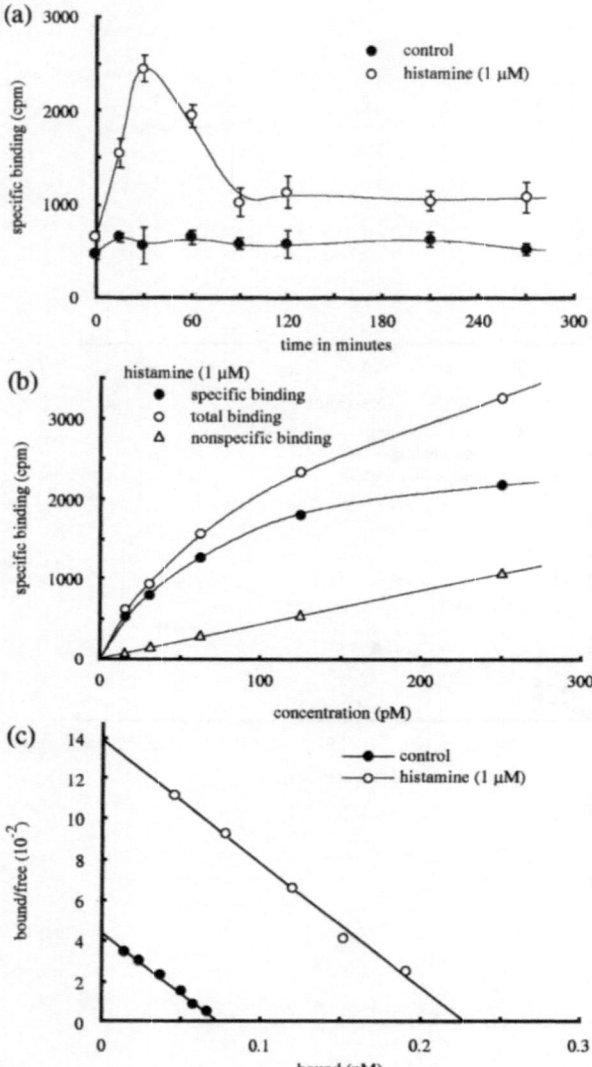

Fig. 25. Specific binding of ^{125}I-rG-CSF to murine neutrophil progenitors treated with histamine (1 μM). (a) Sequential changes in specific binding of ^{125}I-rG-CSF to myeloblasts and promyelocytes treated with histamine for various periods of time. Each point indicates the mean ± S.E.M. of 5 separate experiments. (b) Saturable binding of ^{125}I-rG-CSF to myeloblasts and promyelocytes treated with histamine for 30 min. Each point represents the mean of 5 separate experiments. (c) Scatchard analysis. Two lines run almost parallel to each other. Bmax in histamine treated cells is approximately 3.4 times larger than that of the control cells. Reproduced from Tasaka, K., Doi, M., Nakaya, N. and Mio, M.: Mol. Pharmacol., 45, 837-845 (1994), with permission.

effect was achieved at 5 ng/ml in histamine-pretreated group, while in the group treated with rG-CSF alone, the maximum differentiation was achieved at 10 ng/ml. The results seem to indicate that histamine increases the susceptibility of progenitor cells to rG-CSF.

Fig. 23a shows the changes in the myeloperoxidase activity of murine myeloblasts and promyelocytes treated in both the presence and the absence of histamine, and Fig. 23b shows those incubated with various concentrations of rG-CSF for 24 hr with or without histamine pretreatment for 2 hr. Usually, the myeloperoxidase activity of neutrophil precursors reaches the maximum level at the stage of promyelocytes, and as the differentiation proceeds further, the enzyme activity decreases.

As shown in Fig. 23a, myeloperoxidase activity was not altered by 2 hr of histamine treatment. However, a significant decrease was noticed after 4 hr of incubation and the enzyme activity decreased to nearly half that of the control after 24 hr of incubation. In rG-CSF treated cells, myeloperoxidase activity decreased dose-dependently and rG-CSF treatment at a concentration of 5 ng/ml decreased the enzyme activity less than a half of that of the control. As in the case of morphological changes, the reduction of enzyme activity was more remarkable in the histamine-treated group than in those treated with rG-CSF alone and the enzyme activity decreased to the minimum at a concentration of 2 ng/ml of rG-CSF. The results also suggest that histamine pretreatment augments the action of rG-CSF.

To clarify the reason why the reactivity of progenitor cells to rG-CSF increased after

Table 13. Effects of histamine antagonists on histamine-induced increase in the specific binding of ^{125}I-G-CSF to murine myelyblasts and promyelocytes.

drugs	μM	specific binding (cpm)
control	0	550 ± 13
histamine	0.1	1510 ± 38
	1.0	2347 ± 65
	10.0	2983 ± 245
+ cimetidine	0.01	1928 ± 65
	0.1	1477 ± 123
	1.0	619 ± 71
	10.0	534 ± 48
+ ranitidine	1.0	1074 ± 26
	10.0	551 ± 103
+ pyrilamine	0.01	2659 ± 242
	0.1	2606 ± 39
	1.0	2503 ± 104
	10.0	2370 ± 91

After antihistamines were pretreated for 30 min, histamine (1 μM) was added to the medium and incubation was continued for another 30 min. N = 5. Reproduced from Tasaka, K., Doi, M., Nakaya, N. and Mio, M.: Mol. Pharmacol., 45, 837-845 (1994), with permission.

histamine pretreatment, the binding capacity of rG-CSF to the progenitor cells was determined using ^{125}I-labeled rG-CSF. As shown in Fig. 24a, when the time course of ^{125}I-rG-CSF binding to myeloblasts and promyelocytes was determined, ^{125}I-rG-CSF binding reached the maximum at 10 min and then decreased to a stationary level at 30 min. Thus, in the following experiment, 30 min was taken as a standard time in the binding assay. When saturable binding of ^{125}I-rG-CSF was measured and Scatchard analysis was carried out subsequently, the binding constant (K_D) and the maximum binding capacity (Bmax) were 60.6 ± 2.3 pM and 521.7 ± 30 binding sites/cell, respectively (Figs. 24b and 24c). When the progenitor cells were treated with histamine (1 μM) for various periods of time and rG-CSF binding was determined, the binding reached the maximum at 30 min of histamine treatment and then declined to the stationary level (Fig. 25a). When Scatchard analysis was performed after 30 min of histamine pretreatment, the K_D and Bmax of ^{125}I-rG-CSF binding were 63.2 ± 2.04 pM and 1754.8 ± 119.2 binding sites/cell, respectively (Figs. 25b and 25c). Although the K_D values were almost the same before and after histamine treatment, an approximately 3.4 times increase in Bmax was noticed after histamine treatment.

Table 13 indicates a dose-dependent increase of rG-CSF binding to neutrophil precursor cells after histamine treatment and its inhibition by pretreatment with histamine antagonists. The rG-CSF binding to immature cells treated with histamine significantly increased at concentrations higher than 1×10^{-7}M compared to the non-treated control.

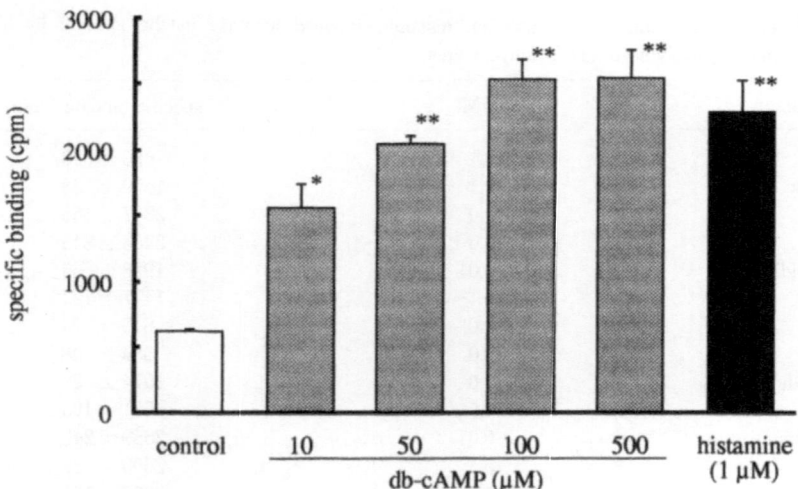

Fig. 26. Effect of db-cAMP on ^{125}I-rG-CSF binding to murine myeloblasts and promyelocytes. Each value represents the mean $\pm$ S.E.M. of 5 separate experiments. * and ** represent the significant difference between the control and db-cAMP-treated groups at $p < 0.05$ and $p < 0.01$, respectively. Reproduced from Tasaka, K., Doi, M., Nakaya, N. and Mio, M.: Mol. Pharmacol., 45, 837-845 (1994), with permission.

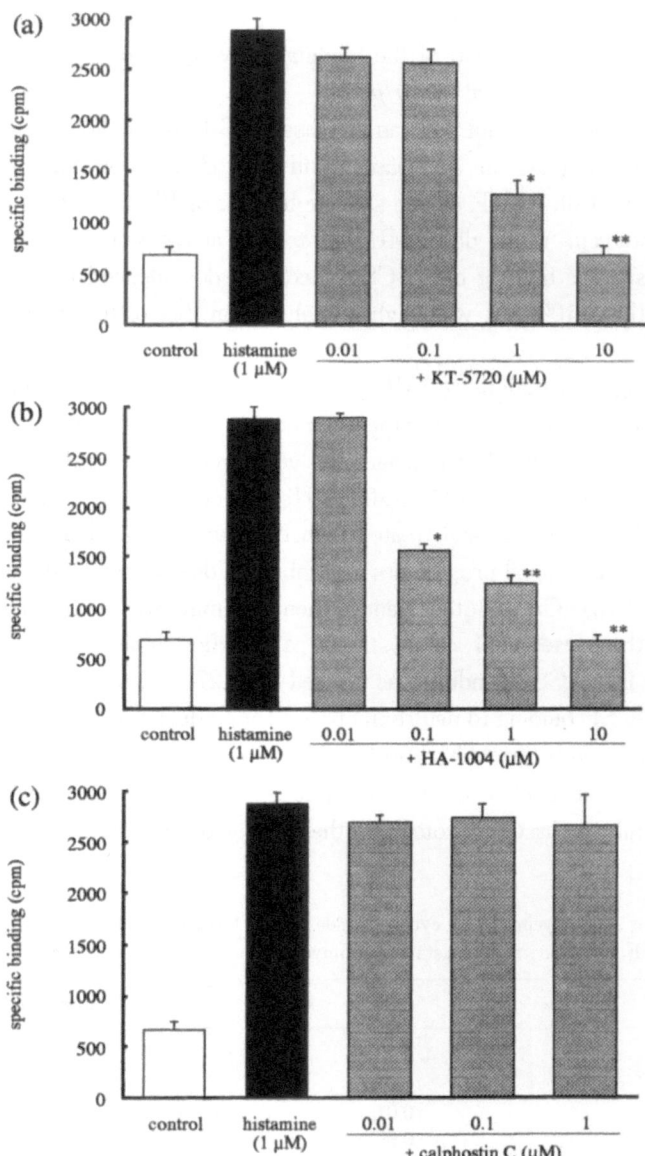

Fig. 27. Effects of protein kinase inhibitors on the histamine-induced increase in ^{125}I-rG-CSF binding to murine myeloblasts and promyelocytes. Effects of KT-5720 (a), HA-1004 (b) and calphostin ℂ (c) are shown in the corresponding figures. These drugs were incubated for 30 min prior to histamine addition (1 μM). Each value represents the mean $\pm$ S.E.M. N = 5. * and ** represent the significant difference between histamine treated cells and protein kinase inhibitor + histamine treated cells at $p < 0.05$ and $p < 0.01$, respectively. Reproduced from Tasaka, K., Doi, M., Nakaya, N. and Mio, M.: Mol. Pharmacol., 45, 837-845 (1994), with permission.

Pretreatment with cimetidine at 1×10^{-6}M was effective in preventing the histamine (1 μM)-induced increase to nearly that of the control level. Also, ranitidine (at 10 μM) induced almost complete inhibition of the binding. However, no inhibition was affected by pyrilamine pretreatment, even at 10 μM.

To investigate whether or not such an increase in rG-CSF binding sites on neutrophil progenitor cells takes place in association with the changes in intracellular cAMP contents, the effect of db-cAMP on the specific binding of ^{125}I-G-CSF to the neutrophil precursors was studied. When db-cAMP was treated at concentrations of 10-100 μM for 30 min, the specific binding of rG-CSF increased dose-dependently (Fig. 26). The effect of db-cAMP at 100 μM was slightly higher than that of histamine detected at 1 μM.

In many biological reactions, cAMP acts in association with protein kinase A to complete its function. To confirm whether or not this is the case, the effect of protein kinase inhibitor was tested. When progenitor cells were pretreated for 30 min with KT-5720 (Kase et al., 1987) or HA-1004 (Hidaka et al., 1984), both were potent inhibitors of protein kinase A, and treated with histamine in the following 30 min, the rG-CSF binding to neutrophil progenitors was inhibited dose-dependently and significantly (Figs. 27a and 27b). On the other hand, when histamine was added similarly in both the absence and the presence of calphostin C, a spacific inhibitor of protein kinase C, no visible change in rG-CSF binding was noticed (Fig. 27c). This seems to indicate that an increase of rG-CSF binding to neutrophil precursors induced by histamine is intimately related in some way with the cAMP-protein kinase A system.

It was supposed that when a histamine-induced increase of rG-CSF binding takes place on neutrophil precursors, protein synthesis may occur concurrently. To confirm

Table 14. Effects of actinomycin D and cycloheximide on the histamine-induced increase in the specific binding of ^{125}I-G-CSF to murine myeloblasts and promyelocytes.

drugs	μM	specific binding (cpm)
control	0	756 ± 21
histamine	1.0	2685 ± 30
+ actinomycin D	0.01	2323 ± 376
	0.001	2778 ± 356
	0.0001	2787 ± 171
+ cycloheximide	1.0	2672 ± 305
	0.1	2488 ± 156
	0.01	2597 ± 72
	0.001	2571 ± 32

The cells were pretreated with actinomycin D or cycloheximide for 30 min, then incubation was continued for another 30 min with histamine (1 μM). N = 5. Reproduced from Tasaka, K., Doi, M., Nakaya, N. and Mio, M.: Mol. Pharmacol., 45, 837-845 (1944), with permission.

whether or not this is the case, histamine treatment $(10^{-6}M)$ was carried out in both the presence and the absence of either cycloheximide or actinomycin D for 2 hr and rG-CSF binding was carried out similarly. As shown in Table 14, the increase of rG-CSF binding to neutrophil precursors caused by histamine was not inhibited in the presence of either protein synthesis inhibitors. Furthermore, ^{3}H-leucine uptake into myeloblasts and promyelocytes was studied in the presence and the absence of histamine (1 μM). As shown in Table 15, histamine had no effect on ^{3}H-leucine uptake. The values determined during 120 min were almost the same between the histamine-treated group and the corresponding control. The results clearly indicate that rG-CSF binding to progenitor cells takes place without concurrent ^{3}H-leucine uptake. This seems to coincide with the finding that neither actinomycin D nor cycloheximide inhibited histamine-induced increase in rG-CSF binding.

When neutrophil progenitors were exposed to α-chymotrypsin and incubated with

Table 15. Effect of histamine on [^{3}H]-leucine uptake into murine myeloblasts and promyelocytes.

time in minutes	[^{3}H]-leucine uptake (cpm)	
	control	histamine
0	3444 ± 105	3606 ± 62
10	4295 ± 77	4135 ± 77
20	6450 ± 379	6281 ± 280
30	7609 ± 591	7600 ± 144
60	6947 ± 124	8408 ± 182
90	8556 ± 133	9150 ± 169
120	10747 ± 425	9961 ± 456

No significant difference was noticed between the control and histamine-treated group. N = 5. Reproduced from Tasaka, K., Doi, M., Nakaya, N. and Mio, M.: Mol. Pharmacol., 45, 837-845 (1994), with permission.

Table 16. Effect of db-cAMP on rG-CSF binding to neutrophil progenitors.

time in minutes	specific binding of ^{125}I-G-CSF (cpm)	
	cycloheximide (1 μM) alone	cycloheximide+db-cAMP (100 μM)
0	134 ± 20	132 ± 12
30	135 ± 17	134 ± 24
60	148 ± 15	163 ± 36
120	165 ± 28	$296 \pm 18**$
240	239 ± 23	$358 \pm 10**$
360	277 ± 19	$429 \pm 15**$

Each value represents the mean $\pm$ S.E.M. of 5 separate experiments. ** represents significant difference in $p < 0.01$. Reproduced from Tasaka, K., Doi, M., Nakaya, N. and Mio, M.: Mol. Pharmacol., 45, 837-845 (1994), with permission.

db-cAMP in the presence of cycloheximide (1 μM), ^{125}I-rG-CSF binding increased remarkably in comparison to the control cells (treated with cycloheximide alone) (Table 16). However, when protease-treated cells were exposed to histamine (1 μM) in the presence of cycloheximide, no such increase in rG-CSF binding was observed (data not shown), suggesting that H_2 receptors were digested or remained as in inactive state.

Since no relation was observed between the increase in rG-CSF receptors and protein synthesis, the translocation of the receptor from the cell interior to the cell surface (externalization) (Deutsch et al., 1982) was considered to be a possible event leading to the receptor increase. To confirm this possibility, the agents acting on the functions of cytoskeletal elements were tested. As shown in Fig. 28, pretreatment with cytochalasin D or colchicine inhibited histamine-induced increase in rG-CSF binding dose-dependently.

It is known that the c-*myc* proto-oncogene is expressed in various immature cells. As shown in Fig. 29a, neutrophil precursor cells expressed the c-*myc* gene. In contrast, matured cells such as neutrophils did not express the oncogene at all. When neutrophil precursor cells were treated for 0-4 days with rG-CSF at a concentration of 5 ng/ml, c-*myc* gene expression in these cells became apparent after 24 hr treatment. However, gene expression gradually decreased thereafter. Also, when progenitor cells were treated with histamine (1 μM), gene expression markedly diminished within a few days (Fig. 29b).

From the experiments indicated above, it became apparent that histamine augments

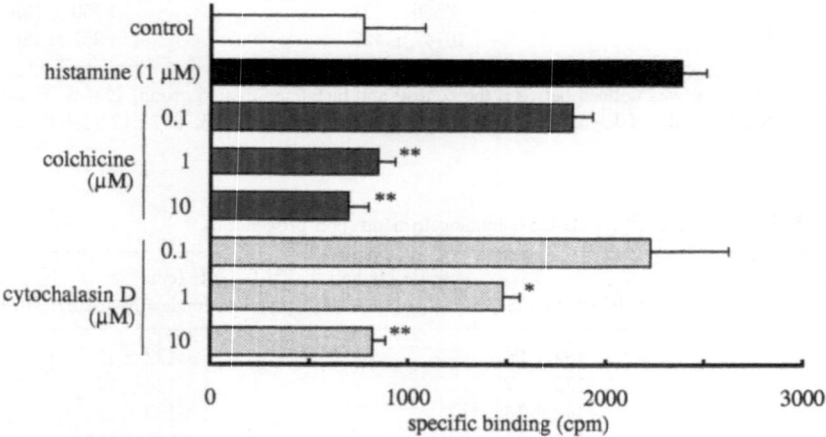

Fig. 28. Effects of colchicine and cytochalasin D on the histamine-induced increase in ^{125}I-rG-CSF binding to murine myeloblast and promyelocytes. * and ** represent the significant difference between histamine-treated cells and test compound + histamine-treated cells at p < 0.05 and p < 0.01, respectively. Reproduced from Tasaka, K., Doi, M., Nakaya, N. and Mio, M.: Mol. Pharmacol., 45, 837-845 (1994), with permission.

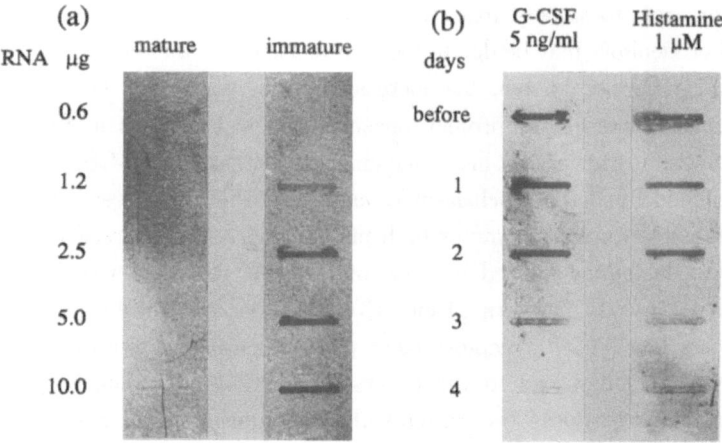

Fig. 29. Effects of rG-CSF and histamine on c-*myc* gene expression in murine myeloblasts and promyelocytes. Immature cells were treated with rG-CSF (5 ng/ml) or histamine (1 μM) for 0-4 days. Total RNA was isolated and c-*myc* gene expression was determined by slot blot analysis using a biotin-labeled v-*myc* probe. Reproduced from Tasaka, K., Doi, M., Nakaya, N. and Mio, M.: Mol. Pharmacol., 45, 837-845 (1994), with permission.

the differentiation of neutrophil precursors induced by rG-CSF mainly through an increase in the number of G-CSF receptors located on the surface of the immature cells. It is known that when G-CSF binds to a receptor, the ligand-receptor complex is transferred into the cell interior by the internalization process (Lotem and Sachs, 1987; Nicola et al., 1988). As a consequence of this, a down-regulation of the receptors takes place (Walker et al., 1985). Furthermore, many CSFs are required not only for cell proliferations but also for cell survival (Dunn, 1987; Platzer et al., 1988; Lin et al., 1989), so that the biosynthesis of receptors may actively take place in these cells. However, contradictory results were obtained: that is, histamine increased the number of G-CSF receptors on neutrophil progenitors without concomitant protein synthesis (Table 14 and 15). The increase in rG-CSF receptors may result from the inhibition of internalization of the receptors or the externalization of receptors (Deutsch et al., 1982) from the inside to outside of the cell surface. In association with this, Deutsch et al (1982) reported that after proteolytic inactivation of cell surface insulin receptors of 3T3-L-1 adipocytes, the cells recovered 20 % of their external insulin binding activity within 2 hr even when protein synthesis was inhibited by cycloheximide. They explained that the pre-existing pool supplies the active insulin receptors to the cell surface (Deutsch et al., 1982). When neutrophil progenitors were treated with α-chymotrypsin in the present experiment, ^{125}I-rG-CSF binding in the presence of cycloheximide was markedly decreased to approximately 21 % of the non-treated control. However, addition of db-cAMP significantly increased the binding even in the presence of cycloheximide. These findings seem to suggest that the histamine-induced increase in G-CSF receptors of neutrophil progenitors may be due to the externalization of the receptors. Although the exact mechanism is not known, the increase in the number of G-CSF receptor is definitely related to the cAMP-protein kinase-A system as shown in Figs 26 and 27. Also, as shown in Fig. 28, a histamine-induced increase in G-CSF receptors was markedly inhibited by both cytochalasin D and colchicine. This seems to suggest that either microtubules, microfilaments or both play the active part of receptor mobilization. Furthermore, a histamine-induced increase of rG-CSF receptors was inhibited by H_2 blockers but not by H_1 blockers (Table 13). This clearly indicates that a histamine-induced increase of rG-CSF receptors takes place in association with the stimulation of H_2 receptors which may lead to the process increasing intracellular cAMP content. When neutrophil progenitors were treated with α-chymotrypsin, histamine treatment did not increase the rG-CSF binding. In this case, H_2 receptors may disappear from the cell surface of neutrophil progenitors, and consequently, no increase in cAMP level was affected.

In the experiments indicated above, human rG-CSF was used in place of murine rG-CSF. It is known that homology between human G-CSF and murine G-CSF is higher than 70 % and the biological activities of both substances are very similar (Nomura et al., 1986). Actually, as shown in Fig. 22b, rG-CSF caused the differentiation of murine myeloblasts and promyelocytes at concentrations higher than 1 ng/ml. However, when the cells were treated with histamine, the cells became more sensitive to rG-CSF and cell differentiation occurred even at 0.5 ng/ml.

A rapid and significant elevation of cAMP was found in murine myeloblasts and promyelocytes treated with either histamine or rG-CSF as shown in Fig. 21. Furthermore, db-cAMP increased rG-CSF binding to neutrophil precursors in a dose-dependent fashion (Fig. 26). Together with these findings, it was assumed that cAMP plays an essential role in the histamine-induced increment of rG-CSF binding. Actually, it was reported that db-cAMP is effective not only in increasing the intracellular cAMP but also in stimulating the differentiation of murine myeloblasts, murine promyelocytes and human promyelocytic leukemia cells (HL-60) (Koeffler, 1983). It has been shown that KT-5720 is a specific and potent inhibitor of protein kinase A (Kase et al., 1987). When KT-5720 was preincubated simultaneously with histamine, the histamine-induced increase of rG-CSF binding was completely inhibited. However, calphostin C, a specific inhibitor of protein kinase C (Kase et al., 1987), showed no inhibitory effect on the histamine-induced increase in G-CSF receptors. The finding clearly indicates that the increase in rG-CSF receptors may be intimately related to the cAMP-protein kinase A system.

C-*myc* is a proto-oncogene which is expressed in some tumor cells (Spotts and Hann, 1990). The product synthesized from this gene is an initiation factor of the transcription as well as a factor controlling the translation (Liebermann and Hoffman-Liebermann, 1989). There is a report that c-*myc* expression occurred most remarkably in promyelocytes among the neutrophil lineage cells in humans (Gowda et al., 1986). This may mean that proliferation of neutrophil progenitors takes place most actively at the stage of promyelocytes. In the present study, both rG-CSF and histamine reduced c-*myc* expression in myeloblasts and promyelocytes. It is known that c-*myc* expression is augmented by the activation of protein kinase C (Salehi et al., 1988). Cyclic AMP exerts the inhibitory influence on protein kinase C (Knight and Scrutton, 1984; 28. Ohta et al., 1987). As shown in Fig. 21, both histamine and rG-CSF increased cAMP contents of neutrophil precursor cells. In fact, McCachren et al. (1986) found that db-cAMP reduces the expression of c-*myc* during HL-60 differentiation.

In conclusion, it was considered that histamine increases the rG-CSF-induced differentiation of murine neutrophil progenitors mainly through an increase in the binding sites of rG-CSF of these cells in association with receptor externalization. However, the decrease in c-*myc* expression caused by histamine treatment may well be another reason. It seems quite likely that cAMP plays crucial roles in these two events.

References

Adachi K., Aoyagi T., Iizuka H., Halprin K.M. and Levine V. (1980) Cyclic GMP system in the epidermis. *Curr. Probl. Dermatol.*, 10: 39-65

Akiyama Y., Mukai T., Kamei C. and Tasaka K. (1990) Histamine lipolysis I: Changes in the free fatty acid levels of dog plasma after intravenous infusion of histamine. *Meth. Find. Exp. Clin. Pharmacol.*, 12: 315-324

Bainton D.F., Ullyot J.L. and Farquhar M.G. (1971) The development of neutrophilic polymorphonuclear leukocytes in human bone marrow. *J. Exp. Med.*, 134: 907-934

Black J.W., Duncan W.A.M., Durant G.J., Gannelin C.R. and Parsons M.E.: Definition and antagonism of histamine H_2-receptors. *Nature*, 236: 385-390

Burgess A.W. and Metcalf D. (1980) The nature and action of granulocyte-macrophage colony stimulating factors. *Blood*, 56: 947-958

Burtin C. (1986) Mast cells and tumour growth. *Ann. Inst. Pasteur Immunol.*, 137D: 289-294

Burtin C., Scheinmann P., Salomon J.C., Lespinats G., Frayssinet C., Lebel B. and Canu P. (1981) Increased tissue histamine in tumour-bearing mice and rats. *Br. J. Cancer*, 43: 684-688

Byron J.W. (1976) Bone-marrow toxicity of metiamide. *Lancet*, 2: 1350

Byron J.W. (1980) Pharmacodynamic basis for the interaction of cimetidine with the bone marrow stem cells (CFU-S). *Exp. Hematol.*, 8: 256-263

Catini C., Gheri G., Giampaoli M. and Miliani A. (1984) Histamine uptake by leukocytes in vitro. *Basic Appl. Histochem.*, 28: 329-336

Chaplinski T.J. and Niedel J.E. (1986) Cyclic AMP levels and cellular kinetics during maturation of human promyelocytic leukemia cell. *J. Leukocyte Biol.*, 39: 323-331

Clark R.A.F., Gallin J.I. and Kaplan A. (1975) The selective eosinophil chemotactic activity of histamine. *J. Exp. Med.*, 142: 1462-1476

Code C.F. (1976) Suppression of histamine leucocytosis by metiamide. *J. Physiol.* (Lond.) 254: 31P-32P

Collins S.J., Ruscetti F.W., Gallagher R.E. and Gallo R.C. (1978) Terminal differentiation of human promyelocytic leukemia cells induced by dimethylsulfoxide and other polar compounds. *Proc. Natl. Acad. Sci. USA*, 75: 2458-2462

de Galoscy C. and van Ypersele de Strihou C. (1979) Pancytopenia with cimetidine. *Ann. Intern. Med.*, 90: 274

Deutsch P.J., Rosen O.M. and C.S. Rubin (1982) Identification and characterization of a latent pool of insulin receptors in 3T3-L1 adipocytes. *J. Biol. Chem.*, 257: 5350-5358

Dexter T.M., Allen T.D. and Lajtha L.G. (1977) Conditions controlling the proliferation of haemopoietic stem cells *in vitro*. *J. Cell. Physiol.*, 91: 335-344

Dexter T.M. and Testa N.G. (1976) Differentiation and proliferation of hemopoietic cells in culture. *Methods Cell Biol.* 14: 387-405

Dinarello C.A. (1988) Interleukin-1. *Ann. N.Y. Acad. Sci.*, 546: 122-132

Dinarello C.A., Ikejima T., Warner S.J.C., Orencole S.F., Lonnemann G., Cannon J.G. and Libby P. (1987) Interleukin 1 induces interleukin 1. *J. Immunol.*, 139: 1902-1910

Dunn A.R. (1987) The role of growth factors in normal and neoplastic haemopoiesis. *Ann. N. Y. Acad. Sci.* 511: 1-9

Evans S.W., Rennick D. and Farrar W.L. (1987) Identification of a signal-transduction pathway shared by haematopoietic growth factors with diverse biological specificity. *Biochem. J.*, 244: 683-691

Feldman E.J. and Isenberg J.I. (1976) Effects of metiamide on gastric acid hypersecretion, steatorrhea and bone-marrow function in a patient with systemic mastocytosis. *N. Engl. J. Med.*, 295: 1178-1179

Fibbe W.E., Damme J.V., Billiau A., Goselink H.M., Voogt P.J., Eeden G.V., Ralph P., Altrock B.W. and Falkenburg J.H.F. (1988) Interleukin 1 induces human marrow stromal cells in long-term culture to produce granulocyte colony-stimulating factor and macrophage colony-stimulating factor. *Blood* 71: 430-435

Fleming W.A. and McNeill T.A. (1976) Cellular responsiveness to stimulation *in vitro*. Increased responsiveness to colony stimulating factor of bone marrow colony forming cells treated with surface active agents and cyclic 3', 5'AMP. *J. Cell Physiol.* 88: 323-330

Forrest J.A.H., Shearman D.J.C., Spence R. and Celestin L.R. (1975) Neutropenia associated with metiamide. *Lancet*, 1: 392-393

Gespach C., Saal F., Cost H. and Abita J.P. (1982) Identification and characterization of surface receptors for histamine in the human promyelocytic leukemia cell line HL-60. Comparison with human peripheral neutrophils. *Mol. Pharmacol.*, 22: 547-553

Gespach C., Marrec N. and Belitrand N. (1985a) Relationship between ^{3}H-histamine uptake and H_2-receptors in the human promyelocytic leukemia cell line HL-60. *Agents Actions*, 16: 279-283

Gespach C., Cost H. and Abita J.-P. (1985b) Histamine H_2 receptor activity during the differentiation of the human monocytic-like cell line U-937. Comparison with prostaglandins and isoproterenol. *FEBS Lett.*, 184: 207-213

Gespach C., Courillon-Mallet A., Launay J.M., Cost H. and Abita J.-P. (1986b) Histamine H_2 receptor activity and histamine metabolism in human U-937 monocyte-like cells and human peripheral monocytes. *Agents Actions*, 18: 124-128

Gowda S.D., Koler R.D. and Bagby G.C. Jr. (1986) Regulation of c-*myc* expression during growth and differentiation of normal and leukemic human myeloid progenitor cells. *J. Clin. Invest.*, 77: 271-278

Graham H.T., Lowry O.H., Wheelwright F., Lenz M.A. and Parish H.H. Jr. (1955) Distribution of histamine among leukocytes and platelets. *Blood*, 10: 467-481

Harper R.A., Flaxman B.A. and Chopra D.P. (1974) Mitotic response of normal and psoriatic keratinocytes in vitro to compounds known to affect intracellular cAMP. *J. Invest. Dermatol.*, 62: 384-387

Hatamochi A., Fujiwara K. and Ueki H. (1985) Effects of histamine on collagen synthesis by cultured fibroblasts derived from guinea pig skin. *Arch. Dermatol. Res.*, 277: 60-64

Hay L.J., Vacro R.L., Code C.F. and Wangensteen O.H. (1942) The experimental production of gastric and duodenal ulcers in laboratory animals by the intramuscular injection of histamine in beeswax. *Surg. Gynecol. Obstet.*, 75: 170-182

Hidaka H., Inagaki M., Kawamoto S. and Sakai Y. (1984) Isoquinolinesulfonamides, novel and potent inhibitors of cyclic nucleotide dependent protein kinase and protein kinase C. *Biochemistry*, 23: 5036-5041

Idvall J. (1979) Cimetidine-associated thrombocytopenia. *Lancet*, 2: 159

Iizuka H., Adachi K., Halprin K.M. and Levine V. (1978) Cyclic AMP accumulation in psoriatic skin: differential responses to histamine. AMP and epinephrine by the uninvolved and involved epidermis. *J. Invest. Dermatol.*, 70: 250-253

Imaizumi, M. and Breitman, T.R. (1988) Changes in c-*myc*, c-*fms* and N-*ras* proto-oncogene expression associated with retinoic acid-induced monocytic differentiation of human leukemia. *Cancer Res.*, 48: 6733-6738

Johnson M.McI., Black A.E., Hughes A.S.B. and Clarke S.W. (1977) Leucopenia with cimetidine. *Lancet*, 2: 1226-1227

Kase H., Iwahashi K., Nakanishi S., Matsuda Y., Yamada K., Takahashi M., Murakata C., Sato A. and Kaneko M. (1987) K-252 compounds, novel and potent inhibitors of protein kinase C and cyclic nucleotide-dependent protein kinases. *Biochem. Biophys. Res. Commun.*, 142: 436-440

Kiss Z., Deli E., Vogler W.R. and Kuo J.F. (1987) Anti-leukemic agent alkyl-lysophospholipid regulates phosphorylation of distinct proteins in HL-60 and K-562 cells and differentiation of HL-60 cells prompted by phorbol ester. *Biochem. Biophys. Res. Commun.*, 42: 661-666

Klotz S.A. and Kay B.F. (1978) Cimetidine and agranulocytosis. *Ann. Intern. Med.*, 88: 579-580

Knight D.E. and Scrutton M.C. (1984) Cyclic nucleotides control a system which regulates Ca^{2+} sensitivity of platelet secretion. *Nature*, 309: 66-68

Kobayasi Y., Appella E., Yamada M., Copeland T.D., Oppenheim J.J. and Matsushima K. (1988) Phosphorylation of intracellular receptors of human IL-1. *J. Immunol.*, 140: 2279-2287

Koeffler H.P. (1983) Induction of differentiation of human acute myelogenious leukemia cells: Therapeutic implications. *Blood*, 62: 709-721

Lanotte M., Metcalf D. and Dexter T.M. (1982) Production of monocyte/macrophage colony-stimulating factor by preadipocyte cell lines derived from murine marrow stroma. *J. Cell.*

Physiol., 112: 123-127

Liebermann D.A. and Hoffman-Liebermann B. (1989) Proto-oncogene expression and dissection of the myeloid growth to differentiation developmental cascade. *Oncogene*, 4: 583-592

Lin H.-S., Lokekshwar B.L. and Hsu J.R. (1989) Both granulocyte-macrophage CSF and macrophage CSF control the proliferation and survival of the same subset of alveolar macrophages. *J. Immunol.*, 142: 515-519

Lingberg S. and Tönqvist Å. (1966) The inhibitory effect of aminoguanidine on histamine catabolism in human pregnancy. *Acta Obstet. Gynecol. Scand.*, 45: 131-139

Lotem J. and Sachs L. (1987) Regulation of cell-surface receptors for hematopoietic differentiation-inducing protein MGI-2 on normal and leukemic myeloid cells. *Int. J. Cancer*, 40: 532-539

Marks R.M., Roche W.R., Czerniecki M., Penny R. and Nelson D.S. (1986) Mast cell granules cause proliferation of human microvascular endothelial cells. *Lab. Invest.*, 55: 289-294

Maslinski C.Z., Kierska D., Sasiak K. and Adamas B. (1984) Histamine and its catabolism in tumor-bearing rat and mouse. *Agents Actions*, 14: 497-500

McCachren S.S. Jr., Nichols J., Kaufman R.E. and Niedel J.E. (1986) Dibutyric cyclic adenosine monophosphate reduces expression of c-*myc* during HL-60 differentiation. *Blood*, 68: 412-416

Mochizuki D.Y., Eisenman J.R., Conlon P.J., Larsen A.D. and Tushinski R.J. (1987) Interleukin 1 regulates hematopoietic activity, a role previously ascribed to hemopoietin 1. *Proc. Natl. Acad. Sci. USA* 84: 5267-5271

Nakao S., Matsushima K. and Young N. (1989) Decreased interleukin 1 production in aplastic anaemia. *Br. J. Haematol.*, 71: 431-436

Nakaya N. and Tasaka K. (1988a) The influence of histamine on precursors of granulocytic leukocytes in murine bone marrow. *Life Sci.*, 42: 999-1010

Nakaya N. and Tasaka K. (1988b) Histamine incorporation into murine myeloblasts and promyelocytes. Formation of a histamine transport system. *Biochem. Pharmacol.*, 37: 4523-4530

Nicola N.A., Peterson L., Hilton D.J. and Metcalf D. (1988) Cellular processing of murine colony-stimulating factor (multi-CSF, GM-CSF, G-CSF) receptors by normal hemopoietic cells and cell lines. *Growth Factors*, 1, 41-49

Nomura H., Imazeki I., Oheda M., Kubota N., Tamura M., Ono M., Ueyama Y. and Asano S. (1986) Purification and characterization of human granulocyte colony-stimulating factor (G-CSF). *EMBO J.*, 5: 871-876

Nonaka T., Mio M., Doi M. and Tasaka K. (1992) Histamine-induced differentiation of HL-60 cells. The role of cAMP and protein kinase A. *Biochem. Pharmacol.*, 44: 1115-1121

Ohta Y., Akiyama T., Nishida E. and Sakai H. (1987) Protein kinase C and cAMP-dependent protein kinase induce opposite effects on actin polymerizability. *FEBS Lett*, 222: 305-310

Platzer E., Simon S. and Kalden J.R. (1988) Human granulocyte colony stimulating factor: Effects of human long-term bone marrow cultures. *Blood Cells*, 14: 463-469

Plet A., Evain D. and Anderson W.B. (1982) Effect of retinoic acid treatment of F9 embryonal carcinoma cells on the activity and distribution of cAMP dependent protein kinase. *J. Biol. Chem.*, 257: 889-893 kinase. *J. Biol. Chem.*, 257: 889-893

Radermecker M. and Maldague M.-P. (1981) Depression of neutrophil chemotaxis in atopic individuals. An H_2 histamine receptor response. *Int. Arch. Allergy Appl Immunol.*, 65: 144-152

Rapoport A.P., Abboud C.N. and DiPersio J.F. (1992) Granulocyte-macrophage colony-stimulating factor (GM-CSF) and granulocyte colony-stimulating factor (G-CSF): Receptor biology, signal transduction, and neutrophil activation. *Blood Rev.*, 6: 43-57

Rocklin R.E. and Haberek-Davidson A. (1981) Histamine activates suppressor cells in vitro using a coculture technique. *J. Clin. Immunol.*, 1: 73-79

Rocklin R.E., Greineder D. and Melmon K.L. (1979) Histamine-induced suppressor factor (HSF): further studies on the nature of the stimulus and the cell which produces it. *Cell Immunol.*, 44: 404-415

Rosenwasser L.J. and Dinarello C.A. (1981) Ability of human leukocytic pyrogen to enhance hemagglutinin induced murine thmocyte proliferation. *Cell Immunol.*, 63: 134-142

Salehi Z., Taylor J.D. and Niedel J.E. (1988) Dioctanoylglycerol and phorbol esters regulate transcription of c-*myc* in human promyelocytic leukemia cells. *J. Biol Chem.*, 263: 1898-1903

Sawutz D.G., Kalinyak K., Whitsett J.A., Johnson C.L. (1984) Histamine H_2-receptor desensitization in HL-60 human promyelocytic leukemia cells. *J. Pharmacol Exp. Ther.*, 231: 1-7

Schwartz J.-C., Arrang J.-M., Bouthenet M.-L., Garbarg M., Pollard H. and Raut M. (1991) Histamine receptors in brain, in 'Handbook of Experimental Pharmacology, vol. 97. Histamine and histamine antagonists' (ed. Uväns B.), Springer-Verlag, Berlin, pp. 191-242

Spotts G.D. and Hann S.R. (1990) Enhanced translocation and increased turnover of c-*myc* proteins occur during differentiation of murine erythroleukemia cells. *Mol. Cell. Biol.*, 10: 3952-3964

Tasaka K. (1991) Histamine and the Blood. in 'Handbook of Experimental Pharmacology, vol. 97. Histamine and histamine antagonists' (ed. Uvnäs B.), Springer-Verlag, Berlin, pp. 473-510

Tasaka K. and Code C.F. (1964) Histamine leukocytosis. *Fed. Proc.* 23: 471

Tasaka K. and Nakaya N. (1987) The relationship between incorporation of histamine and differentiation of neutrophil progenitors in murine bone marrow. *Agents Actions*, 20: 320-323

Tasaka K., Nakaya N. and Code C.F. (1992a) Histamine leukocytosis. I. Effect of histamine on peripheral leukocyte counts. *Meth. Find. Exp. Clin. Pharmacol.*, 14: 667-675

Tasaka K., Shorter R.G. and Code C.F. (1992b) Histamine leukocytosis. II. Source of histamine leukocytosis. *Meth. Find. Exp. Clin. Pharmacol.*, 14: 799-804

Tasaka K., Mio M., Shimazawa M. and Nakaya N. (1993) Histamine-induced production of interleukin-1α from murine bone marrow stromal cells and its inhibition by H_2 blockers. *Mol. Pharmacol.*, 43: 365-371

Tasaka K., Doi M., Nakaya N. and Mio M. (1994) Reinforcement effect of histamine on the differentiation of murine myeloblasts and promyelocytes: Externalization of G-CSF receptors induced by histamine. *Mol. Pharmacol.*, 45: 837-845

Voorhees J.J., Duell E.A., Bass L.J., Powell J.A. and Harrell E.R. (1972) The cyclic AMP system in normal and psoriatic epidermis. *J. Invest. Dermatol.*, 59: 114-120

Walker F., Nicola N.A., Metcalf D. and Burgess A.W. (1985) Hierarchical down-modulation of hemopoietic growth factor receptors. *Cell*, 43: 269-276

Zsebo K.M., Yuschenkoff V.N., Schiffer S., Chang D., McCall E., Dinarello C.A., Brown M.A., Altrock B. and Bagby G.C. (1988) Vascular endothelial cells and granulopoiesis: Interleukin-1 stimulates release of G-CSF and GM-CSF. *Blood* 71: 99-103

Zucali J.R., Dinarello C.A., Oblon D.J., Gross M.A., Anderson L. and Weiner R.S. (1986) Interleukin 1 stimulates fibroblasts to produce granulocyte-macrophage colony-stimulating activity and prostaglandin E_2. *J. Clin. Invest.*, 77: 1857-1863

Chapter 9

Pharmacology of newly developed H_1 antagonists: antiallergic profile of H_1 antagonists

1. Introduction

In 1937, the first antihistaminic compound was found by Bovet and Staub in one of a series of amines with a phenolic ether function. Although this compound, 2-isopropyl-5-methylphenoxyethyldiethylamine, protected guinea pigs not only from several lethal doses of histamine but also from the symptoms of anaphylactic shock, this drug was too toxic for clinical use. As a result of the following search for a histamine antagonist, pyrilamine maleate was found as one of the most specific and effective histamine antagonists in the early 1940's. By the early 1950's, many compounds having antihistaminic property had been described (Douglas, 1985). However, all these classical histamine antagonists (H_1 antagonists) caused central side effects such as sedation, lassitude and drowsiness (Wyngaarden and Seevers, 1951). In addition, concurrent ingestion of alcohol and other CNS depressants produced an additive effect in impairing CNS actions, such as disturbing motor skills (Garrison, 1990).

Recently, non-sedative H_1 antagonists have been developed, and it has become evident that these drugs are not only effective as highly potent and specific H_1 antagonists but also they are very effective in treating allergic symptoms such as allergic rhinitis, atopic dermatitis, acute urticaria and bronchial asthma, though the CNS depressant effects are almost negligible. As one of the most important mechanisms, it became apparent that these H_1 blocking agents exert a potent inhibition on histamine release from mast cells. On the basis of histamine release inhibition, these drugs are capable of inhibiting intracellular Ca^{2+} mobilization as well as Ca^{2+} influx when sensitized mast cells are exposed to antigen.

In this chapter, the effects of several newer H_1 antagonists on antiallergic property and the CNS action were compared with those of classical H_1 antagonists. In addition, the characteristics of the antiallergic properties of each drug are mentioned in relation to the histamine release inhibition.

2. Determination of H_1 antagonistic activity (isolated guinea-pig ileum)

It is well known that histamine is capable of contracting the smooth muscle of various tissues in a wide variety of species. Among them, guinea pig ileum is the most sensitive to histamine, more so than the other species or tissues, therefore it had been widely employed in the bioassay for histamine determinations. At the present time, however, histamine determinations are mostly carried out by either the fluorometric method (Shore et al., 1959; Siraganian, 1974) or the radioimmunoassay method (Guesdon et al., 1986). However, the values of IC_{50} and pA_2 are useful parameters even today in assessing the strength of the H_1 blocking activity of H_1 antagonists.

Diphenhydramine and pyrilamine potently inhibited the histamine-induced contraction of guinea pig ileum, while chlorpheniramine was less potent than diphenhydramine and pyrilamine. Ketotifen, epinastine and levocabastine showed potent anti-histaminic activity, while the effects of terfenadine and loratadine were relatively weak. The order of the H_1 blocking potency observed in the present study was almost the same as those reported previously. For instance, Fügner et al. (1988) reported that epinastine was more potent than astemizole and terfenadine in inhibiting the histamine-induced contraction. Fukuda et al. (1984) described that the effect of ketotifen was greater than that of chlorpheniramine and diphenhydramine. The data obtained mainly by us on the IC_{50} of newer H_1 antagonists are collected and summarized in Table 1.

Table 1. Effects of certain H_1 antagonists on the contraction of guinea pig ileum induced by histamine

Drugs	IC_{50} values (M)
Diphenhydramine	$5.34 \times 10^{-8} (4.60-6.17)$
Pyrilamine	$4.85 \times 10^{-9} (3.92-6.03)$
Chlorpheniramine	$1.43 \times 10^{-7} (1.05-1.96)$
Ketotifen	$5.49 \times 10^{-9} (3.25-22.5)$
Mequitazine	$1.82 \times 10^{-7} (1.36-2.43)$
Epinastine	$2.40 \times 10^{-8} (0.09-6.40)$
Astemizole	$3.23 \times 10^{-7} (1.44-5.67)$
Azelastine	$2.79 \times 10^{-9} (0.31-21.6)$
Terfenadine	$1.48 \times 10^{-6} (1.03-2.21)$
Oxatomide	$3.06 \times 10^{-7} (2.62-3.55)$
Levocabastine	$2.29 \times 10^{-8} (1.31-4.56)$
Loratadine	$5.49 \times 10^{-6} (3.92-9.03)$

Reproduced from Tasaka, K. and Akagi, M.: Arzneim.-Forsch., 29, 488-493 (1979); Kamei, C., Mio, M., Izushi, K., Kitazumi, K., Tsujimoto, S., Fujisawa, K., Adachi, Y. and Tasaka, K.: Arzneim.-Forsch., 41, 932-936 (1991a); Kamei, C., Mio, M., Kitazumi, K., Tsujimoto, S., Yoshida, T., Adachi, Y. and Tasaka, K.: Immunopharmacol. Immunotoxicol., 14, 207-218 (1992b); Tasaka, K., Kamei, C., Akagi, M., Mio, M., Shirasaka, T. and Chokki, M.: Arzneim.-Forsch., 43, 1331-1337 (1993), with permission.

3. Effect on histamine release and SRS release

3.1. Histamine release from rat peritoneal mast cells

Recently, it is accepted that various H_1 antagonists are capable of inhibiting histamine release from mast cells and this may be one of the reasons for the antiallergic activities of H_1 antagonists.

In 1957, Tasaka reported that pyrilamine and diphenhydramine were effective in inhibiting histamine release from the abdominal skin of rats and dogs. Actually, these H_1 antagonists showed dual effects on histamine release; when the specimens were pretreated with lower concentrations (ranging from 0.1 to 0.001 mg/ml) of these drugs, histamine release from chopped dog skin induced by sinomenine, a potent histamine liberator, was inhibited dose-dependently (Yamasaki and Tasaka, 1957). The maximum inhibition was achieved at a concentration of 0.05 mg/ml in both drugs. Histamine release inhibition induced by pyrilamine and diphenhydramine was exhibited not only in vitro but also in vivo. Arunlakshana (1953) reported that when the specimen including diphenhydramine was autoclaved, diphenhydramine was exclusively destroyed without affecting the coexisting histamine in the medium. By employing this procedure, she reported that antihistamines prevent histamine release. However, the method of autoclaving the sample is not the proper way to destroy antihistamines selectively. In the experiment carried out according to this method, enough antihistamine was not destroyed to perform the bioassay of histamine. However, at concentrations higher than 2 mg/ml, these drugs caused histamine release in a dose-dependent manner (*in vitro*). Actually, Pellerat and Murat (1946) reported that when 2339RP and 2786RP were injected intravenously in human beings, the level of blood histamine increased; both 2339RP and 2786RP were on the way to being developed as useful antihistamines.

In accordance with the advent of newer H_1 antagonists, research concerning the inhibitory effect of H_1 antagonists on histamine release was resumed actively. Tasaka and his associates reported a series of papers on histamine release inhibition by new H_1 antagonists. First of all, it was reported that terfenadine is effective in inhibiting the histamine release from rat peritoneal mast cells.

Rat peritoneal mast cells were collected and purified to a level higher than 95 % by Percoll density gradient centrifugation (Nemeth and Röhlich, 1980). The mast cells were preincubated with test drugs for 20 min at 37 °C, and thereafter, compound 48/80 or A-23187 were added to make a final concentration of 0.35 μg/ml and 0.1 μM, respectively, and the incubation was continued for another 10 min. The histamine releasing process was terminated by chilling the test tube in an ice bath.

Astemizole, oxatomide, terfenadine and mequitazine caused a potent inhibition of histamine release induced by compound 48/80. Epinastine also caused an inhibition of histamine release induced by compound 48/80, but the potency was less than that of loratadine. Ketotifen and levocabastine had almost no inhibitory effect on histamine release from rat peritoneal mast cells induced by compound 48/80.

Fügner et al. (1988) reported that epinastine inhibited anaphylactic histamine release from rat peritoneal mast cells with protective IC_{50} of $98 \mu g/ml$ ($342.9 \mu M$). Lau and Pearce (1985) described how astemizole inhibited the histamine release from rat peritoneal mast cells induced by antigen with IC_{50} of $18.5 \mu M$, while levocabastine is not capable of inhibiting the histamine release ($IC_{50} > 100 \mu M$). Martin and Römer (1978) reported that ketotifen showed a concentration-dependent inhibition of histamine release induced by compound 48/80; however, a high concentration was needed. IC_{50} of ketotifen was 120 $\mu g/ml$ ($387.8 \mu M$). These results are almost identical to the present data summarized in Table 2.

Table 2. Inhibitory effects of newer H_1 antagonists on histamine release from rat peritoneal mast cells induced by compound 48/80, A-23187 and antigen

Drugs	IC_{50} values (μM)		
	48/80 (0.35 $\mu g/ml$)	A-23187 (0.1 μM)	Antigen
Ketotifen	432	317	
Epinastine	169	165	81.3
Astemizole	5.02		7.24
Oxatomide	2.58		
Terfenadine	5.90	7.80	
Levocabastine	> 500	> 500	
Loratadine	46.6	> 500	
Mequitazine	9.24	29.6	
MY-1250	186.3		

48/80: compound 48/80. Reproduced from Tasaka, K., Mio, M. and Okamoto, M.: Ann. Allergy, 56, 464-469 (1986a); Akagi, M., Mio, M., Miyoshi, K. and Tasaka, K.: Immunopharmacol. Immunotoxicol., 9, 257-279 (1987); Tasaka, K., Akagi, M., Mio, M., Okamoto, M., Miyoshi, K. and Izushi, K.: Jpn. Pharmacol. Ther., 16, 2465-2480 (1988); Tasaka, K., Akagi, M., Izushi, K. and Mio, M.: Meth. Find. Exp. Clin. Pharmacol., 12, 531-539 (1990a); Tasaka, K., Akagi, M., Izushi, K. and Aoki, I.: Pharmacometrics, 39, 365-373 (1990d); Tasaka, K., Akagi, M., Mio, M., Izushi, K. and Aoki, I.: Arzneim.-Forsch., 40, 1092-1097 (1990e); Tasaka, K., Kamei, C., Akagi, M., Mio, M., Shirasaka, T. and Chokki, M.: Arzneim.-Forsch., 43, 1331-1337 (1993), with permission.

3.2. Histamine or SRS release from the lung pieces of sensitized guinea pigs

It is well known that when lung pieces of actively sensitized guinea pigs are exposed to antigen, histamine and SRS are released into the medium (Tasaka et al., 1990d). Therefore, this model of anaphylaxis has been widely used to study the mechanism of allergic reactions and to estimate drugs' effect.

The animals were immunized with egg albumin as an antigen according to the method of Mota (1964). Two weeks after immunization, the animals were killed and the excised lung was cut into pieces approximately $1 mm^3$. The lung pieces, weighing 100 mg wet weight each, were suspended in 1.9 ml of Tyrode solution and preincubated for 10 min at 37°C. Subsequently, 0.1 ml of egg albumin solution ($50 \mu g/ml$) was added and incubation was continued for 20 min (Akagi et al., 1987). The released amount of histamine was determined by means of a fluorometric assay (Shore, 1959), and expressed as the percentage of the total histamine content in the tissues. The released amount of SRS was measured biologically using guinea pig ileum in the presence of pyrilamine (10^{-6} M) and atropine (10^{-5} M) and expressed as dose equivalent to leukotriene C_4 (LTC_4). Recently, the measurement of leukotrienes was carried out by means of radioimmunoassay. Specific antibody can be obtained from commercial source. In *ex vivo* experiments, test drugs were administered by p.o. 1 hr before exsanguination; in *in vitro* experiments, lung pieces were pretreated with the test drugs for 15 min at 37°C prior to antigen challenge.

Oxatomide (*in vitro*), astemizole (both *in vitro* and *ex vivo*) and epinastine (*ex vivo*) caused a potent inhibition of histamine release in actively sensitized guinea pig lung preparations. Levocabastine (*in vitro*) and loratadine (*in vitro*) also caused a significant inhibition, whereas ketotifen (both *in vitro* and *ex vivo*) and MY-1250 (*in vitro*) were almost completely ineffective in preventing histamine release from the lung pieces of sensitized guinea pigs. Astemizole also inhibited SRS release from sensitized guinea pig lung preparations.

Table 3. Inhibitory effects of newer H_1-antagonists on histamine release from the lung pieces of sensitized guinea pigs

Drugs	ID_{50} values		
	Histamine release		SRS release
	(*in vitro*, μM)	(*ex vivo*, mg/kg)	(*in vitro*, μM)
Ketotifen	430	> 20	
Epinastine	75.9	4.69	
Astemizole	9.27	8.18	3.25
Levocabastine	39.5		
Oxatomide	8.76		
Terfenadine		15.6	
Loratadine	21.8		
Mequitazine			
MY-1250	268		

Reproduced from Akagi, M., Mio, M., Miyoshi, K. and Tasaka, K.: Immunopharmacol. Immunotoxicol., 9, 257-279 (1987); Tasaka, K., Akagi, M., Mio, M., Miyoshi, K. and Nakaya, N.: Int. Archs Allergy appl. Immun., 83, 348-353 (1987); Tasaka, K., Akagi, M., Izushi, K. and Mio, M.: Meth. Find. Exp. Clin. Pharmacol., 12, 531-539 (1990a); Tasaka, K., Akagi, M., Izushi, K. and Aoki, I.: Pharmacometrics, 39, 365-373 (1990d); Kamei, C., Mio. M., Izushi, K., Yoshii, N., Fujisawa, K. and Tasaka, K.: Immunopharmacol. Immunotoxicol., 13, 341-356 (1991b); Tasaka, K., Kamei, C., Akagi, M., Mio, M., Shirasaka, T. and Chokki, M.: Arzneim.-Forsch., 43, 1331-1337 (1993), with permission.

Ohmori et al. (1982) reported that oxatomide inhibited histamine release from the lung pieces of sensitized guinea pigs with an ED_{50} of about 1 μM. Fukuda et al. (1984) explained that although chlorpheniramine caused about 50 to 60 % inhibition of histamine release at 1000 and 3000 μM, ketotifen did not provide any inhibitory effect at a concentration range of 100 to 10000 μM. In human lung preparations, Church and Gradidge (1980) reported that many H_1 antagonists inhibited histamine release, and the effect of oxatomide was greater than that of ketotifen. These results are almost identical to those shown in Table 3.

4. Effect on ^{45}Ca uptake and intracellular Ca^{2+} mobilization

4.1. ^{45}Ca uptake into rat peritoneal mast cells

It is known that an increase in the intracellular Ca^{2+} level is a key event in triggering histamine release from rat mast cells. Previously we have shown that mobilization of Ca^{2+} takes place from intracellular stores of rat peritoneal mast cells after treatment with compound 48/80 (Tasaka et al., 1986b). Also it has been reported that some Ca antagonists effectively inhibit antigen-induced bronchospasm and exercise-induced asthma in humans (Ritchie et al., 1984; Patel, 1981). Therefore, effects of newer H_1 antagonists on ^{45}Ca uptake and intracellular Ca^{2+} mobilization was studied.

The measurement of ^{45}Ca uptake into rat peritoneal mast cells was performed using a modification of the method of Spataro and Bosmann (1976). Peritoneal mast cells were incubated in PBS containing with ^{45}Ca (^{45}CaCl$_2$, 1.35 μCi/2.5 $\times$ 10^5 cells) at 37°C for 1 min. Thereafter, test compounds were added and the incubation was continued for 10 min. After addition of compound 48/80 (0.5 μg/ml), the incubation was continued for further 5 min. Subsequently, the cells were washed twice with ice cold PBS and solubilized with 0.2 ml of 10 % Triton X-100. Five ml of a liquid scintillator (Atomlight, New England Nuclear) was added and radioactivity was measured by a liquid scintillation counter (Aloka, LSA-700).

Oxatomide at concentrations higher than 0.1 μM significantly inhibited ^{45}Ca uptake in a dose-dependent fashion and the effective concentration range was similar to that observed in the case of histamine release inhibition (Table 4). Oxatomide alone did not affect the spontaneous influx of ^{45}Ca into normal mast cells. Verapamil was much less effective in preventing ^{45}Ca uptake; significant inhibition was elicited at 100 μM. Terfenadine and MY-1250 also inhibited the increasing ^{45}Ca-influx induced by compound 48/80. This range is almost identical to that of the concentrations which effectively inhibited histamine release (Table 4).

As shown in Table 4, oxatomide elicited the most potent inhibition on the compound 48/80-induced uptake of ^{45}Ca. The effect was much more potent than that of verapamil, a representative Ca antagonist. These findings strongly suggests that oxatomide is a Ca antagonist highly selective toward rat mast cells.

Table 4. Inhibitory effects of newer H_1 antagonists on ^{45}Ca uptake into rat peritoneal mast cells induced by compound 48/80 (0.5 $\mu g/ml$)

Drugs	Concentrations (μM)	% Inhibitions	$ED_{50}s$ (μM)
Oxatomide	0.1	43.3 ± 5.2**	
	1	49.5 ± 0.2**	0.31
	10	90.7 ± 0.2**	
Astemizole	0.1	22.1 ± 5.2	
	1	39.0 ± 3.9**	4.28
	5	50.6 ± 4.7**	
	10	57.1 ± 2.6**	
Terfenadine	2	2.1 ± 0.7	
	5	51.4 ± 0.6**	6.44
	10	67.2 ± 6.5**	
MY-1250	1	7.5 ± 1.5	
	10	44.3 ± 0.3**	75.1
	100	44.7 ± 2.8**	
Tazalest	0.1	5.3 ± 0.8	
	1	11.9 ± 0.7	
	10	11.9 ± 1.3	> 100
	100	22.8 ± 1.2*	
Loratadine	1	9.8 ± 3.6	
	10	40.8 ± 13.1*	22.6
	100	66.6 ± 6.0**	
Verapamil	1	0.8 ± 5.2	
	10	2.7 ± 6.2	> 100
	100	30.9 ± 0.2**	

*, **: Significantly different from the control with $P < 0.05$ and $P < 0.01$, respectively. Reproduced from Tasaka, K., Akagi, M., Mio, M., Miyoshi, K. and Nakaya, N.: Int. Archs Allergy appl. Immun., 83, 348-353 (1987); Akagi, M., Mio, M., Miyoshi, K. and Tasaka, K.: Immunopharmacol. Immunotoxicol., 9, 257-279 (1987); Tasaka, K., Akagi, M., Izushi, K. and Mio, M.: Meth. Find. Exp. Clin. Pharmacol., 12, 531-539 (1990a); Kamei, C., Mio, M., Izushi, K., Yoshii, N., Fujisawa, K. and Tasaka, K.: Immunopharmacol. Immunotoxicol., 13, 341-356 (1991b), with permission.

4.2. Fluorescence increase in quin 2-loaded rat mast cells induced by compound 48/80

It is well known that an increase in the intracellular Ca^{2+} concentration is essential in triggering histamine release from mast cells. It has also been reported that an increase in the intracellular Ca^{2+} level of mast cells takes place during histamine release in a Ca-free medium (White et al., 1984; Beaven et al., 1984). In these studies, an increase in fluorescence intensity in a quin 2-loaded mast cell suspension or 2H3 cell suspension has been exhibited. However, in experiments to measure intracellular Ca^{2+} concentrations of mast cells suspended in a solution using quin 2-Ca complex as a signal, it is indispensable to stir the cells without interruption during the measurement, because mast cells are heavy and otherwise sink to the bottom of the cuvette. It can be easily assumed that mast cells are probably damaged during such mechanical agitation. Furthermore, light scattering of the excitation beam or emitted fluorescence in a heavy cell suspension may well be the reason for errors in fluorescence measurements. To avoid such problems, the changes in fluorescence intensity in association with the alteration of intracellular Ca^{2+} were measured in an individual cell in the following experiments.

Intracellular concentration of free Ca^{2+} was measured according to the method described previously (Tasaka et al., 1986a). In brief, rat peritoneal mast cells washed with Ca-free PBS supplemented with 1 mM of EGTA were suspended in 5 μM of quin 2-AM, and incubated at 37°C for 30 min. The cells were then suspended in Ca-free PBS containing various concentrations of test compounds and incubated at 37°C for 5 min, and subsequently, stimulated with compound 48/80 (0.2 μg/ml) for 1 min. Fluorescence derived from quin 2 chelated with intracellular free Ca was measured using individual cells (n = 200 at minimum) by means of a fluorescence microscope (Nikon XF) connecting to a photon counter (Nikon, P-1).

When mast cells loaded with quin 2 were exposed to compound 48/80 (0.2 μg/ml) for 1 min at 37°C, their fluorescence intensity increased significantly, exhibiting a value of 149.6 ± 8.7 (n = 250, P < 0.01), as compared with the control value 100 (± 6.8, n = 260). Since this experiment was carried out in a Ca-free medium, an increase in fluorescence intensity reflects the amount of Ca^{2+} released from the intracellular Ca store.

It is known that compound 48/80 elicits histamine release from mast cells even in a Ca-free medium. It has been shown that Ca^{2+} is released from the intracellular stores after exposure to compound 48/80 in a Ca-free medium, using fluorescence image analysis of quin 2-loaded mast cells (Tasaka et al., 1986a; Tasaka et al., 1986b).

Table 5. Inhibitory effects of newer H_1 antagonists and Ca antagonists on the fluorescence intensity increase in quin 2-loaded mast cells induced by compound 48/80 (0.2 μg/ml)

Drugs	IC$_{50}$ values (M)
Ketotifen	Ca. 10^{-4}
Epinastine	3.1×10^{-5}
Oxatomide	5.0×10^{-8}
Astemizole	3.1×10^{-8}
Mequitazine	4.8×10^{-6}
MY-1250	5.0×10^{-6}
Verapamil	2.7×10^{-7}
D-600	3.7×10^{-6}

However, when the quin 2-loaded mast cells were pretreated with both oxatomide and astemizole, the fluorescence increase was remarkably suppressed as shown in Table 5. This indicates that one of the inhibitory mechanisms of oxatomide and astemizole involving histamine release is the inhibition of Ca^{2+} release from the intracellular store. Epinastine, mequitazine and MY-1250 also inhibited Ca^{2+} release from the intracellular Ca store, though the extent of inhibition was slightly less than that elicited by oxatomide or astemizole.

5. The role of cyclic nucleotides for inhibition of histamine release

5.1. cAMP and cGMP contents in the lung pieces of actively sensitized guinea pigs

It has been reported that a significant reduction in cAMP content in rat mast cells takes place in association with the histamine release and that when rat mast cells were pretreated with certain antiallergic agents, it is capable to increase the cAMP content; as a consequence of this, histamine release from mast cells was significantly inhibited (Tasaka et al., 1986a). In addition, an elevation of the intracellular cAMP level may correlated with the inhibitory effect of antiallergic drugs (Loeffler et al., 1971; Hayashi et al., 1976). Therefore, the effect of newer H_1 antagonists on cyclic nucleotides was studied using the lung pieces of guinea pigs.

Using egg albumin as an antigen, animals were sensitized according to the method of Mota (1964). The animals were exsanguinated 2 weeks later and the lung was excised. Thereafter, the lung was cut into small pieces (approximately 100 mg) and homogenized in 2 ml of 5 % trichloroacetic acid by means of a Polytron. After centrifugation at $1,000 \times g$ for 15 min, the supernatant was extracted 4 times with water-saturated ether. The cAMP and cGMP contents in 0.1 ml of supernatant were measured using a radioimmunoassay kit, respectively. The samples were pretreated with test drugs for 15 min prior to antigen addition at 37 °C.

In the control group (without drug treatment), after exposure to antigen (antigen-treated group), the cAMP or cGMP contents in the lung pieces of actively sensitized guinea pigs decreased or increased, respectively compared with that of the antigen-untreated groups. The decrease in cAMP contents after exposure to antigen was inhibited in the cases pretreated with either drug used in the experiments. Oxatomide and levocabastine were more effective than ketotifen in preventing the cAMP decrease. It has been reported that cAMP effectively prevents Ca^{2+} release from the intracelluar Ca store of rat mast cells (Tasaka et al., 1986a). Because of this, preventing cAMP reduction may be an effective way of preventing mast cell activation by suppressing the Ca dependent processes.

5.2. Adenylate cyclase and phosphodiesterase activities in rat mast cells or guinea pig lung preparations

As shown in Table 6, some newer H_1 antagonists increased the cAMP level in the lung pieces of guinea pigs (Tasaka et al., 1987). Therefore, adenylate cyclase and phosphodiesterase activities were measured.

Adenylate cyclase activity in rat peritoneal mast cells (*in vitro*) and rat lung sample (*ex vivo*) was measured according to the method of Kebabian et al. (1972) with some modifications. *In vitro* study, rat mast cells, purified to a level higher than 95 %, were suspended in PBS containing various concentrations of the test drugs and incubated for 5 min at 37°C. Subsequently, the cells were disrupted and the cell debris was removed by centrifugation. In *ex vivo*, the excised lungs were cut into pieces and these were homogenized in amount 25 times the sample volume in 5 mM HEPES buffer (pH 7.4) containing 2 mM EGTA, followed by centrifugation. These supernatants were suspended in assay medium (mM: HEPES 5, $MgSO_4$ 2, theophylline 10 and EGTA 0.2). After addition of ATP, the reaction was initiated and subsequent incubation was performed. After the reaction was terminated by placing the assay tube in an ice bath, the mixture was centrifuged to remove insoluble materials. The amount of cAMP in the supernatant was measured using a radioimmunoassay kit. The results were expressed in terms of percent changes as compared with the values determined in the control group (Tasaka et al., 1988; Kamei et al., 1992a).

Table 6. Effects of newer H_1 antagonists on cyclic nucleotide contents in sensitized guinea pig lung

Drugs	Concentrations (μM)	% Changes cAMP	cGMP
Ketotifen	0.1	108.5**	91.5
	1	144.4**	89.5
	10	157.3**	78.8*
	100	138.0**	63.5**
Oxatomide	0.01	114.6**	107.7
	0.1	148.1**	79.6*
	1	149.0**	70.4**
	10	159.4**	94.0
Levocabastine	1	158.8	
	10	190.2*	
	100	217.6**	
MY-1215	1	138.7	
	10	225.9**	
	100	238.7**	

*, ** : Significantly different from control group with P < 0.05 and P < 0.01, respectively. Reproduced from Tasaka, K., Akagi, M., Mio, M., Miyoshi, K. and Nakaya, N.: Int. Archs Allergy appl. Immun., 83, 348-353 (1987); Kamei, C., Mio, M., Izushi, K., Yoshii, N., Fujisawa, K. and Tasaka, K.: Immunopharmacol. Immunotoxicol., 13, 341-356 (1991b); Tasaka, K., Kamei, C., Akagi, M., Mio, M., Shirasaka, T. and Chokki, M.: Arzneim.-Forsch., 43, 1331-1337 (1993), with permission.

Phosphodiesterase activity in rat peritoneal mast cells (*in vitro*) and rat lung sample (*ex vivo*) was measured using the method of Pöch (1971) with some modifications. The isolated mast cells or lung homogenized samples were suspended in a cold buffered magnesium solution and centrifuged at 4°C. To determine the enzyme activity, 0.2 ml of supernatant was used and the drug was added and incubated for 10 min at 37°C. After addition of ^{3}H-cAMP and 5'-AMP, incubation was continued at 37°C, and hydrolysis of cAMP was terminated by adding $ZnSO_4$ (0.17 M) to the test tube. After 10 min, $Ba(OH)_2$ was added to precipitate 5'-AMP. After centrifugation, a scintillator (Atomlight) was added, and samples were counted using a liquid scintillation counter (Aloka, LSC-700). The value determined in the control sample was taken as 100%. The effect of the test drug on the enzyme activity was expressed in terms of percent changes as compared with the control group (Tasaka et al., 1988; Kamei et al., 1992a).

Table 7. Effects of newer H_1 antagonists on adenylate cyclase and phosphodiesterase activities in rat mast cells (in vitro) and lung preparations (ex vivo)

In vitro

Drugs	Concentrations (M)	% Changes	
		Adenylate cyclase	Phosphodiesterase
Control	—	100.0	100.0
Azelastine	10^{-6}	102.6	133.3
	10^{-5}	115.1*	105.1
	10^{-4}	132.7**	59.1**
Epinastine	10^{-6}	104.9	103.0
	10^{-5}	109.8*	102.7
	10^{-4}	116.0**	95.0**
	10^{-3}	139.3**	
Isoproterenol	10^{-6}	115.0**	
	10^{-5}	121.0**	
Theophylline	10^{-4}		74.3

Ex vivo

Drugs	Doses (mg/kg, p.o.)	% Changes	
		Adenylate cyclase	Phosphodiesterase
Control	—	100.0	100.0
Ketotifen	5	113.8	100.0
	10	114.0	98.2
	20	124.1*	101.8
Terfenadine	5	150.0**	94.7
	10	162.1**	98.2
	20	172.4**	100.0

*, **: Significantly different from control group with P < 0.05 and P < 0.01, respectively. Reproduced from Akagi, M., Mio, M., Tasaka, K. and Kiniwa, S.: Pharmacometrics, 26, 191-198 (1983); Tasaka, K., Akagi, M., Mio, M., Okamoto, M., Miyoshi, K. and Izushi, K.: Jpn. Pharmacol. Ther., 16, 2465-2480 (1988); Kamei, C., Akagi, M., Mio, M., Kitazumi, K., Izushi, K., Masaki, S. and Tasaka, K.: Immunopharmacol. Immunotoxicol., 14, 191-205 (1992a), with permission.

It became apparent that azelastine (*in vitro*), epinastine (*in vitro*), ketotifen (*ex vivo*) and terfenadine (*ex vivo*) caused an activation of adenylate cyclase activity, whereas no significant inhibition of phosphodiesterase activity was observed (Table 7). These results indicate that an activation of adenylate cyclase is responsible for an elevation of cAMP content induced by H_1 antagonists.

5.3. Adenylate cyclase activity in purified enzyme

Adenylate cyclase activity was measured using a purified enzyme in the medium containing 25 mM sodium HEPES, 2 mM EDTA, 2 mM $MgCl_2$, 0.5 mM ATP, 0.25 mM IBMX, 0.1 mM sodium ascorbate and 1 μg/ml BSA. The reaction was initiated by the addition of the plasma membrane fraction (adenylate cyclase, extracted from S49 cells, Ross et al., 1977) and preincubated for 5 min. A crude Gs protein (extracted from rat brain) and test compounds were added to the reaction mixture and they were incubated for 20 min. Reaction was terminated by adding of ice-cold 10 % TCA. The cAMP content was determined by means of a cAMP radioimmunoassay kit.

Astemizole (1, 10 μM) and terfenadine (10 μM) caused a significant activation of adenylate cyclase activity (Table 8).

6. Effect on bronchoconstriction in guinea pigs

6.1. Antigen- or histamine-induced response

The bronchoconstriction induced by antigen or histamine is widely used to determine the efficacy of anti-asthmatic drugs which are able to alleviate the dyspnea caused by bronchospasm (Akagi and Tasaka, 1988).

Table 8. Effects of newer H_1 antagonists on adenylate cyclase activity in purified enzyme

Drugs	Concentrations (μM)	Adenylate cyclase activity (pmole/min/mg protein)
Control	—	3.64 ± 0.62
Oxatomide	1	3.87 ± 0.25
	10	4.40 ± 0.37
Astemizole	1	5.03 ± 0.83 *
	10	7.26 ± 0.86 **
Ketotifen	10	3.63 ± 0.75
	100	3.65 ± 0.72
Cromolyn	10	4.33 ± 0.47
sodium	100	4.13 ± 0.45
Terfenadine	1	4.35 ± 0.73
	10	4.98 ± 0.61 *

*, ** : Significantly different from control group with $P < 0.05$ and $P < 0.01$, respectively.

Male Hartley guinea pigs weighing approximately 450 g were sensitized according to the method of Mota (1964). Egg albumin was used as an antigen. Two weeks later, the animals were anesthetized by pentobarbital administration (30 mg/kg, i.p.) and a cannula was inserted into the trachea. Airway resistance was measured according to the method previously described (Akagi and Tasaka, 1988). The jugular vein was cannulated in order to administer the antigen (0.2 mg/kg, i.v.) or histamine (5 μg/kg, i.v.), while test drugs were given orally 1 hr before antigen challenge.

Ketotifen, astemizole and loratadine caused a potent inhibition of antigen-induced bronchoconstriction, while the effects of epinastine and terfenadine were less potent than those of ketotifen and astemizole. Also, ketotifen, azelastine and epinastine exhibited a potent inhibition of histamine-induced bronchoconstriction. However, the effects of astemizole and loratadine were inferior to those of ketotifen, azelastine and epinastine (Table 9). Similar results were observed when antihistaminic activity was measured using guinea pig ileum as shown in Table 1. The antihistaminic activities of ketotifen and epinastine were more potent than those of loratadine and astemizole.

Wauwe et al. (1981) reported that astemizole was effective in inhibiting histamine-induced bronchoconstriction in guinea pigs with ED_{50} of 0.05 mg/kg. Fügner et al. (1988) also reported that epinastine, ketotifen and promethazine are able to protect bronchoconstriction induced by histamine aerosol, and that the appropriate ED_{50} values were 0.047, 0.019 and 0.94 mg/kg, respectively. These results were almost identical to those shown in the present study. The inhibitory effect of ketotifen and epinastine on histamine-induced bronchoconstriction was more remarkable than that shown in antigen-induced bronchoconstriction, whereas the effects of astemizole and loratadine on both antigen- and histamine-induced bronchoconstrictions were almost identical. In conclusion, it became clear that the newer H_1 antagonists used in the present study may be effective in the treatment of asthma.

6.2. PAF-induced response

It is known that PAF caused not only hypotension but also anaphylactoid reactions after intravenous injection (Vargaftig et al., 1980; McManus et al., 1980). Vargaftig et al. (1980) proposed that bronchoconstriction due to PAF was platelet dependent as it was suppressed by immune platelet depletion, and PAF is the most powerful bronchoconstrictor agent. Based on these findings, the effects of newer H_1 antagonists on PAF-induced bronchoconstriction were investigated in addition to antigen-induced bronchoconstriction.

The animals were anesthetized with pentobarbital sodium (30 mg/kg, i.p.), and a cannula was inserted into the trachea. Airway resistance was measured according to the method described previously (Akagi and Tasaka, 1988). In short, jugular vein was cannulated in order to administer PAF (150 ng/kg, i.v.), while test drugs were given orally 1 hr before histamine injection.

Pretreatment with ketotifen, azelastine and epinastine was more effective in inhibiting the PAF-induced bronchoconstriction than that of promethazine (Table 9). Criscuoli et al. (1986) described how ketotifen inhibited PAF-induced bronchoconstriction by 40-50 % at a dose of 3 μmol/kg (0.93 μg/kg). These results are almost the same as those seen in the present study. These results seem to indicate that ketotifen, epinastine and azelastine are effective in inhibiting not only antigen- or histamine- but also PAF-induced bronchoconstrictions.

7. Effect on cutaneous reactions in rats

7.1. Homologous PCA

Anaphylactic reactions are the results of complex mechanisms which are triggered by the interaction between antigen and antibodies fixed to mast cells or circulating basophils (Martin and Römer 1978). The antianaphylactic actions of newer H_1 antagonists were tested in rats by means of both the homologous and heterologous PCA test.

Wistar rats were immunized according to the method of Tada and Okumura (1971). A preparation of TNP-As was used as an antigen and the titer of the antiserum was determined using male Wistar rats, weighing about 150 g which have received intradermal injection of diluted antisera (1:40, 1:80). Forty-eight hr later, 0.5 mg of TNP-As dissolved in 0.5 ml of 2 % Evans blue solution was injected intravenously. Thirty min after the antigen challenge, the animals had been sacrificed, and the blueing area was measured. The test drugs were given orally 1 hr before the antigen challenge.

Table 9. Effects of certain H_1 antagonists on bronchoconstriction in guinea pigs

Drugs	ID$_{50}$ values (mg/kg, p.o.)		
	Antigen-induced	Histamine-induced	PAF-induced
Promethazine		0.26	4.08
Ketotifen	0.43	0.01	0.31
Azelastine		0.05	0.53
Epinastine	2.30	0.03	0.12
Terfenadine	1.70		
Astemizole	0.86	0.24	
Loratadine	0.20	0.34	
Apafant			0.04

Reproduced from Akagi, M., Mio, M., Miyoshi, K. and Tasaka, K.: Immunopharmacol. Immunotoxicol., 9, 257-279 (1987); Tasaka, K., Akagi, M., Izushi, K. and Mio, M.: Meth. Find. Exp. Clin. Pharmacol., 12, 531-539 (1990a); Tasaka, K., Akagi, M., Izushi, K. and Aoki, I.: Pharmacometrics, 39, 365-373 (1990d); Kamei, C., Mio, M., Izushi, K., Kitazumi, K., Tsujimoto, S., Fujisawa, K., Adachi, Y. and Tasaka, K.: Arzneim.-Forsch., 41, 932-936 (1991a); Tasaka, K., Kamei, C. and Nakamura, S.: Arzneim.-Forsch., 44, 327-329 (1994a), with permission.

As shown in Table 10, ketotifen, epinastine and astemizole caused an inhibition of homologous PCA. The effects of terfenadine were less than those of ketotifen, epinastine and astemizole. Martin and Römer (1978) also reported that ketotifen had a marked inhibitory effect on the antigen-antibody reaction in rats (ID_{50} = 0.3 mg/kg, i.v. and 5.1 mg/kg, p.o.). The result observed by Martin and Römer was somewhat less potent than that shown in the present study. On the other hand, Chand et al. (1985) reported that ID_{50}s of ketotifen and astemizole in the IgE-mediated 72 hr PCA were 2.0 and 1.6 mg/kg, p.o., respectively. The results are almost identical to those shown in the present study (Table 10).

7.2. Heterologous PCA

IgE-rich anti-egg albumin mice serum was prepared according to the method of Levine and Vaz (1970), and the titer of the antiserum was determined in rats by means of 4 hr heterologous PCA. Shaved, normal rats received intradermal injections of diluted antisera (1:200, 1:400, 1:800). Four hr later, 0.5 ml of a mixture of 1 % egg albumin and 2 % Evans blue solution were injected intravenously. Subsequent procedures were similar to that described in the homologous PCA.

As shown in Table 10, ketotifen, epinastine and astemizole inhibited heterologous PCA at the same dose level as in the case of homologous PCA. Terfenadine was less potent than either ketotifen, epinastine or astemizole.

7.3. Histamine-induced skin response

It is generally known that histamine is one of the most important mediators in an anaphylactic reaction of the skin in many species and humans (Martin and Römer, 1978). Therefore, the effects of newer H_1 antagonists on histamine-induced cutaneous reactions were also studied.

Table 10. Effects of newer H_1 antagonists on cutaneous reactions in rats

Drugs	ID_{50} values (mg/kg, p.o.)		
	Homologous PCA	Heterologous PCA	Histamine response
Ketotifen	1.31	1.59	1.12
Epinastine	2.30	2.94	3.71
Astemizole	1.07	1.35	
Levocabastine			4.18
Terfenadine	4.32	6.31	

Reproduced from Akagi, M., Mio, M., Miyoshi, K. and Tasaka, K.: Immunopharmacol. Immunotoxicol., 9, 257-279 (1987); Kamei, C., Mio, M., Izushi, K., Kitazumi, K., Tsujimoto, S., Fujisawa, K., Adachi, Y. and Tasaka, K.: Arzneim.-Forsch., 41, 932-936 (1991a); Kamei, C., Mio, M., Kitazumi, K., Tsujimoto, S., Yoshida, T., Adachi, Y. and Tasaka, K.: Immunopharmacol. Immunotoxicol., 14, 207-218 (1992b), with permission.

The rats were shaved and the test drugs were given orally. One hr later, 2 % Evans blue was injected intravenously (0.5 ml/100 g). Immediately after that, histamine (10^{-4} g/ml, 0.1 ml) was injected intradermally into the dorsal skin. Thirty min after that, the animals had been sacrificed, and the blueing area was measured.

Ketotifen markedly inhibited the dye leakage due to an increased vascular permeability of the skin after histamine injection. Epinastine and levocabastine also caused an inhibition of histamine-induced cutaneous reactions, but the potencies were less than that of ketotifen. Wauwe et al. (1981) reported that ketotifen inhibited histamine-induced skin reactions with ID_{50} of 1.17 mg/kg. Fügner et al. (1988) described that the ID_{50}s of epinastine and terfenadine in inhibiting histamine-induced skin reaction were 2.2 mg/kg and 3.5 mg/kg, respectively. These results are almost the same as those seen in the present study. As shown in Table 10, in the cases of ketotifen and epinastine, ID_{50}s of homologous and heterologous PCA were almost identical to those seen in histamine-induced skin reactions. These findings may suggest that histamine is the most important mediator of PCA reactions.

8. Effect on experimental allergic rhinitis

8.1. Antigen- or histamine-induced rhinitis

Allergic rhinitis is one of the common allergic diseases and classical H_1 antagonists have long been used as a first choice in its treatment (Bende and Pipkorn, 1987). Although they are often effective in treating allergic rhinitis, a high incidence of CNS depressant side effects have hindered the use of these classic H_1 antagonists. In this item, the effects of some newer H_1 antagonists on experimental allergic rhinitis will be discussed.

In the experiment of allergic rhinitis, rats (4-5 weeks old, body weight: 150-160 g) were immunized by injecting DNP-Ascaris extract (1 mg) coupled with 10^{10} killed B. Pertussis and aluminium hydroxide gel (2 mg) into four footpads. Five days later, they were boostered with 0.5 mg of DNP-As alone in the back muscle (Tada and Okumura, 1971). Eight days after the immunization, the rats were used in the experiments. The antibody titer was 1:128. The provocation of experimental rhinitis was performed according to a modified method of Kojima et al. (1986). In short, the animals were anesthetized with urethan (1.5 g/kg, i.p.) and a trachea cannula was inserted to keep the ventilation. Thereafter, a polyethylene tubing (outside diameter, 1.09 mm, Intramedic, PE-20) was inserted into the upper trachea and the tip was placed in the nasal cavity passing through the pharynx and choana, and the tubing was ligated to the trachea together with gullet. The other end of the polyethylene tubing was connected to a perfusion pump, and saline was perfused at a rate of 0.25 ml/min, and the effluent flowing from the nostrils was collected. Subsequently, 1 % Evans blue was injected intravenously (0.5 ml/100 g), and the effluent was collected every 10 min. In the following, antigen or histamine was perfused, and the collection of effluent was continued for 30 min; the effluent collected within first 10 min was named Period 1, and those collected within the second and the last 10 min were named as Period 2 and 3, respectively. The test drugs were administered orally (0.1 ml/100 g) or topically (10 μl/each nostril) before antigen or histamine perfusion. In oral administration, the test drugs were administered 2 hr before antigen. The lavage fluid was centrifuged at $1,200 \times g$ for 10 min and the amount of Evans blue in the supernatant was determined at 620 nm. The drug effect was evaluated by measuring the amount of dye leakage in Period 1 (Kamei et al., 1992b).

Ketotifen, astemizole and loratadine caused an inhibitory effect on dye leakage into the nasal cavity induced by antigen. The potency of levocabastine was less effective than those of ketotifen and loratadine. Ketotifen, epinastine and azelastine were effective in inhibiting the leakage into the nasal cavity induced by histamine. The potency of levocabastine was less effective than those of ketotifen, epinastine and azelastine when administered orally. On the other hand, when levocabastine was applied topically, it was more potent than ketotifen in inhibiting the dye leakage induced by both antigen- and histamine-induced dye leakage (Table 11).

9. Effect on experimental allergic conjunctivitis

9.1. Antigen- or histamine-induced conjunctivitis

Pharmacological treatment of allergic conjunctivitis is at present mainly limited to vasoconstricting agents and antihistamines. An attempt was made to show how some H_1 antagonists are effective in alleviating allergic conjunctivitis.

Table 11. ID_{50} values for newer H_1 antagonists on the dye leakage into the nasal cavity induced by antigen or histamine perfusion

Drugs	ID_{50} values (mg/kg, p.o.)	
	Antigen	Histamine
Epinastine		0.85 (0.22-2.40)
Ketotifen	2.87 (0.69-14.1)	1.01 (0.53-2.05)
Azelastine		0.94 (0.45-2.08)
Levocabastine	5.86 (3.30-21.0)	3.33 (1.21-11.1)
Loratadine	1.01 (0.43-2.03)	1.74 (1.21-3.03)

Drugs	ID_{50} values (topical application) (μg)	
	Antigen	Histamine
Ketotifen	63.7 (42.0-144.6)	51.0 (28.7-96.1)
Levocabastine	6.52 (5.23-9.03)	4.82 (3.05-7.04)

From Kamei, C., Mio, M., Kitazumi, K., Tsujimoto, S., Yoshida, T., Adachi, Y. and Tasaka, K.: Immunopharmacol. Immunotoxicol., 14, 207-218 (1992b); Tasaka, K., Kamei, C. and Nakamura, S.: Arzneim.-Forsch., 44, 337-341 (1994b), with permission.

Fifteen min after drug application, histamine solution (25 μl) was instilled to bilateral conjunctivae, and the degree of inflammation was determined by a scoring system. In most cases, the vehicle was instilled to the right eye and the test drugs were applied to the left eye. In the case of antigen-induced conjunctivitis, the animals were immunized with egg albumin (1 mg) as an antigen according to the method of Mota (1964). Two weeks after immunization, 25 μl of egg albumin solution (20 mg/ml) was applied to the eyes 15 min after the drug administration. The scoring system used to clarify the severity of conjunctivitis is as follows: 0 = no symptoms; 1 = slight hyperemia; 2 = severe hyperemia; 3 = severe hyperemia and slight edema; and 4 = severe edema.

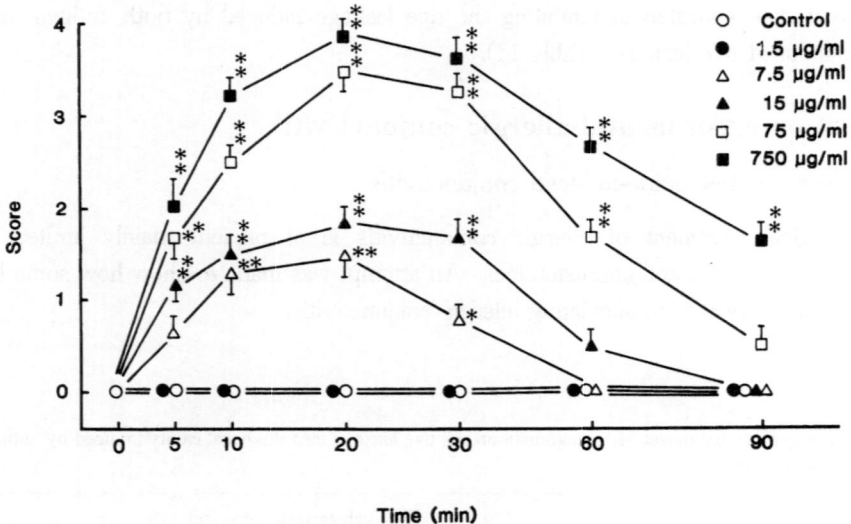

Fig. 1. Sequential changes in the severity of histamine-induced conjunctivitis elicited in guinea pigs. Each point indicates the mean ± S.E.M. *, **: Significantly different from the control group at P < 0.05 and P < 0.01, respectively by Mann-Whitney U test. From Kamei, C., Izushi, K. and Tasaka, K.: J. Pharmacobio-Dyn., 14, 467-473 (1991c), with permission.

Table 12. Effects of certain H_1 antagonists on antigen- and histamine-induced conjunctivitis in guinea pigs

Drugs	ED_{50} values (μg/ml)	
	Antigen	Histamine
Chlorpheniramine	14.2 (0.12-33.1)	11.2 (3.71-36.5)
Ketotifen	4.12 (0.19-16.0)	2.82 (0.03-11.4)
Levocabastine	4.14 (0.14-17.2)	7.67 (3.93-12.1)
Amlexanox	2767	> 2500
KW-4679	3.90 (0.28-13.4)	2.44 (0.59-5.55)

Histamine caused no conjunctivitis at concentrations lower than 1.5 μg/ml, however, at concentrations higher than 7.5 μg/ml, dose-related degrees in inflammation were observed. At a concentration of 7.5 μg/ml, histamine induced significant signs associated with conjunctivitis from 10 to 30 min after application. Severe and long-lasting inflammation was observed at concentrations ranging from 75 to 750 μg/ml (Fig. 1). Ketotifen, levocabastine and KW-4679 caused a potent inhibition of antigen-induced conjunctivitis. Chlorpheniramine, ketotifen, levocabastine and KW-4679 caused an inhibition of histamine-induced conjunctivitis, while no inhibition was observed with amlexanox (Table 12).

9.2. Antigen-induced histamine release from the conjunctiva

As described previously, histamine plays a key role in the conjunctival inflammation. Furthermore, it is very well known that histamine release takes place in antigen-antibody reaction. Based on these findings, it is rational to determine the extent of histamine release from the conjunctiva in antigen-antibody reaction as a possible source of histamine.

Fifteen min after antigen application, guinea pig conjunctiva was carefully excised, weighed and washed twice with saline. The tissue were homogenized with 0.4 N perchloric acid and placed in an ice-bath for 1 hr. After centrifugation at 1,200 × g for 10 min at 4°C, histamine contents in the supernatant were determined by HPLC (Kamei et al., 1993). The percentage of histamine release was calculated by the following equation: $c = (a - b_1$ or $b_2)/a \times 100$. c: % histamine release; a: histamine content of the conjunctiva in immunized animals; b_1: histamine content of the right conjunctiva (instilled by the vehicle) in immunized animals; b_2: histamine content of the left conjunctiva (antigen challenged) in immunized animals.

Both levocabastine and amlexanox caused an inhibition of antigen-induced histamine release. Although ketotifen inhibited both antigen- and histamine-induced conjunctivitis, a more potent effect was observed in histamine-induced conjunctivitis, indicating that ketotifen can be classified as a potent H_1 antagonist but that its antiallergic property is not striking. On the other hand, levocabastine has a somewhat more potent inhibitory effect on antigen-induced conjunctivitis than on histamine-induced conjunctivitis, suggest-

Table 13. Effects of certain H_1 antagonists on antigen-induced histamine release in guinea pig conjunctivae

Drugs	Concentrations (mg/ml)	% Histamine release	
		Control	Drug
Chlorpheniramine	0.1	44.0 ± 4.2	42.9 ± 6.9
Ketotifen	0.1	48.3 ± 1.5	41.8 ± 2.9
Levocabastine	0.5	45.7 ± 5.3	36.1 ± 7.7 *
Amlexanox	2.5	49.1 ± 7.4	27.8 ± 3.7 *

* : Significantly different from control group with P < 0.05. Reproduced from Kamei, C., Izushi, K. and Tasaka, K.: J. Pharmacobio-Dyn., 14, 467-473 (1991c), with permission.

ing that the drug may cause histamine release inhibition from the conjunctivae (Table 13). Based on these findings, the effects of levocabastine on antigen-induced histamine release from the conjunctivae was studied. Consequently, it became apparent that levocabastine and amlexanox are capable of inhibiting histamine release from the conjunctiva and furthermore, can prevent an increase in the vascular permeability elicited by released histamine after antigen instillation. It can be mentioned that levocabastine may be more efficacious than the other drugs in the treatment of allergic conjunctivitis.

10. Effect on LTs- or PAF-induced contractions of isolated guinea pig trachea

It is known that both LTs and PAF are the important mediators of allergic diseases especially in asthma, and it has been reported that oxatomide exerts anti-LTs and anti-PAF activities as determined with bronchospasm in guinea pigs (Ohmori et al., 1985). In addition, anti-PAF activity of thiazinamium was also reported when it was determined with bronchospasm in guinea pigs and rabbit platelet aggregation (Lewis et al., 1984). Antagonistic effects of newer H_1 antagonists on LTs and PAF were compared.

The trachea was excised from guinea pigs, and tracheal chains were mounted in an organ bath containing 7 ml of Tyrode's solution. Dose-response curves for either LTC_4 or PAF were constructed on each tracheal chain. Contractions were recorded isotonically and the contractile heights induced by either LTC_4 (LTD_4) or PAF at a fixed concentration of 0.05 nM were taken as 100 % in each case. The contractions induced by LTC_4 (LTD_4) or PAF after pretreatment with the test drugs were expressed as percentages of the control.

Astemizole and epinastine caused a potent inhibition of the LTC_4-induced contraction of guinea pig trachea, while ketotifen showed no such effect even at a concentration of 10^{-6} M. Astemizole and mequitazine exhibited a more potent inhibition on the contraction induced by PAF than that observed with ketotifen (Table 14).

Table 14. Effects of newer H_1 antagonists on LTs- or PAF-induced contraction of isolated guinea pig trachea

Drugs	IC_{50} values (M)	
	LTC_4 (LTD_4)	PAF
Ketotifen	$> 10^{-6}$	4.33×10^{-5}
Astemizole	5.80×10^{-7}	2.74×10^{-7}
Epinastine	3.41×10^{-7}	
Mequitazine	(4.45×10^{-8})	8.62×10^{-8}

Reproduced from Tasaka, K., Akagi, M., Mio, M., Izushi, K. and Aoki, I.: Arzneim.-Forsch., 40, 1092-1097 (1990e); Tasaka, K., Akagi, M., Izushi, K. and Mio, M.: Meth. Find. Exp. Clin. Pharmacol., 12, 531-539 (1990a); Tasaka, K., Akagi, M., Izushi, K. and Aoki, I.: Pharmacometrics, 39, 365-373 (1990d), with permission.

11. Effect on rabbit platelet aggregation induced by PAF

As shown in Table 15, in experimental animals, PAF is a very potent trigger of acute bronchoconstriction (Venuti, 1985), which is highly platelet-dependent (Vargaftig et al., 1980). Based on these findings, the effects of newer H_1 antagonists on platelet aggregation induced by PAF was studied.

Blood was removed from rabbits by cardiac puncture, and 3.15 % sodium citrate was added to the blood (1 part of sodium citrate to 9 parts of blood) to prevent coagulation. It was then centrifuged at $300 \times$ g for 10 min to obtain platelet rich plasma (PRP). Platelet poor plasma (PPP) was obtained by centrifuged blood at $1,500 \times$ g for 10 min. Platelet aggregation studies were carried out by means of an aggregometer (NBS hema tracer 601) using PRP which was adjusted to a platelet count of $500,000/\mu l$ with PPP. For drug studies, $225 \ \mu l$ of PRP was incubated with a drug or vehicle ($2.5 \ \mu l$) for 5 min at 37°C before the addition of PAF ($5 \times 10^{-8}M$) (Kamei et al., 1992b).

As shown in Table 15, ketotifen and epinastine inhibited PAF-induced platelet aggregation. By contrast, apafant was much more potent than these two drugs. Criscuoli et al. (1986) found that the IC_{50} of ketotifen required to inhibit PAF-induced platelet aggregation was $1.75 \times 10^{-4}M$. This value is essentially the same as that found in the present study.

12. Effect on the central nervous system

12.1. Locomotor activity in mice

It has been reported that diphenhydramine and promethazine induce a marked sedation in mice as judged by the decrease of their spontaneous movement, but the induction of sleep was not observed even at high doses (Weidmann and Petersen, 1953). Therefore, the influence on locomotor activity of mice was investigated for estimating the sedative effects of newer H_1 antagonists.

Table 15. IC_{50} values for the inhibition of newer H_1 antagonists on the rabbit platelet aggregation induced by PAF

Drugs	IC_{50} values (M)
Ketotifen	2.18×10^{-4}
Epinastine	2.62×10^{-4}
Levocabastine	$> 10^{-4}$
Apafant	1.56×10^{-7}

Reproduced from Kamei, C., Mio, M., Kitazumi, K., Tsujimoto, S., Yoshida, T., Adachi, Y. and Tasaka, K.: Immunopharmacol. Immunotoxicol., 14, 207-218 (1992b); Tasaka, K., Kamei, C., Akagi, M., Mio, M., Shirasaka, T. and Chokki, M.: Arzneim.-Forsch., 43, 1331-1337 (1993), with permission.

Locomotor activity (exploratory behavior) was measured for 3 min after each mouse was placed in a Hall's open-field apparatus (Hall, 1936). All drugs were administered orally 1 hr before the test.

As shown in Table 16, classic H_1 antagonists, diphenhydramine and promethazine, showed a potent depressant effect on locomotor activity. Ketotifen and azelastine induced a relatively potent decrease in locomotor activity in mice. On the other hand, levocabastine, mequitazine and astemizole showed weak depressant effects on locomotor activity. These findings are in agreement with the findings reported previously. For instance, Wauwe et al. (1981) reported that astemizole caused no decrease in locomotor activity even at a dose of 160 mg/kg, p.o. On the other hand, Kaneko et al. (1981) demonstrated that orally administered azelastine (10 and 40 mg/kg) resulted in a decrease in spontaneous locomotor activity. In addition, Martin and Römer (1978) described that motor activity measured by means of the climbing test was inhibited by ketotifen at doses

Table 16. Comparative effects (ED_{50}, mg/kg) of certain H_1 antagonists in some behavioral tests

Drugs	Locomotor activity	Thiopental sleep	Acetic acid writhing	Anti-MES	Anti-avoidance
Diphenhydramine	17.9 (7.99–40.1)	15.9 (4.18–60.5)	6.54 (2.26–9.56)	23.0 (18.8–28.4)	11.0 (3.67–33.0)
Promethazine	26.5 (10.2–68.6)	26.9 (17.0–42.5)	15.4 (8.06–29.4)		8.20 (5.16–13.0)
Ketotifen	20.0 (14.0–28.6)	20.0 (5.00–80.0)	16.5 (2.84–30.6)	> 50.0	32.5 (19.7–53.6)
Azelastine	24.4 (13.6–43.9)	58.9 (56.7–61.4)	11.5 (9.26–15.0)	> 50.0	21.0 (9.55–46.2)
Mequitazine	56.0 (22.8–138)	128.0 (81.7–201)	48.2 (23.5–98.8)		55.6 (33.7–91.7)
Epinastine	26.5 (11.9–59.1)	73.5 (28.4–190)		> 50.0	23.0 (14.6–49.0)
Astemizole	> 50.0	58.9 (56.7–61.4)	39.9 (36.8–43.7)	> 50.0	74.0 (26.4–207)
Oxatomide					60.0 (23.1–156)
Terfenadine					17.5 (10.9–28.0)
Levocabastine	38.1 (24.7–130)	97.4 (52.0–781)	17.3 (1.30–37.1)	> 50.0	111 (101–124)
Emedastine					24.9 (12.7–95.4)

Reproduced from Tasaka K., Kamei, C., Tsujimoto, S., Chung, Y.H. and Nakano, S.: Pharmacometrics, 37, 509-516 (1989b); Tasaka, K., Kamei, C., Nakano, S., Tsujimoto, S. and Chung, Y. H.: Pharmacometrics, 38, 53-62 (1989c); Tasaka, K., Kamei, C., Chung, Y.H., Tsujimoto, S. and Mukai, T.: Pharmacometrics, 39, 197-206 (1990b); Tasaka, K., Kamei, C., Tsujimoto, S., Yoshida, T. and Aoki, I.: Arzneim.-Forsch., 40, 1295-1299 (1990c), with permission.

of 30 and 100 mg/kg, p.o. Therefore, it is apparent that ketotifen and azelastine decreased exploratory behavior in mice.

12.2. Thiopental-induced sleep in mice

It is well known that diphenhydramine and chlorpheniramine prolong the sleep-producing effects of pentobarbital (Winter, 1948). Further, he proposed that the potentiation of barbiturate in the presence of classical H_1 antagonists correlated well with the incidences of sedation in patients receiving these drugs.

When thiopental sodium was injected intravenously (40 mg/kg) 1 hr after oral administration of the test drugs, the time required to restore the righting reflex was measured as the sleeping time. Loss of the righting reflex was defined as the animal's inability to right themselves after being placed on their backs (Kamei et al., 1986).

Classic H_1 antagonists, diphenhydramine and promethazine caused a marked prolongation of thiopental sleep as in the case reported by Winter (1948) (Table 16). Ketotifen also caused remarkable prolongation, while newly developed H_1 antagonists, such as azelastine, mequitazine, epinastine, astemizole and levocabastine induced moderate prolongation of thiopental-induced sleep. It is also reported that newer H_1 antagonists, especially astemizole and terfenadine exhibit little sedative side effects in clinical use (Garrison, 1990).

12.3. Acetic acid-induced writhing in mice

It is generally recognized that writhing syndromes induced by acetic acid or phenylquinone can be inhibited by various kinds of drugs including analgesics. Although it is realized that chemically induced writhing does not simply reflect pain sensation, this method is commonly used as a method for estimating the CNS depressing effect of the drugs.

Writhing response was induced by intraperitoneal injection of a 0.6 % solution of acetic acid (0.1 ml/10 g) according to the method of Koster et al. (1959). Test drugs were administered orally 50 min before the injection of acetic acid. The number of writhing movement was counted 30 min after injection of acetic acid. Ketotifen, azelastine and levocabastine caused a potent inhibition of acetic acid-induced writhing.

The potency of these drugs was almost the same as that of promethazine (Table 16). The inhibitory effects of mequitazine and astemizole were relatively weak. As for azelastine, Kaneko et al. (1981) described that the drug produced an inhibition corresponding to the extent exhibited by diphenhydramine. This was also the case in the present study. Although the interpretation of how drugs act to inhibit this characteristic abdominal constriction is not simple, both astemizole and mequitazine provide very weak suppression. Since some H_1 antagonists provide potent local anesthetic activity, this can be an explanation as to why H_1 antagonists inhibit chemically-induced writhing. For instance, it is known that diphenhydramine and promethazine possess potent local anesthetic actions (Leavitt and Code, 1947; Naranjo and Naranjo, 1958).

12.4. Anti-convulsant action in mice

It has been reported that high doses of classic H_1 antagonists caused a seizure or convulsions in infants and children (Wyngaarden and Seevers, 1951). However, in animal study, Swinyard et al. (1950) reported that diphenhydramine at doses of 125 mg/kg (p.o.) and 11.0 mg/kg (i.p.) showed a marked protective effect on the maximal electroshock seizure (MES). To investigate the same is the case in newer H_1 antagonists, the effect of H_1 antagonists on MES was studied.

MES was elicited according to the method of Swinyard et al. (1950). The ear electrodes were used to deliver an electroshock of 12 mA (475 V) for 0.2 sec. The duration of tonic extension (TE) was measured 1 hr after drug administration. Diphenhydramine caused a dose-related shortening of TE seizures reported by Swinyard et al. (1950), however, other H_1 antagonists showed no depressant effects (Table 16).

It was described that there is a certain relationship between an inhibition of TE seizure and sleepiness produced by drugs. Kreindler et al. (1958) and Bergmann et al. (1963) reported that the bulbopontomesencephalic reticular formation is responsible for the manifestation of the TE seizure induced by maximal electroshock. It is known that mesencephalic reticular formation is intimately related to sleep and wakefulness. Therefore, it became clear that the newer H_1 antagonists may have no inhibitory influences on the mesencephalic reticular formation.

12.5. Step-through active avoidance response in rats

It is well known that several classic H_1 antagonists disrupt a variety of conditioned response in rats (Winter and Flataker, 1951). In this study, the effects of newer H_1 antagonists on one way active avoidance response were studied in comparison with those of classic H_1 antagonists.

The apparatus is divided into two compartments. One with lighted room (19 × 15 × 14 cm) and the other is dark one (33 × 20.5 × 14 cm), and these are separated by a sliding door (5 × 5 cm). The dark compartment has a grid floor consisting of 3 mm copper rods, spaced 1.0 cm between each; and electrical shock (40 V, 0.2 mA) can be applied to the rods. The lighted compartment, painted white color, has a flat floor. The animals were placed in the dark compartment and unless the rats moved into the lighted room within 5 sec, the sliding door was closed and unescapable electroshock was delivered for 5 sec. Drug studies were performed using the rats well trained to move from the dark to lighted room less than 2.0 sec (Kamei et al., 1990).

Classical H_1 antagonists, diphenhydramine and promethazine caused a potent inhibition of the active avoidance response. On the other hand, mequitazine, astemizole, oxatomide and levocabastine showed a weak inhibitory effect on the avoidance response. Ketotifen, azelastine, epinastine, terfenadine and emedastine also caused an inhibitory effect. However, the extent of inhibition was much less than those of classic H_1 antagonists (Table 16). In two-way conditioned avoidance response in rats, it was found that diphenhydramine, pyrilamine and promethazine also showed a potent inhibitory effect on the avoidance response, while chlorpheniramine was less effective (Tasaka et al., 1985). It is assumed that an inhibition of the active avoidance response means a disturbance of the retrieval of the memory, so that, it seems likely that newer H_1 antagonists cause little or no disturbance of the retrieval of memory, in sharp contrast to the case of classical H_1 antagonists.

12.6. EEG spectral power in rats

It has been frequently shown that when classical H_1 antagonists were administered, an increase in the amount of sleep-like EEG activity was noticed (Tasaka et al., 1986c;

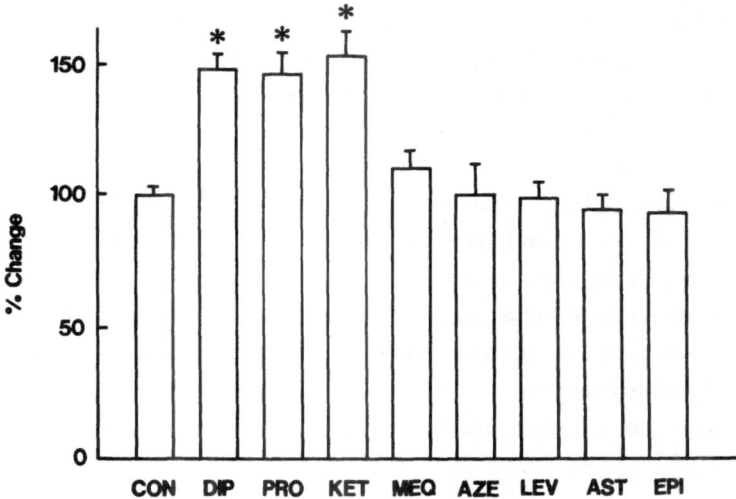

Fig. 2. Effects of certain H_1 antagonists on EEG spectral power recorded at the frontal cortex in rats. All drugs were administered orally at the dose of 20 mg/kg, p.o. CON; Control, DIP: Diphenhydramine, PRO: Promethazine, KET: Ketotifen, AZE: Azelastine, MEQ: Mequitazine, EPI: Epinastine, AST: Astemizole, LEV: Levocabastine. ∗ : Significantly different from the control group with $P < 0.05$. Reproduced from Tasaka K., Kamei, C., Tsujimoto, S., Chung, Y.H. and , Nakano, S.: Pharmacometrics, 37, 509-516 (1989b); Tasaka, K., Kamei, C., Nakano, S., Tsujimoto, S. and Chung, Y. H.: Pharmacometrics, 38, 53-62 (1989c); Tasaka, K., Kamei, C., Chung, Y.H., Tsujimoto, S. and Mukai, T.: Pharmacometrics, 39, 197-206 (1990b); Tasaka, K., Kamei, C., Tsujimoto, S., Yoshida, T. and Aoki, I.: Arzneim.-Forsch., 40, 1295-1299 (1990c), with permission.

Kamei et al., 1981). Both promethazine and chlorpheniramine also appear to induce a drowsy EEG pattern in animals and patients (Goldstein et al., 1968: Kamei et al., 1981). This study was carried out to investigate the changes in EEG spectral power after administration of several H₁ antagonists; special attention was placed on whether or not newer H₁ antagonists cause a drowsy EEG pattern.

Under pentobarbital anesthesia, chronic electrodes were implanted for EEG recording in the frontal cortex (FCOR), occipital cortex (OCOR), hippocampus (HPC) and amygdala (AMG) according to the atlas of de Groot (1959). Stainless steel screws were used as cortical and reference electrodes. EEG was converted into degital values by mean of multichannel A-D converter with 20 msec (τ) of interval (Tasaka et al., 1989a). EEG power processing was performed by the FFT method in real time. The FFT programming was carried out according to the algorithm of Cooley and Tukey (1965). On most occasions, the EEG powers were averaged for 2 min and they were divided into the 4 frequency bands, delta band (0-6 Hz), theta band (6-10 Hz), alpha band (10-16 Hz) and beta band (16-25 Hz). The effect of test drugs was expressed as percent changes in the power densities to that determined in the control periods.

The effects of H₁ antagonists on EEG spectral power (alpha band) recorded at the FCOR is shown in Fig. 2. Ketotifen (20 mg/kg, p.o.) caused an increase in the densities of spectral power in the delta band. Similar findings were observed after administration of diphenhydramine (20 mg/kg, p.o.) and promethazine (20 mg/kg, p.o.). On the other hand, azelastine, mequitazine, epinastine, astemizole and levocabastine caused no significant changes in the power densities even at a dose of 20 mg/kg, p.o. In the other areas such as OCOR, HPC and AMG, almost the same effects were observed, especially in the low frequency band.

It is known that diphenhydramine caused a drowsy EEG pattern (Tasaka et al., 1986c; Kamei et al., 1981), and ketotifen was also reported to demonstrate an increase in slow-wave sleep in dogs (Awouters et al., 1983). As shown in Fig. 2, all newer H₁ antagonists had no remakable effect on EEG spectral power when they were administered at a dose of 20 mg/kg (p.o.). It seems likely, therefore that an increase in EEG spectral power is closely related with the sleepiness or sedation when these drugs were applied in clinical uses. It can be expected that these drugs may not induce sleepiness in clinical uses.

13. Conclusion

Both oxatomide and astemizole caused a potent inhibition on histamine release from rat peritoneal mast cells induced by compound 48/80 and the lung pieces of sensitized guinea pigs. The inhibitory effect of oxatomide on histamine release may be caused by a combination of (1) prevention of Ca^{2+} uptake, which is highly selective toward mast cells, and (2) inhibition of Ca^{2+} release from the intracellular Ca^{2+} store. Therefore, the antiallergic mechanism of oxatomide and astemizole may be an inhibition of signal transduction from the mast cell membrane to the intracellular systems. In addition, astemizole exhibited little or no CNS depressant activity though it provided potent anti-LTs or anti-PAF activity.

Levocabastine is a drug with weak CNS effects, and potent anti-histaminic effects. Based on these findings, the antiallergic effect of levocabastine is mainly dependent on its potent antihistaminic activity. In addition, levocabastine provides a relatively potent inhibition of histamine release from the lung pieces of actively sensitized guinea pigs exposed to antigen, and simultaneously the drug prevented a decrease in cAMP contents. The most remarkable feature of levocabastine is its potent inhibition of experimental rhinitis and conjunctivitis when applied topically, i.e., the drug caused an inhibition of the dye leakage induced not only by histamine and antigen but also by substance P.

Terfenadine has a potent inhibition on histamine and SRS-A release from sensitized guinea pig lung samples, and on histamine release from rat peritoneal mast cells. The inhibitory effects of terfenadine on mediator release from mast cells are in some way related to its antiallergic effects, and that an elevated cAMP content and prevention of Ca^{2+} uptake may be effective in enhancing mediator release inhibition. In addition, terfenadine dose-dependently inhibited rat homologous PCA and experimentally-induced asthma in guinea pigs. On the other hand, the drug caused no appreciable changes in power densities in EEGs recorded in the FCOR, OCOR and AMG, indicating that the drug showed no CNS depressant activity.

Epinastine caused an inhibition of histamine release induced by the antigen-antibody reaction and compound 48/80. The drug was effective in inhibiting not only Ca^{2+} uptake into lung mast cells in actively sensitized guinea pigs but also Ca^{2+} release from the intracellular Ca^{2+} store of rat peritoneal mast cells exposed to both compound 48/80 and substance P. Epinastine has potent anti-histamine, anti-PAF and anti-LTs effects. Consequently, these effects may significantly contribute to its antiallergic activity. The characteristic property is that the drug had a potent inhibition on histamine-induced experimental rhinitis and PAF- or histamine-induced bronchospasms.

The antiallergic activity of mequitazine may be expected by antagonizing histamine, LTs, and PAF, as well as by inhibiting histamine release induced by either compound 48/80 or A23187. In the histamine release inhibition process, mequitazine may act not only to inhibit the Ca^{2+} release from the intracellular Ca^{2+} store of mast cells but also to stabilize the lipid bilayer of the cell membrane. In addition, this drug showed no obvious CNS depressant effects on such areas as locomotor activity, thiopental-induced sleep or active avoidance response.

References

Akagi, M., Mio, M., Tasaka, K. and Kiniwa, S.: Mechanism of histamine release inhibition induced by azelastine. Pharmacometrics, 26, 191-198 (1983)

Akagi, M., Mio, M., Miyoshi, K. and Tasaka, K.: Antiallergic effects of terfenadine on immediate type hypersensitivity reactions. Immunopharmacol. Immunotoxicol., 9, 257-279 (1987)

Akagi, M. and Tasaka, K.: Analysis of bronchomotor tone in anesthetized guinea pigs by impedance plethysmography: a simple method for the evaluation of bronchodilator action. Meth. Find. Exp. Clin. Pharmacol., 10, 143-150 (1988)

Arunlakshana, O.: Histamine release by antihistamines. J. Physiol., 119, 47-48P (1953)

Awouters, F.H.L., Niemegeers, C. J. E. and Janssen, P.A.J.: Pharmacology of the specific histamine H_1-antagonist astemizole. Arzneim.-Forsch., 33, 381-388 (1983)

Beaven, M.A., Rogers, J., Moore, J.P., Hesketh, T.R., Smith, G.A. and Metcalfe, J.C.: The mechanism of the calcium signal and correlation with histamine release in 2H3 cells. J. Biol. Chem., 259, 7129-7136 (1984)

Bende, M. and Pipkorn, U.: Topical levocabastine, a selective H_1 antagonist, in seasonal allergic rhinoconjunctivitis. Allergy, 42, 512-515 (1987)

Bergmann, F., Costin, A. and Gutman, J.: A low threshold convulsive area in the rabbit's mesencephalon. EEG Clin. Neurophysiol., 15, 683-690 (1963)

Bovet, D. and Staub, A.-M.: Action protectrice des éthers phénoliques au cours de l'intoxication histaminique. C.R. Soc. Biol., 124, 547-549 (1937)

Chand, N., Harrison, J.E., Rooney, S.M., Sofia, R.D. and Diamantis, W.: Inhibition of passive cutaneous anaphylaxis (PCA) by azelastine: Dissociation of its antiallergic activities from antihistaminic and antiserotonin properties. Int. J. Immunopharmacol., 7, 833-838 (1985)

Church, M. K. and Gradidge, C.F.: Inhibition of histamine release from human lung in vitro by antihistamines and related drugs. Br. J. Pharmacol., 69, 663-667 (1980)

Cooley, J.W. and Tukey, J.S.: An algorithm for machine calculation of complex Fourier series. Math. Comput., 19, 267-301 (1965)

Criscuoli, M., Subissi, A., Daffonchio, L. and Omini, C.: LG 30435, a new bronchodilator/antiallergic agent, inhibits PAF-acether induced platelet aggregation and bronchoconstriction. Agents Actions, 19, 246-250 (1986)

de Groot, J.: The rat forebrain in stereotaxic coordinates, Verh. K. Ned. Acad. Wet. Naturkund., 52, 1-40 (1959)

Douglas, W.W.: Histamine and 5-hydroxytryptamine (serotonin) and their antagonist. In: Gilman, A.G., Goodman, L.S., Rall, T.W., Murad, F. (ed) Goodman and Gilman's The Pharmacological Basis of Therapeutics, 7th ed. Macmillan Publishing Co., New York, 524-627, 1985

Fukuda, T., Morimoto, Y., Iemura, R., Kawashima, T., Tsukamoto, G. and Ito, K.: Effect of 1-(2-ethoxyethyl)-2-(4-methyl-1-homopiperazinyl)-benzimidazole difumarate (KB-2413), a new antiallergic, on chemical mediators. Arzneim.-Forsch., 34, 801-805 (1984)

Fügner, A., Bechtel, W.D., Kuhn, F.J. and Mierau, J.: In vitro and in vivo studies of the non-sedating antihistamine epinastine. Arzneim.-Forsch., 38, 1446-1453 (1988)

Garrison, J.C.: Histamine, bradykinin, 5-hydroxytryptamine, and their antagonists, In: Gilman, A.G., Rall, T.W., Nies, A.S. and Taylor, P. (ed) Goodman and Gilman's The Pharmacological Basis of Therapeutics, 8th ed. Pergamon Press, New York, pp.575-599, 1990

Goldstein, L., Murphree, H.B., Pfeiffer, C.C: Comparative study of EEG effects of antihistamines in normal volunteers. J. Clin. Pharmacol., 8, 42-53 (1968)

Guesdon, J.L., Chevrier, D., Mazié, J.C., David, B. and Avrameas, S.: Monoclonal anti-histamine antibody, preparation, characterization and application to enzyme immunoassay of histamine. J. Immunol. Methods, 87, 69-78 (1986)

Hall, C.S.: Emotional behavior in the rat . III. The relationship between emotionality and ambulatory activity. J. Comp. Psychol., 22, 345-352 (1936)

Hayashi, H., Ichikawa, A., Saito, T. and Tomita, K.: Inhibitory role of cyclic adenosine 3', 5'-monophosphate in histamine release from rat peritoneal mast cells in vitro. Biochem. Pharmacol., 25, 1907-1913 (1976)

Kamei, C., Kiniwa S., Ikegami, N. and Tasaka, K.: Effect of 4-(p-chlorobenzyl)-2[N-methylperhydroazepinyl-(4)]-1-(2H)-phthalazinone hydrochloride (azelastine) on EEGs and behavior in rats. Jpn. J. Clin. Pharmacol. Ther., 12, 297-310 (1981)

Kamei, C., Akahori, H. and Tasaka, K.: Influence of histamine and related compounds on the hypnotic effect of thiopental in mice. J. Pharmacobio-Dyn., 9, 112-116 (1986)

Kamei, C., Chung, Y.H. and Tasaka, K.: Influence of certain H_1-blockers on the step-through active avoidance response in rats. Psychopharmacology, 102, 312-318 (1990)

Kamei, C., Mio, M., Izushi, K., Kitazumi, K., Tsujimoto, S., Fujisawa, K., Adachi, Y. and Tasaka, K.: Antiallergic effects of major metabolites of astemizole in rats and guinea pigs. Arzneim.-Forsch., 41, 932-936 (1991a)

Kamei, C., Mio, M., Izushi, K., Yoshii, N., Fujisawa, K., and Tasaka, K.: Inhibitory effect of MY-1250 on histamine release from rat peritoneal mast cells and guinea pig lung fragments: The elucidation of the mechanism. Immunopharmacol. Immunotoxicol., 13, 341-356 (1991b)

Kamei, C., Izushi, K. and Tasaka, K.: Inhibitory effect of levocabastine on experimental allergic conjunctivitis in guinea pigs. J. Pharmacobio-Dyn., 14, 467-473 (1991c)

Kamei, C., Akagi, M., Mio, M., Kitazumi, K., Izushi, K., Masaki, S. and Tasaka, K.: Antiallergic effect of epinastine (WAL 801 CL) on immediate hypersensitivity reactions: (I) Elucidation of the mechanism for histamine release inhibition. Immunopharmacol. Immunotoxicol., 14, 191-205 (1992a)

Kamei, C., Mio, M., Kitazumi, K., Tsujimoto, S., Yoshida, T., Adachi, Y. and Tasaka, K.: Antiallergic effect of epinastine (WAl 801 CL) on immediate hypersensitivity reactions: (II) Antagonistic effect of epinastine on chemical mediators, mainly antihistaminic and anti-PAF effects. Immunopharmacol. Immunotoxicol., 14, 207-218 (1992b)

Kamei, C., Okumura, Y. and Tasaka, K.: Influence of histamine depletion on learning and memory recollection in rats. Psychopharmacology, 111, 376-382 (1993)

Kaneko, T., Kitahara, A., Ozaki, S., Takizawa, K. and Yamatsu, K.: Effects of azelastine hydrochloride, a novel anti-allergic drug, on the central nervous system. Arzneim.-Forsch., 31, 1206-1212 (1981)

Kebabian, J.W., Petzold, G.L. and Greegard, P.: Dopamine-sensitive adenylate cyclase in caudate nucleus of rat brain, and its similarity to the "Dopamine Receptor". Proc. Natl. Acad. Sci., 69, 2145-2149 (1972)

Koster, R., Anderson, M. and de Beer, E.J.: Acetic acid for analgesic screening. Fed. Proc., 18, 412 (1959)

Kojima, M., Tsutsumi, N., Abe, M., Komatsu, H., Ujiie, A., Naito, J. and Nakazawa, M.: Experimental allergic rhinitis in rats and the influence of tranilast and antihistaminics on it. Jpn. J. Allergol., 35, 180-187 (1986)

Kreindler, A., Zuckermann, E., Steriade, M. and Chimion, D.: Electroclinical features of convulsions induced by stimulation of brain stem. J. Neurophysiol., 21, 430-436 (1958)

Lau, H.Y.A. and Pearce, F.L.: Dual effect of antihistamines on rat peritoneal mast cells: induction and inhibition of histamine release. Agents Actions, 16, 176-178 (1985)

Leavitt, M.D. and Code, C.F.: Anesthetic action of beta-dimethylaminoethyl benzhydryl ether hydrochloride (benadryl) in the skin of human beings. Proc. Soc. exp. Biol. (N.Y.), 65, 33-38 (1947)

Levine, B.B. and Vaz, N.M.: Effect of combinations of inbred strain, antigen, and antigen dose on immune responsiveness and reagin production in the mouse. Int. Arch. Allergy, 39, 156-171 (1970)

Lewis, A.J., Dervinis, A. and Chang, J.: The effects of antiallergic and bronchodilator drugs on platelet-activating factor (PAF-acether) induced bronchospasm and platelet aggregation. Agents Actions, 15, 636-642 (1984)

Loeffler, L.J., Lovenberg, W. and Sjoerdsma, A.: Effects of dibutyryl-3', 5'-cyclic adenosine monophosphate, phosphodiesterase inhibitors and prostaglandin E_1 on compound 48/80-induced histamine release from rat peritoneal mast cells in vitro. Biochem. Pharmacol., 20, 2287-2297 (1971)

McManus, L.M., Hanahan, D.J., Demopoulos, C.A. and Pinckard, R.N.: Pathobiology of the intravenous infusion of acetyl glyceryl ether phosphorylcholine (AGEPC), a synthetic platelet-activating factor (PAF), in the rabbit. J. Immunol., 124, 2919-2924 (1980)

Martin, U. and Römer, D.: The pharmacological properties of a new, orally active antianaphylactic compound: Ketotifen, a benzocycloheptathiophene. Arzneim.-Forsch., 28, 770-782 (1978)

Mota, I.: The mechanism of anaphylaxis. I. Production and biological properties of 'mast cell sensitizing' antibody. Immunology, 7, 681-699 (1964)

Naranjo, P. and Naranjo, E.B. De: Local anesthetic activity of some antihistamines and its relationship with the antihistaminic and anticholinergic activities. Arch. int. Pharmacodyn., 113, 313-335 (1958)

Nemeth, A. and Röhlich, P.: Rapid separation of rat peritoneal mast cells with Percoll. Eur. J. Cell Biol., 20, 272-275 (1980)

Ohmori, K., Ishii, H., Kubota, T., Shuto. K. and Nakamizo, N.: Inhibitory effects of oxatomide on several activities of SRS-A and synthetic leukotrienes in guinea pigs and rats. Arch. int. Pharmacodyn., 275, 139-150 (1985)

Ohmori, K., Ishii, H., Shuto, K. and Nakamizo, N.: Pharmacological studies on oxatomide: (4) Effect on the histamine release from rat isolated peritoneal exudate cells (PEC) and lung slices. Folia pharmacol. japon., 80, 441-449 (1982)

Patel, K.R.: The effect of calcium antagonist, nifedipine in exercise-induced asthma. Clin. Allergy, 11, 429-432 (1981)

Pellerat, J. and Murat, M.: Action des antihistaminiques de synthese sur l'histaminemie. C.R. Soc. Biol. (Paris), 140, 297 (1946)

Pöch, G.: Assay of phosphodiesterase with radioactively labeled cyclic 3', 5'-AMP as substrate. Naunyn-Schmiedebergs Arch. Pharmak., 268, 272-299 (1971)

Ritchie, D.M., Sierchio, J.N., Bishop, C.M., Hedli, C.C., Levinson, S.L. and Capetola, R. J.: Evaluation of calcium entry blockers in several models of immediate hypersensitivity. J.

Pharmacol. Exp. Ther., 229, 690-695 (1984)

Ross, E.M., Maguire, M.E., Sturgill, T.W., Biltonen, R.L. and Gilman, A.G.: Relationship between the β-adrenergic receptor and adenylate cyclase. J. Biol. Chem., 252, 5761-5775 (1977)

Siraganian, R.P.: An automated continuous-flow system for the extraction and fluorometric analysis of histamine. Anal. Biochem., 57, 383-394 (1974)

Spataro, A.C. and Bosmann, H.B.: Mechanism of action of disodium cromoglycate-mast cell calcium ion influx after a histamine-releasing stimulus. Biochem. Pharmacol., 25, 505-510 (1976)

Shore, P.A., Burkhalter, A. and Cohn, V.H.: A method for the fluorometric assay of histamine in tissues. J. Pharmacol. Exp. Ther., 127, 182-187 (1959)

Swinyard, E.A., Jolley, J.M. and Goodman, L.S.: Anticonvulsant properties of benadryl and pyribenzamine. Proc. Soc. Exp. Biol. Med., 75, 239-242 (1950)

Tada, T. and Okumura, K.: Regulation of homocytotropic antibody formation in the rat. I. Feed-back regulation by passively administered antibody. J. Immunol., 106, 1002-1011 (1971)

Tasaka, K.: Histamine release and its inhibition by antihistamines. Folia pharmacol, japon., 53, 1029-1035 (1957)

Tasaka, K. and Akagi, M.: Anti-allergic properties of a new histamine antagonist, 4-(p-Chlorobenzyl)-2-[N-methyl-perhydroazepinyl-(4)]-1-(2H)-phthalazinone hydrochloride (Azelastine). Arzneim.-Forsch., 29, 488-493 (1979)

Tasaka, K., Kamei, C., Akahori, H. and Kitazumi, K.: The effects of histamine and some related compounds on conditioned avoidance response in rats. Life Sci., 37, 2005-2014 (1985)

Tasaka, K., Mio, M. and Okamoto, M.: Intracellular calcium release induced by histamine releasers and its inhibition by some antiallergic drugs. Ann. Allergy, 56, 464-469 (1986a)

Tasaka, K., Mio, M., Okamoto, M.: Changes in intracellular Ca^{2+} distribution of rat peritoneal mast cells before and after histamine release. Agents Actions, 18, 61-64 (1986b)

Tasaka, K., Kamei, C., Katayama, S., Kitazumi, K., Akahori, H. and Hokonohara, T.: Comparative study of various H_1-blockers on neuropharmacological and behavioral effects including 1-(2-ethoxyethyl)-2-(4-methyl-1-homopiperazinyl) benzimidazole difumarate (KB-2413), a new antiallergic agent. Arch. int. Pharmacodyn., 280, 275-291 (1986c)

Tasaka, K., Akagi, M., Mio, M., Miyoshi, K. and Nakaya, N.: Inhibitory effects of oxatomide on intracellular Ca mobilization, Ca uptake and histamine release, using rat peritoneal mast cells. Int. Archs Allergy appl. Immun., 83, 348-353 (1987)

Tasaka, K., Akagi, M., Mio, M., Okamoto, M., Miyoshi, K. and Izushi, K.: Antiallergic effects of terfenadine on immediate type hypersensitivity reactions and the mechanisms of those actions. Jpn. Pharmacol. Ther., 16, 2465-2480 (1988)

Tasaka, K., Chung, Y.H., Sawada, K. and Mio, M.: Excitatory effect of histamine on the arousal system and its inhibition by H_1 blockers. Brain Res. Bull., 22, 271-275 (1989a)

Tasaka, K., Kamei, C., Tsujimoto, S., Chung, Y.H. and Nakano, S.: Effects of mequitazine and some other H_1-blockers on the central nervous system. Pharmacometrics, 37, 509-516 (1989b)

Tasaka, K., Kamei, C., Nakano, S., Tsujimoto, S. and Chung, Y.H.: Effect of epinastine, a new antiallergic agent, on the central nervous system. Pharmacometrics, 38, 53-62

(1989c)

Tasaka, K., Akagi, M., Izushi, K. and Mio, M.: Antiallergic effects of astemizole on immediate type hypersensitivity reactions. Meth. Find. Exp. Clin. Pharmacol., 12, 531-539 (1990a)

Tasaka, K., Kamei, C., Chung, Y.H., Tsujimoto, S. and Mukai, T.: Influence of astemizole, a non-sedative H_1-blocker on the central nervous system. Pharmacometrics., 39, 197-206 (1990b)

Tasaka, K., Kamei, C., Tsujimoto, S., Yoshida, T. and Aoki, I.: Central effect of the potent long-acting H_1-antihistamine levocabastine. Arzneim.-Forsch., 40, 1295-1299 (1990c)

Tasaka, K., Akagi, M., Izushi, K. and Aoki, I.: Antiallergic effect of epinastine: the elucidation of the mechanism. Pharmacometrics, 39, 365-373 (1990d)

Tasaka, K., Akagi, M., Mio, M., Izushi, K. and Aoki, I.: Anti-platelet activating factor, anti-leukotriene D_4 and some other antiallergic activities of mequitazine. Arzneim.-Forsch., 40, 1092-1097 (1990e)

Tasaka, K., Kamei, C., Akagi, M., Mio, M., Shirasaka, T. and Chokki, M.: Antiallergic profile of the novel H_1-antihistaminic compound levocabastine. Arzneim.-Forsch., 43, 1331-1337 (1993)

Tasaka, K., Kamei, C. and Nakamura, S.: Inhibitory effect of epinastine on bronchoconstriction induced by histamine, platelet activating factor and serotonin in guinea pigs and rats. Arzneim.-Forsch., 44, 327-329 (1994a)

Tasaka, K., Kamei, C. and Nakamura, S.: Effects of antiallergic agents including levocabastine on experimental rhinitis in rats. Arzneim.-Forsch., 44, 337-341 (1994b)

Vargaftig, B.B., Lefort, J., Chignard, M. and Benveniste, J.: Platelet-activating factor induces a platelet-dependent bronchoconstriction unrelated to the formation of prostaglandin derivatives. European J. Pharmacol., 65, 185-192 (1980)

Venuti, M.C.: Platelet-activating factor: multifaceted biochemical and physiological mediator. Ann. Rep. Med. Chem., 20, 193-202 (1985)

Wauwe, J. Van., Awouters, F., Niemegeers, C.J.E., Janssens, F., Neuten, J.M. Van and Janssen, P.A.J.: In vivo pharmacology of astemizole, a new type of H_1-antihistaminic compound. Arch. int. Pharmacodyn., 251, 39-51 (1981)

Weidmann, H. and Petersen, P.V.: A new group of potent sedatives. J. Pharmacol. Exp. Ther., 108, 201-216 (1953)

White, J.R., Ishizaka, T., Ishizaka, K. and Sha'afi, R.I.: Direct demonstration of increased intracellular concentration of free calcium as measured by quin-2 in stimulated rat peritoneal mast cell. Proc. Natl. Acad. Sci. U.S.A., 81, 3978-3982 (1984)

Winter, C.A. The potentiating effect of antihistaminic drugs upon the sedative action of barbiturates. J. Pharmacol. Exp. Ther., 94, 7-11 (1948)

Winter, C.A. and Flataker, L.: The effect of antihistaminic drugs upon the performance of trained rats. J. Pharmacol. Exp. Ther., 101, 156-162 (1951)

Wyngaarden, J.B. and Seevers, M.H.: The toxic effects of antihistaminic drugs. J. Amer. Med. Ass., 145, 277-282 (1951)

Yamasaki, H. and Tasaka, K.: Dual action of antihistamines on histamine release. Acta Med. Okayama, 11, 290-299 (1957)

Subject index

A

A23187   296

α-fluoromethylhistidine   39, 40, 41, 42, 43, 44, 45, 48, 83

acetic acid (-induced) writhing   314, 315

acetylcholine   31, 33, 37, 58, 59, 93

acquisition   27, 28, 29, 45, 47, 59, 60

ACTH   69, 70, 72, 73, 75, 76, 77, 79, 80, 81, 82, 83, 84, 85, 86, 87, 88, 91, 93

actin   169, 174

actinomycin D   282

active avoidance response   28, 29, 30, 37, 40, 41, 45, 47, 51, 52, 53, 54, 55, 56, 59, 60, 316

adenylate cyclase   302, 303, 304

adrenal gland   89

adrenal nerve   89, 90, 91

adrenocortical cell   92, 93

age-related change   29

amino acid sequence   219, 221, 226

aminoguanidine   247, 248

amlexanox   310, 311

anti-avoidance   314

anti-convulsant action   24

antiallergic drug   126, 204

antigen-antibody reaction   296

apafant   306, 313

apyrase   122

arecoline   31, 32

arousal system   1

astemizole   27, 52, 53, 55, 56, 60, 62, 294, 296, 297, 299, 300, 304, 306, 307, 312, 314, 317

ATP   116, 139, 140

ATP depletion   123

ATP release   130

ATP-dependent Ca uptake   116

β, γ-methylene ATP   140

atropine   58, 59

azelastine   27, 52, 53, 55, 56, 60, 62, 294, 303, 306, 309, 314, 317

B

beeswax   248

β-escin   128, 129, 140, 146

bone marrow cell   252

bone marrow stromal cell   264

bronchoconstriction   304, 306

C

c-myc   284, 285

Ca^{2+} 97, 134, 149

^{45}Ca binding   118

Ca^{2+}-induced histamine release   133, 134, 135, 140, 148, 149, 150

Ca blocker (Ca antagonist)   124, 125, 141, 300

Ca^{2+} release from endoplasmic reticulum   101, 102, 104, 105, 106, 107, 111, 113, 117, 119, 125, 184

Ca uptake   103, 104, 105, 109, 116,

185, 218, 235, 298, 299

calmidazolium 134, 207

calmodulin 207

calmodulin inhibitor 128, 134, 207

calphostin C 143, 150, 151, 196, 281

cAMP 97, 108, 109, 119, 128, 135, 154, 158, 236, 260, 268, 301, 302

cDNA 219

central nervous system (CNS) 1, 27, 69

cerebral cortex 49, 50,

cGMP 108, 109, 119, 301, 302

CGRP 204, 208

CFU-c 255

cell differentiation 252, 259

cell proliferation 258

chlorpheniramine 52, 53, 55, 56, 60, 61, 294, 310, 311

cimetidine 11, 20, 21, 27, 38, 78, 79, 88, 249, 252, 261, 267, 268, 272, 279

colchicine 135, 185, 186, 187, 204, 206, 284

compound 48/80 97, 99, 100, 101, 102, 104, 105, 107, 112, 113, 124, 154, 174, 184, 209, 233, 296, 299, 300

concanavalin A 233

conjunctivitis 309, 310

contraction of guinea pig ileum 294

corticosteroid 70, 73, 75, 77, 80

corticosterone 72, 74, 76, 78, 79, 80, 82, 83, 84, 85, 86, 87, 88, 91

corticosteroid release 69

corticotropin-releasing hormone (CRF) 80

cortisol 70, 71, 73, 75, 77, 80, 81

CTP 140

cyclic nucleotide 108, 109, 119, 301, 302

cycloheximide 282

cytochalasin D 135, 185, 186, 204, 206, 284

cytoskeleton 169

D

D-600 (gallopamil) 105, 124, 125, 141, 300

dantrolene sodium 124, 125

degranulation 170, 173, 175, 176, 178, 213

delta wave 5, 11

dibutyryl cAMP (db-cAMP) 99, 100, 154, 155, 156, 157, 206, 280, 283

diltiazem 124, 125

dimaprit 266

diphenhydramine 10, 27, 53, 58, 60, 61, 294, 314, 317

DSCG (cromolyn sodium) 100, 126, 204, 205, 209, 304

E

EEG 1, 3, 4, 27

EEG arousal response 27

EEG power spectrum 2, 4

EEG spectral power 4, 5, 7, 8, 9, 10, 11, 12, 16, 20, 21, 317

EEG synchronization 2, 16, 21, 22

electrocoagulation 17

emedastine 314

endogenous histamine 39, 42

endoplasmic reticulum 110, 114, 116, 119, 121, 123, 184

epinastine 53, 294, 296, 297, 300, 302, 306, 307, 309, 312, 313, 314,

317

epinephrine   93

evoked potential   89, 90

F

famotidine   261, 267, 268, 270, 272

fast Fourier transform (FFT)   2

flunarizine   124, 125, 141

fluorescence microscopy   169, 181, 182, 187, 197

frontal cortex (FCOR)   4, 5, 7, 8, 9, 10, 11, 12, 16, 20

G

gallopamil   141

glucose-6-phosphatase   114

granulocyte-colony stimulating factor (G-CSF)   273, 274

granulocyte-colony stimulating factor (G-CSF) receptor   273

granulocyte-colony stimulating factor (G-CSF) binding   283

GTP   123, 140

GTP-γ-S   123, 127, 130, 131, 208

guinea pig major basic protein (GMBP)   215

guinea pig major basic protein-1 (GMBP1)   216, 225

guinea pig major basic protein-2 (GMBP2)   216, 225

H

H-7   143

H-8   155, 156

H_1 agonist   36, 75

H_1 antagonist   1, 10, 27, 38, 51, 52, 53, 54, 55, 56, 57, 59, 60, 77, 79, 293, 294, 295, 296, 297, 298, 299, 300, 302, 303, 304, 306, 307,

308, 311, 312, 313, 314, 317

H_1 receptor   27, 59, 69, 75

H_2 agonist   36, 75, 79

H_2 antagonist   10, 38, 77, 79

H_2 receptor   27, 69, 75, 78, 249

H_3 agonist   12, 39, 80

H_3 antagonist   12, 39, 80

H_3 receptor   69

HA1004   272, 281

Harr plot   224

hexamethonium   90

hippocampal lesion   45, 46, 47, 48, 49, 50

hippocampectomy   48

hippocampus   44, 45, 46, 49, 50

histamine   1, 7, 8, 11, 12, 16, 19, 20, 21, 27, 34, 35, 38, 41, 45, 47, 48, 49, 50, 57, 58, 69, 70, 71, 72, 73, 77, 79, 81, 82, 83, 85, 90, 91, 93, 245, 264, 273, 281, 307, 309, 310

histamine content   42, 43, 44, 48, 49, 50, 83, 84

histamine incorporation   253

histamine release   97, 99, 100, 103, 112, 124, 128, 129, 130, 131, 134, 135, 139, 140, 141, 142, 143, 146, 148, 149, 154, 155, 169, 184, 203, 212, 215, 227, 233, 295, 296, 297, 301, 311

histidine   8, 34, 35, 47, 48, 49, 50

histone   212

HL-60 cell   258

hopatenate calcium   33, 34

hypophysectomized animals   81

hypothalamus   44, 49, 82, 83, 84, 86, 87, 88, 89, 90

hypothalamic lesion   85, 86

I

idebenone   33, 34
impromidine   37
indeloxazine   33, 34
inositol 1, 4, 5-trisphospahte (IP$_3$) pro-
    duction   112, 132, 155, 186,
    234, 235
inositol 1, 4, 5-trisphosphate (IP$_3$)
    111, 112, 114, 117, 118, 122, 130,
    131, 132, 134, 135
^{3}H-IP$_3$ binding   118
inositol 4, 5-bisphosphate (IP$_2$)   112,
    117, 132
inositol 5-monophosphate (IP$_1$)   112,
    117, 132
interleukin-1α 235, 270, 271, 272
interleukin-2   233, 235, 236, 238,
    239
interleukin-3   235
interleukin-4   235
interleukin-5   235
intracellular Ca^{2+} concentration   101,
    102, 113, 180, 181, 186, 204, 206
intracellular Ca^{2+} release   99, 101,
    111
intracellular Ca blocker   124, 125
intracellular Ca mobilization   103,
    105, 110, 298
isoproterenol   303
ITP   140

K

K-252b   143
ketotifen   52, 53, 55, 56, 60, 62,
    104, 105, 108, 109, 126, 204, 205,
    209, 294, 296, 297, 300, 302, 303,
    304, 306, 307, 309, 310, 311, 312,
    313, 314, 317
KT-5720   261, 262, 269, 272, 279

KW-4679   310

L

lactate dehydrogenase (LDH) release
    130
lanthanum chloride   116
learning   27, 39, 45
leucine uptake   237, 283
leukocytosis   245
leukotrienes (LTs)   312
levocabastine   53, 294, 296, 297,
    302, 307, 309, 310, 311, 313, 314,
    317
lipocortin-1   233, 238
local degranulation   179
locomotor activity   40, 41, 47, 313,
    314
loratadine   294, 296, 297, 299, 306,
    309
lumicolchicine   185

M

major basic protein (MBP)   215
mass degranulation   213, 214
mast cell   97, 99, 100, 101, 104,
    105, 106, 107, 111, 112, 113, 114,
    121, 124, 128, 141, 154, 155, 156,
    157, 169, 184, 203, 212, 215, 227,
    233, 295, 296, 298, 299, 300
mast cell permeabilization   122, 128
maximal electroshock seizure (MES)
    314, 316
membrane enveloped degranulation
    213, 214
memory   27, 39
memory deficit   41, 44, 45, 57, 59
memory process   27
memory retention   27, 29, 30, 45,
    47

memory retrieval   38, 45, 50, 59

mequitazine   53, 100, 204, 205, 294, 296, 297, 300, 312, 314, 317

metabolic inhibitor   141

2-methylhistamine   10, 36, 37, 57, 58, 75, 266

4-methylhistamine   10, 37, 57, 58, 75, 77, 265, 266

microfilament   97, 106

microfilament-associated degranulation 106

microtubule   112, 184, 187

mitochondria   114

murine mast cell   111, 112

MY-1250   296, 297, 299, 300, 302

myeloperoxidase   259

N

N-ethylmaleimide   141

nefiracetam   33, 34

neomycin   131

neurotensin   205, 208

neurotransmitter   92, 93

neutrophil   247

neutrophil precursor   252

nifedipine   124, 125

norepinephrine   93

nucleotide   140

nucleus medialis centralis thalami (CM) 16

nulceus ventralis thalami (VE)   4, 5, 7, 8, 9, 10, 11, 12, 16

O

OAG   142

ouabain   141

oxatomide   27, 52, 53, 55, 56, 60, 62, 104, 105, 106, 107, 108, 109, 126, 204, 205, 209, 294, 296, 297, 299, 300, 302, 304, 314

P

PAF (platelet-activating factor)   305, 306, 312, 313

passive avoidance response   27, 28

passive cutaneous reaction (PCA) 306, 307

permeabilization   128

permeabilized mast cell   128, 134, 135, 139, 140, 141, 143, 146, 148, 149, 150, 151

phorbol ester   142

4α-phorbol 12, 13-didecanoate (4α-PDD)   142

4β-phorbol 13-monoacetate (PAC) 142

phosphatidylinositol (PI) turnover 111

phosphodiesterase   302, 303

physostigmine   31, 32

plasma histamine concentration   246

platelet aggregation   313

polymerase chain reaction (PCR)   219

promethazine   27, 53, 60, 61, 306, 314, 317

protein kinase A   154, 157, 236, 260, 262, 271, 281

protein kinase A inhibitor   155, 271

protein kinase C   139, 143, 146, 148, 149, 151, 152, 196, 211

protein kinase C inhibitor   143, 281

protein phosphorylation   154, 156, 157, 192, 193, 240, 263

protein synthesis   238

Prozime-10   70

pyramidal cell   47

pyrilamine   10, 20, 21, 27, 38, 52, 53, 55, 56, 57, 58, 59, 60, 61, 78,

79, 88, 247, 261, 279, 294

Q

quin 2 (quin 2/AM)   97, 98, 99, 100, 101, 102, 105, 106, 107, 113, 180, 205, 218, 300

R

(R)-α-methylhistamine   12, 39, 80, 88

ranitidine   11, 79, 261, 267, 268, 270, 272, 279

receptor externalization   273, 285

recovery of memory deficit   31

RF (reticular fomation) stimulation   4, 5, 6, 7, 8, 9, 10, 16, 20, 22

rhinitis   308

rhodamine-phalloidin   170, 180

S

scanning electron microscopy (SEM)   121, 131, 170

Scatchard analysis   118, 208, 277, 278

Sequential changes in histamine release   112

smg p21B   154, 156, 157

sodium vanadate   116, 141

somatostatin   208

SRS release   295, 296, 297

steroidogenesis   92, 93

striatum   49

substance P   97, 100, 203, 204, 205, 208, 210

substance P-binding   208

succinate dehydrogenase   114

T

tazalest   299

tetramethylrhodamine   169

terfenadine   98, 100, 204, 205, 209, 294, 296, 297, 299, 303, 304, 306, 307, 314

thalamic lesion   19

thalamus   17, 18, 49

theophylline   99, 206, 303

theta wave   5

2-thiazolylethylamine   36, 37

thiopental (-induced) sleep   314, 315

thioperamide   12, 39, 80, 88

thymidine labelled cell   250, 251, 253

thymocyte comitogenic activity   269

TMB-8   124, 125, 131

TPA (12-O-tetradecanoylphorbol-13-acetate)   142

translocation of protein kinase C   146, 148, 149, 151

transmission electron microscopy (TEM)   121, 172

trifluoperazine   207

two-dimensional polyacrylamide gel electrophoresis   157

tyrosine kinase   209

U

UTP   140

V

verapamil   104, 105, 109, 141, 299, 300

video-intensified microscopy   98, 106, 107

vimentin   192

vinblastin   185, 186

W

W-5   134, 207

W-7   134, 207